Governing Technology in the Quest for Sustainability on Earth

Governing Technology in the Quest for Sustainability on Earth explores how human technologies can be managed to ensure the long-term sustainability of our species and of other life forms with which we share this world. It analyses human impact, the discourses of environmentalism and issues of economics, history and science. As these variables are complex, drawing on issues from the social, physical and life sciences as well as the humanities, Dain Bolwell uses an interdisciplinary approach to investigate these concepts and their related public policies.

Exploring three major existing and emerging technologies – chemical herbicides, nuclear-electric power generation, and robotics and artificial intelligence – the book demonstrates the multifaceted and complicated nature of the grand challenges we face and draws out the measures required to effect sustainability in the wider political sphere.

Exploring how we can govern technology most effectively to ensure a long term and sustainable future, this book will be of great interest to students and researchers of environmental studies, science and technology and environmental law and policy.

Dain Bolwell is an associate with the Institute for the Study of Social Change in the Department of Politics and International Relations, University of Tasmania. Dain is a former international consultant, author and editor with particular expertise in labour and development fields. He has lived and worked in the Central Pacific as advisor to the Kiribati Minister for Labour, in East and West Africa for Public Services International and in Cambodia, China and Geneva for the United Nations.

Routledge Studies in Sustainability

Towards the Ethics of a Green Future
The Theory and Practice of Human Rights for Future People
Edited by Marcus Düwell, Gerhard Bos and Naomi van Steenbergen

Sustainability Transitions in South Africa
Edited by Najma Mohamed

Strongly Sustainable Societies
Organising Human Activities on a Hot and Full Earth
Edited by Karl Johan Bonnedahl and Pasi Heikkurinen

Managing Complexity, Earth Systems and Strategies for the Future
Walter R. Erdelen and Jacques G. Richardson

Measuring Intangible Values
Rethinking How to Evaluate Socially Beneficial Actions
Marie Harder and Gemma Burford

Education for Sustainable Human and Environmental Systems
From Theory to Practice
Edited by Will Focht, Michael A. Reiter, Paul A. Barresi and Richard C. Smardon

The Politics of Sustainability in the Arctic
Reconfiguring Identity, Space, and Time
Edited by Ulrik Pram Gad and Jeppe Strandsbjerg

Global Planning Innovations for Urban Sustainability
Edited by Sébastien Darchen and Glen Searle

The Question of Limits
A Historical Perspective on the Environmental Crisis
Christian Marouby

For a full list of titles please visit www.routledge.com/Routledge-Studies-in-Sustainability/book-series/RSSTY

Governing Technology in the Quest for Sustainability on Earth

Dain Bolwell

Routledge
Taylor & Francis Group
LONDON AND NEW YORK

First published 2019
by Routledge
2 Park Square, Milton Park, Abingdon, Oxon OX14 4RN

and by Routledge
52 Vanderbilt Avenue, New York, NY 10017

Routledge is an imprint of the Taylor & Francis Group, an informa business

British Library Cataloguing-in-Publication Data
A catalogue record for this book is available from the British Library

Library of Congress Cataloging-in-Publication Data
Names: Bolwell, Dain, author.
Title: Governing technology in the quest for sustainability on Earth / Dain Bolwell.
Description: Abingdon, Oxon ; New York, NY : Routledge, 2019. |
Includes bibliographical references.
Identifiers: LCCN 2018045738 | ISBN 9781138367739 (hardback) |
ISBN 9780429429651 (e-book)
Subjects: LCSH: Technological innovations–Environmental aspects. |
Sustainability. | Glyphosate–Environmental aspects. | Nuclear energy–Environmental aspects. | Robotics | Artificial intelligence.
Classification: LCC HC79.T4 B6155 2019 | DDC 338.9/27–dc23
LC record available at https://lccn.loc.gov/2018045738

ISBN: 978-1-138-36773-9 (hbk)
ISBN: 978-0-429-42965-1 (ebk)

Typeset in Bembo
by Newgen Publishing UK

MIX
Paper from responsible sources
FSC www.fsc.org FSC® C013056

Printed and bound in Great Britain by
TJ International Ltd, Padstow, Cornwall

Contents

Figures and tables

Figures

Tables

Acknowledgements

Associate Professor Fred Gale and Professor Richard Eccleston of the University of Tasmania have unfailingly and inspirationally provided wise counsel during my research for the dissertation on which this book is based.

I am very grateful to Dr Kerryn Higgs, fellow of the Club of Rome, for her insights on environmental politics. Dr Higgs graciously delivered a well-attended public lecture and seminar at my request through the Institute for the study of Social Change, and further supervised my research report on the Club for an associated qualification.

My friend Kevon Kenna, exploration geophysicist, information technology practitioner and poet, I also particularly thank. He has been a sounding board, tough critic and source of inspiration throughout the past years.

My wife, Adjunct Professor Rosalind Harvey (Queensland University of Technology), has been consistently enthusiastic, supportive and a source of insight and material important to this book.

To those academics and practitioners from around the world who have unfailingly and positively responded to my requests and visits I am indebted. These especially include Professor Will Steffen, Australian National University (ANU) and the Stockholm Resilience Centre (climatology), Professor Tristan Perez, Queensland University of Technology (QUT) (robotics), Professor Clifford Shearing, University of Cape Town, Oxford and the ANU (law), Dr Gwynn MacCarrick, University of Tasmania and Dr Bronwyn Lay, European Graduate School (law), Dr Ken Chomitz, ex-World Bank (economics), Professors Tim Wootton and David Jablonski, University of Chicago (paleontology), Professor Barry Brook (ecomodernism) and Dr Alex Bissember (chemistry), both of the University of Tasmania, Dr Charles Benbrook, Washington State University (agricultural technology), Professor Ted Lefroy, University of Tasmania (ecology), Dr Stephen Duke, US Department of Agriculture (biology), Chris Vein, former White House advisor and World Bank executive (innovation), Dr Pushker Kharecha of the NASA Goddard Centre (nuclear power), Professor Robert Logan, University of Toronto (physics and the future), and Professor Keith Solomon, University of Guelph, (toxicology).

Acronyms

2,4-D	2,4 dichlorophenoxy-acetic acid
2,4,5-T	2,4,5 trichlorophenoxy-acetic acid
4IR	Fourth industrial revolution
AAAI	Association for the Advancement of Artificial Intelligence
ACS	American Chemical Society
AEA	American Economic Association
AFOLU	Agriculture, forestry and other land use
AGI	Artificial general intelligence
AGU	American Geophysical Union
AI	Artificial intelligence
AIS	Automatic identification system (shipping)
AMS	American Meteorological Society
ANS	Adjusted net savings
BJP	Bharatiya Janata Party of India
BWC	Biological Weapons Convention
CFC	Chlorofluorocarbon
CMEPSP	Commission on the Measurement of Economic Performance and Social Progress (France)
CNS	Convention on nuclear safety
CO_2	Carbon dioxide
COP	Conference of the Parties
CSER	Centre for the Study of Existential Risk
CSIRO	Commonwealth Scientific and Industrial Research Organisation (Australia)
CWC	Chemical Weapons Convention
DDT	Dichlorodiphenyltrichloroethane
EF	Ecological footprint
EKC	Environmental Kuznets curve
EPA	Environment Protection Authority (US)
EPTA	European Parliamentary Technological Assessment

EPZ	Export processing zone
EROEI	Energy return on energy invested
FAO	Food and Agriculture Organisation (UN)
FHI	Future of Humanity Institute
FLI	Future of Life Institute
FOI	Freedom of information
G20	Group of Twenty (economic forum of 20 major economies)
GAO	Government Accountability Office (US)
GDP	Gross domestic product
GFC	Global financial crisis
GHG	Greenhouse gas
GNH	Gross National Happiness (Bhutan)
GNP	Gross national product
GPS	Global positioning system
IAEA	International Atomic Energy Agency
ICRP	International Commission on Radiological Protection
ICTSD	International Centre for Trade and Sustainable Development
IEA	International Energy Agency
ILO	International Labour Organisation (UN)
IPAT	Impact-Population-Affluence-Technology
IPBES	Intergovernmental Science Policy Platform on Biodiversity and Ecosystem Services
IPCC	Intergovernmental Panel on Climate Change (UN)
IUU	Illegal, unreported and unregulated (fishing)
JFS	Japan for sustainability
LOC	Level of concern
LED	Light emitting diode
LPI	Living Planet Index (WWF)
MCS	Monitoring, control and surveillance
MDGs	Millennium Development Goals
MIT	Massachusetts Institute of Technology (US)
NTFB	Non-timber forest benefits
OECD	Organisation for Economic Cooperation and Development
PDR	Peoples' Democratic Republic
POPs	Persistent Organic Pollutants
PNAS	Proceedings of the National Academy of Sciences (US)
PR	Public relations
R&D	Research and development
RFMO	Regional fisheries management organisation
SDGs	Sustainable Development Goals
STI	Science, technology and innovation (UNESCO)

SWEIT	Species with energy-intensive technology
TAPIC	Transparency, accountability, participation, integrity, capacity.
TINA	There is no alternative (Margaret Thatcher)
TPP	Trans-Pacific Partnership
TRIPS	Agreement on Trade-Related aspects of Intellectual Property Rights
UK	United Kingdom
UN	United Nations
UNCTAD	United Nations Conference on Trade and Development
UNDP	United Nations Development Program
UNESCO	United Nations Educational, Scientific and Cultural Organisation
UNEP	United Nations Environment Program
UNFCCC	United Nations Framework Convention on Climate Change
UNFF	United Nations Forum on Forests
UNFSA	United Nations Fish Stocks Agreement
UNGA	United Nations General Assembly
UNHPS	United Nations Secretary-General's high-level panel on global sustainability
UNSCEAR	United Nations Scientific Committee on the Effects of Atomic Radiation
US	United States
USSR	Union of Soviet Socialist Republics
WANO	World Association of Nuclear Operators
WCC	World Council of Churches
WEF	World Economic Forum
WHO	World Health Organisation (UN)
WIPO	World Intellectual Property Organisation
WWF	World Wildlife Fund

Notes

As to what may be known, I am attracted to the view of the entomologist E. O. Wilson (1999, p. 291) that: "all tangible phenomena, from the birth of stars to the workings of social institutions, are based on material processes that are ultimately reducible, however long and tortuous the sequences, to the laws of physics". Deconstructionists and post-modernists, are mere gadflies, who are nonetheless useful to keep the 'real' scientists honest (Costanza 2009).

On the other hand, I am not convinced that "the workings of social institutions" in particular are tangible phenomena reducible to physics as Wilson asserts. Rather, I think that Bismarck came close to a truth when he said that the making of laws is like the making of sausages: it is better not to observe the process too closely, although in this he was less invoking a Heisenberg indeterminacy, more its unpleasantness.

And lastly, science is central. But it does not mean that particular scientists are infallible, that methods and results are not influenced by social processes, or that human perceptions of what is important are pure and objective rather than the result of historical and political forces.

Quotations from the works of T. S. Eliot are made with permission from Faber and Faber Ltd publishers.

Reference

Costanza, R 2009, 'Science and Ecological Economics Integrating of the Study of Humans and the Rest of Nature', *Bulletin of Science, Technology & Society*, vol. 29 No. 5, October 2009, pp. 358-373.

1 Introduction

British prime minister Margaret Thatcher and visionary industrialist Aurelio Peccei, co-founder of the Club of Rome, were once in accord. While Mrs Thatcher (1988) spoke of the possibility:

> that with all these enormous changes (population, agricultural, use of fossil fuels) concentrated into such a short period of time, we have unwittingly begun a massive experiment with the system of this planet,

Peccei (1984) had earlier dictated a message from his deathbed that said:

> humanity must learn how to ride the technological tigers we have unleashed; if we do not the Earth will become incapable of supporting our species; there will be no future for humankind.

Two centuries of industrialisation, exploding population growth and economic frenzy based on unchecked technologies have given us smog, polluted waterways, toxic waste, soil erosion, deforestation, the extinction of many species, global warming and rising oceans. Yet those technologies have also given us the means to check disease and extend life, to connect with each other and to access and apply knowledge in ways beyond the wildest imaginings of the Enlightenment. Our present system of extracting, making, consuming and discarding may be unsustainable, but is not inevitable. This book assesses the factors of sustainability and how we might become more adept at riding Peccei's tigers.

Is humanity sustainable?

There can be no greater significance than the issue of sustainability for humanity. While ever-narrowing inquiries typify research within all disciplines, this book attempts a breadth commensurate with the importance of the question. If the answer to the question above is in fact an unqualified 'no', then ultimately there is no point to any other field of human endeavour. Quantum physics, history, philosophy, engineering, the fine arts, love, capital accumulation – in the longer term, all is for nought and for nothing.

This sustainability issue is just as important for the many life forms fast dwindling at our hand, more so now than at any earlier time. Yet besides the possible finding of 'no' to the question, there are also many variations of an affirmative. Perhaps humanity is sustainable if we adopt a particular policy, if we effect different technologies, if we accept a diminished energy or if global capitalism is better controlled. Perhaps humanity is, in the longer term, already on a sustainable trajectory and this current crisis is just a blip, a speed bump on our digital highway to the future.

The words 'sustainable' and 'sustainability' in the second decade of the twenty-first century are ubiquitous throughout the Anglosphere, having leapt to prominence around 1980.[1] Businesses aspire to be sustainable. Public strategies require sustainability. No-one supports endeavours that are said to be unsustainable. While the notion of sustainability is both central and contested in several disciplines, it has been said that understanding the true meaning of sustainability requires the concept of entropy (Eriksson and Andersson 2010). Entropy is the universal tendency to disorder, in which energy, although conserved in accordance with the first law of thermodynamics,[2] becomes dispersed and thus less useable over time as it performs work and transitions from higher states to its lowest state: heat. This means that in the long run humanity must again adapt to using inexhaustible energy resources as we once did, because in a closed system, the non-renewable stock will run out, just as energy (or states of order) will tend to even out over time in thermodynamics. All energy on Earth ultimately derives from the sun.[3]

And while ultimately all matter is stardust, and theoretically anything can be constructed from its basic elements, it requires a great deal of energy to change complex material forms. As a result, the extraction of complex forms of matter from the Earth such as metals, chemicals and organic materials will continue as long as humanity foreseeably exists as a civilisation. But minimisation, conservation and re-use of those materials becomes imperative. Otherwise continued consumption and disposal leads only to a material entropy where grains of nickel, calcium carbonate and humus are dispersed beyond recovery over a featureless desert of grey.

In this way, T. S. Eliot's famous phrase from *The Hollow Men* (1967 [1925]) "this is the way the world ends, not with a bang, but a whimper" has a parallel in the second law of thermodynamics, which states that "any closed system[4] evolves towards a state of thermodynamic equilibrium" – the state of maximum entropy. If we consider the Earth as a closed system, then the whimper towards the end of its Anthropocene epoch will be an entropy in which any remaining human life will possess minimal degraded capabilities.

However, C. P. Snow's *The Two Cultures* (1959) lecture on the two mutually uncomprehending worlds of science and literature supposes that familiarity with Eliot for a scientist is the equivalent of knowledge of the second law of thermodynamics by a literary intellectual. Both are unlikely. Snow deplores this gulf of understanding between the two fields,[5] and especially the lack of appreciation of the importance of applied science in addressing the major concerns of his time: "Industrialisation is the only hope of the poor", he said. Writing during the Cold War, Snow urges a scientific effort of the West to transform the poor societies of "India, Africa, South-East Asia, Latin America and the Middle East" to ameliorate the three menaces that confront humanity: nuclear war, over-population and the gap between rich and poor. While less due to a conscious scientific effort of the West, some of those poor societies have since undergone industrialisation and the threat of nuclear war has arguably diminished. Yet the menaces of over-population and inequality remain. An underlying theme of this book is bridging understandings between cultures and disciplines. The quest for sustainability demands that the sciences and the humanities acknowledge their need for each other.

Consistent with an awareness of the Earth as a linked, closed system rather than a limitless resource (Boulding 1966; Buckminster Fuller 1969; Commoner 1972; Holdren and Ehrlich 1974; Lovelock 1979) that has developed since Snow's identification of these threats, overpopulation and inequality still imperil not only humanity, but all species that form life on Earth. The classic I=PAT equation, meaning human environmental impact is a product of population, affluence and technology, has been used to identify the key relationships that govern sustainability for most of the ensuing six decades. Population alone is not the only basic determinant of our effect on the planet. Unlike other species, our impact is greatly amplified by the factors of affluence and technology. This book uses the equation to structure its analysis of sustainability. However, the relationship, between these factors, originally formulated by Paul Ehrlich in 1968, is not entirely straightforward. Neither is the meaning of each element.

The IPAT relationship remains the most coherent summation of factors that relate to sustainability. However, the meanings of the latter two factors especially – A (affluence) and T (technology) — have changed considerably since its original conception. Also, there is no direct empirical relationship between affluence or technology and environmental impact. Rather, in line with), there is evidence that the environmental impact of at least some pollutants first rises with increasing affluence, then falls as wealth further increases (Kuznets 1955; World Bank 1992). Some technologies benefit the environment. Some cause outright harm.

Moreover, the measurement of affluence itself is problematic. It is usually assumed to be measurable by gross domestic product (GDP), given that production equates to consumption, which is determined by affluence. However,

GDP and similar measures do not quantify all aspects of production and consumption. What is consumed can affect the environment as much as how much is consumed, and consumption of intangibles is increasing.

The interaction of technology with the environment may be positive as well as negative. In the original 1968 I=PAT equation, Ehrlich postulated carbon dioxide emission as a cipher for technology use, as then technological density and industrialisation emitted more carbon dioxide. Certainly this is still essentially true of recently industrialised economies such as China and increasingly India, where coal-fired power stations and secondary industries continue to be built at an astonishing rate and where air pollution is legendary. However newer technologies such as solar cells emit no carbon dioxide in use. Digital technologies facilitate analysis and the acquisition of knowledge and also emit little. Sensors and meteorological computer models help mitigate the effects of weather through accurate prediction. And contraception technologies negatively effect population and thus contribute to a reduced environmental impact.

This book initially explores the nature of human sustainability and its dimensions through the IPAT lens. It examines how perceptions of the variables that compose it have changed, their interrelationships and how the concept might guide policies that lead to a more sustainable world. Because each of its variables relate to different fields, including the social, physical and life sciences as well as the humanities, an interdisciplinary approach is used to investigate this concept and its related public policy.

Now we have the advantage of being able to look backwards in time to the beginnings of the modern environmental awakening, as well as forwards to where our fate may lead. That awakening was directly related to the accelerating impact of new technologies after World War II, as well as to the accumulation of technologies that began with the Industrial Revolution. Half a century since that environmental awakening, innovation and still newer technologies are exploding into diverse realms, based on digital computing, new means of data capture and analysis capabilities on a scale unimaginable only a generation earlier. The relationship between humanity and technology is symbiotic. Our technologies affect us as much as we affect them and the relationship is intensifying. It is to this bond that this book turns as the only practicable policy area whereby sustainability might be effected in the medium term.

Notes

1 According to Google's N-gram viewer, 2017.
2 The first law of thermodynamics is a version of the general law of conservation of energy.
3 This includes fossil fuels that are due to photosynthesis, radioactive materials that came from solar matter, wind that is caused by solar radiation differentials and the tides that result from the orbits of the Earth and the Moon about the Sun.

4 There is an important difference between a 'closed' and an 'isolated' system in thermodynamic theory, which is discussed later in this book.

5 Although DiMaggio (2015) asserts "that the era of the 'two cultures' (Snow, 1959) is over" based on his experience as a social scientist working with computer scientists, finding instead of chasms, only "modest differences in orientation" due to their respective intellectual traditions.

References

Boulding, K 1966, 'The Economics of the Coming Spaceship Earth' in Jarrett, H (ed.), *Environmental Quality in a Growing Economy*, Johns Hopkins University Press, Baltimore, US, pp. 3–14.

Buckminster Fuller, R 1969, *Operating Manual for Spaceship Earth*, Southern Illinois University Press, Carbondale, US.

Commoner, B 1972, *The Closing Circle*, Knopf, New York, US.

DiMaggio, P 2015, 'Adapting Computational Text Analysis to Social Science (and Vice Versa)', *Big Data and Society*, vol. 2, no. 2, December, pp. 1–5.

Eliot, T S 1967 [1925], 'The Hollow Men', in *Selected Poems*, Harvest Books, San Diego, US.

Eriksson R and Andersson, J 2010, *Elements of Ecological Economics*, Routledge, London, UK.

Holdren, J and Ehrlich, P 1974, 'Human Population and the Global Environment', *American Scientist*, vol. 62, no. 3, May–June, pp. 282–292.

Kuznets, S 1955, 'Economic Growth and Income Inequality', Presidential address delivered at the sixty-seventh Annual Meeting of the American Economic Association, Detroit, Michigan, December 29, 1954, *The American Economic Review*, vol. 45, no. 1, March, pp. 1–28.

Lovelock, J 2000 [1979], *Gaia: A New Look at Life on Earth*, Oxford University Press, Oxford, UK.

Peccei, A 1984, 'The Club of Rome: Agenda for the End of the Century', in Malaska, P and Vapaavuor, M (eds), 2005, *The Club of Rome "The Dossiers" 1965–1984*, Finnish Association for the Club of Rome, Helsinki, Finland, pp. 37–43.

Snow, C 1959, *The Two Cultures: The Rede Lecture*, Cambridge University Press, Cambridge, UK.

Thatcher, M 1988, Speech to the Royal Society, 27 September, Fishmongers' Hall, City of London, UK.

World Bank 1992, *World Development Report: Development and the Environment*, Oxford University Press, New York, US.

2 The concept of sustainability

This is the dead land
This is cactus land
Here the stone images
Are raised, here they receive
The supplication of a dead man's hand
Under the twinkle of a fading star

– T. S. Eliot, *The Hollow Men*

Eliot's image of a lifeless Earth looms as one reflects on what has happened to our planet home in recent times. By contrast with this representation, a more hopeful world is alive, vibrant, interconnected: a sustainable one. The concept of sustainability is recent, yet concerns about it have been with us for millennia. It encompasses all resource extraction, production and consumption of goods and services. It includes how waste is treated and our impact on other life forms on which we depend, or which are part of the one biosphere that we share. It encompasses and is yet distracted by 'climate change' as it was once distracted by 'the population explosion'. Sustainability is simply human activity that might continue indefinitely.

The early Christian, Tertullian, observed that humans were "burdensome to the world" nearly 2,000 years ago, and instances of unsustainable practices have increased throughout history. Lead pollution was evident in ancient times. The burning of 'sea coal' in London became increasingly obnoxious during the Middle Ages. Timber harvesting in England and much of Europe became unsustainable during Elizabethan times. The volume of sewage in Victorian cities led to epidemics before the imperative of reform. The Romantics were dismayed at the price nature paid for the Industrial Revolution. In the United States, the Sierra Club and national parks were established in the late nineteenth century in reaction to a fast diminishing natural world. The 1930s Dust Bowl tragedy showed the importance of sustainable agricultural practices. In the 1940s and 1950s, the many deaths from air pollution in Donora, Pennsylvania and in London helped spur more sustainable industrial practice, at least in the United

States and the United Kingdom. Fatalities and deformities from heavy metal accumulation in Japan in the 1960s finally led to better controls over following decades. Asbestos and lead poisoning has been curtailed in rich countries since the late twentieth century and emissions of ozone depleting substances have lately been limited by near universal international law, because their use was unsustainable. Sustainability implies minimal harm to both humans and the biosphere. It is a self-evident imperative where they intersect.

Although human awareness that demanding no more than the environment can supply is historically and geographically age-old, contemporary anxieties about sustainability date from the 1960s when Rachel Carson's *Silent Spring* (1962) and the Ehrlichs' *The Population Bomb* (1968)[1] were published. Carson highlighted the perils of synthetic pesticides while the Ehrlichs trumpeted the threat of runaway population growth. In 1969 a new mandate for the International Union for the Conservation of Nature (IUCN) referred to management of 'air, water, soils, minerals and living species including man, so as to achieve the highest sustainable quality of life' (Adams 2006, p. 2).

However, the Massachusetts Institute of Technology (MIT) report for the newly formed Club of Rome, *The Limits to Growth* (Meadows et al. 1972), first showed the profound effects of continued exponential consumption on a finite planet, using computer modelling. Graham Turner's 2014 article *Is Global Collapse Imminent?* that compares the report's predictions to reality, suggests that the original *Limits* was remarkably accurate in forecasting patterns of population, economic growth and resource use in the four decades since it was written. Pointedly, he concludes that if the unsustainable 'business as usual' trajectory continues, then global collapse within the next decade or two is imminent.

Prompted by the *Limits* projections, the word "sustainability"[2] was first articulated at a World Council of Churches meeting involving scientists, theologians and economists in Bucharest in 1974, at which the "intolerable strain on the Earth's resources" was discussed in the context of science, technology and human development:

> What emerged out of the Bucharest discussion on the role of science and technology in the development of human societies was the articulation of a concept called "sustainability" – the idea that the world's future requires a vision of development that can be sustained in the long run, both environmentally and economically.
>
> (Hallman 2002, para. 8)

These early prophecies of peril resulted in measures aimed at curbing excesses, especially in the industrialised West, as well as the creation of global mechanisms relating to the environment and sustainability. During the 1980s these included the *World Conservation Strategy: Living Resource Conservation for Sustainable*

Development in 1980, the *World Commission on Environment and Development* in 1983, the *Montreal Protocol on Substances that Deplete the Ozone Layer* and the Brundtland report *Our Common Future*, both in 1987. Since that flurry of activity, measures relating to sustainability noticeably faltered. There was the Kyoto protocol of 1997, the stalemate of Copenhagen in 2009 and the diluted extension of Doha in 2012, all of which attempt to forestall climate change by limiting greenhouse gas emissions. The 2015 Paris emissions agreement yet stood out as an important step forward, at least as far as global warming is concerned.

There have also been the 1992 Rio and follow-up Earth Summits (*Rio+5, +10* and *+20*) that concerned conservation, poverty reduction and empowerment. There were the eight Millennium Development Goals from 2000 to 2015 that included a very much 11th-hour goal seven, to *ensure environmental sustainability.*[3] This goal included targets for biodiversity loss, deforestation, fish stocks and water quality as well as emissions. However, the focus of the Millennium goals was elsewhere – on poverty reduction, health and gender equality. More importantly, the goals were targeted at developing countries – there was no goal seven that applied to the industrialised world, nor in practice to China. At least the Sustainable Development Goals that replaced them from 2016 do involve targets for rich countries; thus for the first time wider sustainability is recognised as a global problem for global action.

Empathy and entropy

Beyond these historical aspects of the concept, there are also the anthropological and the planetary. While it has been observed that no social animal is guided by the welfare of the entire species to which it belongs, humans are a potential exception. Originally tribal and territorial, our species first conceived of 'universal orders',[4] when the entire human race could be imagined as a single unit about 3,000 years ago. This meant that everyone was potentially 'us'. There was no longer 'them' (Harari 2014, pp. 171–172). Since then, despite many setbacks, the overall historical trend has been towards greater global unity. Around the end of the last ice age there were thousands of autonomous human worlds on the planet. Five hundred years ago most people lived in the single Afro-Asian landmass with limited cultural and economic interconnections. Now there is only one world. It has been observed that Tasmania, with the arrival of British settlers from Sydney in 1803, was the last autonomous human world to be brought under the dominant Afro-Asian (in this case European) sphere of influence.[5]

This impinges on the notion of moral progress, which means "including ever more people (or beings) in the group of those whose interests are to be respected", illustrated by Hierocles, the second-century Stoic philosopher, who described our relationships as a series of concentric circles radiating from the self, then the immediate family, the neighbourhood, the state and so on. It further links with the notion of the 'land ethic' by which we are bound to respect our biotic community: water, animals, plants and the soil, according to Aldo Leopold, one of the founders of the US environmental movement writing in 1949.

In *The Empathic Civilisation*, the US social theorist Jeremy Rifkin takes this idea further and conceives "a grand paradox". He suggests that the whole of history is a struggle between the polar forces of empathy and entropy. Just as we have extended empathy to all of humankind, the industrialised infrastructure needed to accomplish that interconnectivity is confronted by an "accelerating entropic juggernaut" that is formed by climate change and proliferating weapons of mass destruction. This self-destruction can only be averted by developing what Rivkin terms a "biosphere consciousness", a collective sense of affiliation with the entire biosphere and its systems (Rifkin 2009, p. 21).

Indeed, humanity is now involved in a global conversation about human longevity on Earth, but a satisfactory vision of sustainability has yet to be framed. Sustainability is still often regarded as just one of many issues – rather than the cornerstone that anchors all other concerns. In this vein, one pithy statement about sustainability is from the former British ambassador to the UN, Sir Crispin Tickell (2013), who speaking of sustainable development, described it as "treating the Earth as if we intended to stay".

Alternatively, the question of sustainability can be seen from the viewpoint of the Earth itself, rather than from the position of humanity. James Lovelock is associated with this idea that goes beyond the Anthropocene and dynamic Earth systems approaches.[6] *Gaia* – the Greek goddess of the Earth – envisages a planet that is itself "alive, in the same way that a gene is selfish". According to Lovelock, the Earth is "a self-regulating entity" and able to dispense with threats to life – including humans. The atmosphere is a biological construction, not living, but like a bird's feathers or a wasp's nest, an extension of a living system designed to maintain a chosen environment.

The term 'Gaia' was suggested to Lovelock by his neighbour, William Golding, the 1983 Nobel laureate for literature. The concept reputedly dismayed some biologists because the strong form of the hypothesis presupposes a singular member (the Earth) of a separate species that has evolved without natural selection. However, it became popular with the wider public and scientists who saw it as more of a metaphor. Lovelock developed the idea while working for the Jet Propulsion Laboratory in California, when pondering how to discover life on other planets. He reasoned that all life is anti-entropic. Therefore, evidence for life would be found in signs of anti-entropic processes (Lovelock 2000 [1979]).

Thus sustainability on Earth is linked to a maintenance or increase of life, a conservation or lessening of entropy. This notion has profound consequences. For example, a loss of biodiversity points to a reduction in sustainability. A loss of populations within species points to less sustainability. Any factor on which diverse life forms depend, such as forests or seas or entire ecosystems, if degraded, means that sustainability has been lost.

Industrialisation

Predominating these considerations is the notion that human impact on the planet, now clearly a matter of dire concern, is largely the result of the industrialisation that began only within the past 200-odd years. This industrialisation was and still is based on the extraction and burning of coal, oil and gas. Even the icons of modernity – mass production, grid electricity, the telephone, the car, chemicals, plastics – derive from these fossil substances. The problem is not only that the amounts of these materials are limited, but more immediately that their extraction and use is harmful to the biosphere on which all life depends.

Coal, oil and gas are all forms of stored energy from the sun, the result of ancient plants and forests that grew in the oxygen-rich atmosphere of the carboniferous period, 300–350 million years ago. Since the Industrial Revolution we have been raiding that stored sunlight and now face the consequences. Either the storehouse will be emptied or the environment will be permanently altered, or both.

Over the past thousand years, global average temperatures have varied less than one degree Celsius but show a sharp rise after the 1960s. But while the current widespread concern about 'climate change', or 'global warming' and the earlier term 'greenhouse effect' has become part of global consciousness only within the past 20 years or so, one of its more obvious effects, that of melting ice, was known to popular science much earlier. Rachel Carson's lesser known 1951 classic *The Sea Around Us* describes in some detail the opening up of the Arctic Sea due to reduced in ice cover from the early twentieth century:

> We are witnessing a startling alteration of climate…a definite change in the arctic climate set in about 1900, that it became astonishingly marked about 1930, and that it is now spreading into sub-arctic and temperate regions… In 1932, for example, the Knipowitsch sailed around Franz Joseph Land for the first time in the history of arctic voyaging.
>
> (Carson 1951, p. 103)

Carson also shows that global glacier melt, too, was evident quite some time ago. She lists a "catastrophic" decline in glaciers of the Alps as well as those of Norway, the North Atlantic coast and Alaska during the 1920s. Further:

> the glaciers of several East African high volcanoes have been diminishing since they were first studied in the 1800s – very rapidly since 1920 – and there is glacial shrinkage in the Andes and also in the high mountains of central Asia.
> (Ibid., p. 105)

Yet while the effects of global warming are evident, Carson is wary of pointing to a definitive cause. Changes in ocean currents, tidal movements, natural post-Pleistocene warming, and an increase in solar activity are all countenanced. Anthropogenic causes are not.

Today that cause is clear. It is anthropogenic and due to industrialisation. Its impacts include not only short term disasters such as stronger hurricanes, but also long term increases in salinity, loss of water table, droughts, crop failures, sea level rise and biodiversity decline – all of which imply loss of sustainability. Even the venerable author of *The Population Bomb*, Paul Ehrlich, is more than concerned about climate: "Climate change may be the most serious issue there is, another may be the toxification of the planet" (Ehrlich 2009). Former US President Barack Obama was curt about what this means for fossil fuel extraction warning that: "We're not going to be able to burn it all" (Friedman 2014).

Industrialisation affects sustainability well beyond the dimension of climate change, however. Its effluents may directly destroy life forms and its resource extraction may destroy whole ecosystems. And there are many views on its remedy. John Dryzek in *The Politics of the Earth* (1997) identifies eight different discourses that counter the long-dominant Promethean paradigm of industrialism, by which unlimited growth is both assumed and approved. These discourses he classifies as either 'radical' or 'prosaic' and either 'reformist' or 'imaginative', as outlined in Table 2.1.

Table 2.1 Alternative environmental discourses

	Radical	*Reformist*
Prosaic	Survivalism Survivalist – challenges limitless growth but not societal structure, such as *'Limits to Growth'*	Environmental problem solving by: Experts (administrative rationalism) 'the people' (democratic pragmatism) the market (economic rationalism)
Imaginative	Green radical Green romanticism Green rationalism – includes social ecology, deep ecology, ecofeminism and environmental justice movements	Sustainability Sustainable development Ecological modernisation or 'ecomodernism' -seeks to resolve environmental and economic conflict and ignore limits

Source: Based on Dryzek 1997.

The *survivalist* discourse is radical because it challenges the notion of limitless growth, yet is prosaic because its solutions are within the constraints of industrialism. The three *environmental problem-solving* discourses are both reformist and prosaic because they all accept industrialism and their solutions are adjustments to the status quo. They differ according to agency for control of environmental policies – experts, 'the people' or the market. The two *green radical* discourses – green romanticism and green rationalism – are both radical and imaginative. These discourses reject the structure of industrial society and imagine radically different societies and environmental relationships. They include strands of the social ecology, deep ecology, ecofeminist and environmental justice movements. Lastly, the two *sustainability* discourses are both reformist and imaginative. They seek to dissolve the conflict between environmental and economic values and seek to minimise the notion of limits.

When considering the politics of sustainability, what sort of discourse is acceptable depends on how the concepts are 'framed'. For example, people may relate more to identity and values rather than fact and self-interest. More importantly, the tendency for sustainability to be couched as a bio-physical systems issue rather than a social issue makes it less likely to be addressed with the policy urgency it deserves.

In this vein, Alex Steffen (2009) makes the framing rather simpler. He divides environmental sustainability into three shades of green. *Light greens* see the issue as one of personal responsibility and individual lifestyle, *dark greens* see a radical change in economics and consumption as the only answer to continuing destruction, while his preferred *bright greens* favour technological innovation and regulation as the best way to lighten human impact on the planet. A fourth category are the *greys* who deny there is a problem and promote business as usual. For practical reasons, this book concludes that a form of bright green framing is most apt.

The IPAT relationship

This sort of framing emerged from intense debates within the environmental movement. The classic Impact = Population × Affluence × Technology (I = PAT, or IPAT) relationship is directly relevant to sustainability as it conceives the dimensions of human impact on the biosphere. Where that impact exceeds the capacity of the biosphere, it is not sustainable. Yet the formulation has been interpreted quite differently depending on who has referred to it over its half-century of existence.

The IPAT equation was developed during the late 1960s during debates between the biologist and ecologist Barry Commoner (1917–2012), biologist and demographer Paul Ehrlich (b. 1932), and physicist-environmentalist John Holdren (b. 1944), who was President Obama's White House advisor on science. Ehrlich and Holdren argued that of each of these three factors, the population element is most important. Ehrlich and his wife Anne, who co-authored the 1968 best seller, *The Population Bomb*, had visited India as students and had been shocked at the apparent overpopulation and widespread poverty. It is likely that this experience led them to believe that overpopulation was the main cause of poverty and of environmental impact. Commoner, however, argued that environmental impacts (especially pollution) were caused primarily by changes in production technology following World War II, because these impacts far outweighed any population increase, especially in the United States. Ehrlich and Holdren opposed this and focussed their argument increasingly on the population explosion and its consequences. Ehrlich has continued to hold this view since:

> While the basis of climate change was known towards the end of the nineteenth century everybody who didn't need to take off their shoes to count to twenty, knew that population is part of the problem fifty years ago.
>
> (Ehrlich 2009)

Yet it was Ehrlich and Holdren who first linked population, consumption and technological impact in 1971. They wrote that:

> Pollution can be said to be the result of multiplying three factors; population size, per capita consumption, and an 'environmental impact' index that measures, in part, how wisely we apply the technology that goes with consumption.

Commoner attempted to re-interpret this more precisely in 1972 as:

> *pollution* = *(population)* × *(production per capita)* × *(pollution emission per production).*

This version is at least logically valid, since the above the line and below the line 'populations' (or 'per capita') and 'productions' cancel each other out, leaving *pollution* = *pollution emission*. This was then re-written by Ehrlich as the now familiar I = PAT equation. However, in 1974 Ehrlich and Holdren again re-interpreted this as:

> *environmental disruption* = *population* × *consumption per person* × *damage per unit of consumption.*

Ehrlich and Holdren saw the variables as interdependent, whereas Commoner viewed them as independent of each other. The former pair wanted population as the central culprit, thus population affected both affluence and technology in producing a final impact. Whereas Commoner believed the new post-war technologies to be the culprit, irrespective of population and affluence. In this respect, he advanced considerable evidence that new technologies had environmental impacts orders of magnitude beyond the other two factors.

The 'I' in the equation (or environmental impact) is at first regarded as simply *pollution*, whereas later it became *environmental disruption*, a much wider concept that, for example, may include land clearing, loss of habitat and species extinction as well as pollution alone. Further, technology (T) is equated with *"pollution"*, or *"pollution per unit of production"*, or else *"damage per unit of consumption"*, which tends to assume the damaging and heavily polluting industrial technologies of the mid-twentieth century. The possibility that some technologies might reduce environmental impact is not countenanced. Lastly, affluence is equated with either consumption or production depending on the formulation, but its impact depends on how it is measured.

While the IPAT formula was originally developed by life-scientists and a physicist[7] as distinct from other groups, issues surrounding it have also been investigated by social scientists, in parallel, but separately and "often antagonistically" (Dietz & Rosa 1994, p. 277). For example, in *The Wealth of Nature,* US historian Donald Worster (1993) outlines examples of unsustainability based on material "revolutionary forces" such as demography, technology and energy embodied in the "effortless industrialism" that supplies almost limitless goods to the affluent. These 'forces' are close to the IPAT concept but describe a more detailed modern world. Still, Worster suggests the real cause of environmental destruction in modern America is culture – driven by an attitude of human innocence in an Eden where all is fruit for the picking.

In 2000, the Australian engineer, Sharon Beder, advanced the IPAT formula as:

I: environmental impact = *P: number of people* × *A: resource use per person* ×
T: environmental impact per unit of resource used.

This is again slightly different and focuses on resource use rather than pollution emission as the key factor.

The Kaya identity is closely related to the I=PAT equation. But while the I = PAT equation is more general and describes a more abstract 'impact', the Kaya identity describes the impact of human activity specifically on carbon dioxide (CO_2) emissions. It was developed by Japanese energy economist Yoichi Kaya in the book *Environment, Energy, and Economy: Strategies for Sustainability* arising from the 1993 Tokyo *Conference on Global Environment, Energy, and Economic* Development (Kaya and Yokobori 1997). The Kaya identity differs from IPAT, having four rather than three variables:

> *global CO_2 emissions = global population × gross world product per global population × gross energy consumption per gross world product × global CO_2 emissions per gross energy consumption.*

Like the I = PAT equation, whereby (in some versions) pollution ultimately equals pollution, variables above and below the divisor lines cancel each other out, so that its mathematical validity is demonstrable. One limitation of this equation is that it does not account for the direct release of carbon dioxide by deforestation through burning, nor the loss of the carbon sink also due to forest destruction, because it assumes carbon dioxide is directly linked to energy consumption. While carbon dioxide output rose noticeably with the Industrial Revolution and its increasing consumption of fossil fuels, this was inter-related with temperate deforestation across Europe, North America and Australasia during the same period, which also saw considerable growth in world population and production.

In his article, *Energy, Economic Growth and Environmental Sustainability,* Steven Sorrell (2010) discusses the IPAT identity in relation to energy and resource efficiency. Again the terms are subtly twisted. In particular, technology (T) becomes 'performance' or 'efficiency':

> Over the long term, continued economic growth can only be reconciled with environmental sustainability if implausibly large improvements in energy and resource efficiency can be achieved. This point is easy to demonstrate with the I = P*A*T equation, which represents total environmental impact (I) as the product of population (P), affluence or income level (A) and technological performance or efficiency (T)...The decoupling strategy seeks reductions in T that will more than offset the increases in P and A, thereby lowering I.

Elizabeth Kolbert (2011) used the I = PAT equation to visualise the current state of human impact on the planet over time, relative to both 1900 and 1950. If conceptually valid, the acceleration of the impact since 1950 is staggering. However, while still considerable, the impact is arguably much less than portrayed in the illustration that accompanies her article because 'population' is double-counted in both 'affluence' and 'technology'.

A more recent article discussing the 'Anthropocene' concept by Brondizio et al. (2016) suggests that because it represents a "state change" in the "interdependent social-ecological" Earth system, the concept is different "from earlier ideas of human pressures" due to "a combination of population growth and economic and technical change, having an impact on natural systems, whether local or global". Yet, whether the Anthropocene concept does in fact differ much from what is a reformulation of IPAT, may be less important than the authors' call for more collaboration across disciplines to help reach sustainability.

The feminist critic Patricia Hynes believes that the IPAT identity is so entrenched that it is "like a mental boxing ring", that both its advocates and its critics debate within it, rather than from a critical outside position. Its combination of simplicity and universality makes the relationship particularly hard to escape from. In Hynes' view, the relationship lacks any dimension of environmental or social justice, especially for the lowest quintile of humanity whose only goal is daily survival. She goes on to suggest that environmental justice means that women's health should be an end in itself, rather than a means to reach population goals. The general education of women and girls, as well as the education of men and boys in peace studies, non-violence to women and environmental management are some of its public policy implications.

The development economist Jeffrey Sachs (2008) does begin to look at justice issues, while at the same time factoring in some of the elements of the IPAT relationship. He outlines four main causes of the 'unsustainability crisis' – human pressure on ecosystems and climate, rapidly rising population especially in areas least able to cope, the poverty of one sixth of humanity unrelieved by global economic growth, and paralysis of global problem-solving due to cynical, defeatist attitudes and inadequate institutions. Sachs was influential in the creation of the Millennium Development Goals (MDGs). However, it is intriguing that he does not mention technology in relation to sustainability. It is as if he is concerned with the 'I' (impact), the 'P' (population) and the 'A' (poverty in this case), but not at all with the potential of the 'T' (technology) as either a positive or negative factor.

Yet the original concept lingers at high levels. The ghost of the IPAT relationship lurks within the Intergovernmental Panel on Climate Change (IPCC), which warned that if extra efforts to cut greenhouse gas emissions were not put in place, "global emission growth is expected to persist driven by growth in global population and economic activities (high confidence)" (IPCC 2014d).

The IPAT relationship remains central to issues of sustainability. As an overarching concept it can constrain debates but it also disciplines them and it has inherent flexibility. It tells us, for example, that to reduce human impact we need to moderate population, reconsider the nature of affluence and develop technologies that enhance rather than degrade life systems. It is at least a well-trodden start line, a place to begin. For these reasons the next chapters of this book follow its formulation.

Images

Formulae can be uninspiring, however. The potency of a concept frequently depends on imagery. It is often asserted that the first photographs of the entire planet by the Apollo moon missions in the late 1960s led to a change of consciousness, a new awareness of the limited and fragile nature of the biosphere that we occupy. Certainly the new 'blue marble' image of the entire planet was used to support environmental causes and a new environmental awareness became widespread at around the same time.[8] However, this was predicted nearly two decades before. In 1950 the Cambridge astronomer Fred Hoyle predicted that:

> Once a photograph of the Earth, taken from outside, is available, we shall, in an emotional sense, acquire an additional dimension… a new idea as powerful as any in history will be let loose.

Nonetheless, the 2013–2014 President of the American Association for the Advancement of Science, Phillip Sharp, believes that the power of the blue marble is less than Hoyle had hoped. In his annual presidential address of 2014 he observed that:

> our awareness of the global nature of major problems facing our planet is relatively new and demands global responses for which neither the scientific community nor the general public is well-prepared.

The issue now is more how to act on that awareness, rather than contesting the degree of impact our species is having on the biosphere. In that vein, Sharp is more positive. A microbiologist, he thinks that "ecosystem engineering" will reduce damage and restore ecosystem functions through analysis of microorganisms using DNA sequencing. This is a prominent example of techno-optimism, the view that technology can be used to save rather than destroy our environment. More broadly, he suggests that we need to converge across disciplines (life, physical, social and engineering sciences) as well as between research and its implementation through entrepreneurship.

This echoes the economist Kenneth Boulding, who in anticipation of CP Snow in 1956 drew attention to the increasing development of 'knowledge silos', whereby physicists only talk to physicists and economists to economists. He feared that this mutually unintelligible "assemblage of walled-in hermits" would slow the growth of knowledge. However, he is encouraged that there were then an increasing number of 'hybrid disciplines' such as social psychology and astrophysics. Interestingly, he also draws attention to the possible re-birth of political economy, which had died prematurely in the mid nineteenth century.

The polymath inventor Buckminster Fuller was of a similar frame of mind when writing his 1969 *Operating Manual for Spaceship Earth*:

> our failures are a consequence of many factors, but possibly one of the most important is the fact that society operates on the theory that specialisation

is the key to success, not realising that specialisation precludes comprehensive thinking.

While Sharp and his predecessors appear to presume that capitalism is a friend of the environment and sustainability through its emphasis on technological innovation, Boulding and Snow have reservations as the overspecialisation it engenders means sustainability will be made difficult to achieve. By contrast, Marshall McLuhan and Robert Logan asserted in 1976 that Snow was naïve about increasing specialisation because, "in the electric age there can be no more monopolies of knowledge".

Kenneth Boulding (1966) was one of the first to identify the Earth as a "spaceship", meaning a self-contained economy with physical limits as opposed to an endless frontier that could be exploited indefinitely without ecological consequence. In this way, it is relevant to consider the defining characteristics of artificial spaceships that are designed to transport people through the cosmos. Even the early orbital and lunar spaceships had to provide resources and environments that supported human life, including air, water, food and means of waste disposal. Longer voyages such as the Mars mission, possibly around the year 2030, will require intricate systems for recycling of fluids and waste.

In science fiction, starships are commonplace, whether of alien or human construction. The most sophisticated iteration of the *Star Trek* concept – the exploration of an intelligence-populated universe by a technologically advanced and idealistic humanity – was achieved in the 1990s *Star Trek Voyager* television series (Berman, Piller & Taylor 1995–2001). The starship crew are seeking to return to Earth from a distant quadrant of the galaxy, through a space littered with minefields diplomatic, military and epistemological. There are three relevant observations here: (i) the starship possesses technologies that not only enable it to travel at multiples of light speed and to create instantly any (relatively small) material item,[9] but allow it to cruise indefinitely – or sustainably – throughout the cosmos, (ii) its on-board artificial intelligence (AI) is not only immortal,[10] but develops as a character and is in many ways more human than those he supports, and (iii) this advanced ship is a product of a government federation (albeit of a military nature), rather than a means of exploration provided through capitalist enterprise.

It is intriguing that such a successful series, made by the most triumphant of entrepreneurial societies, views our exploration of deep space in this way, the same way that our relationship with space began in both the former Soviet Union (USSR) and the United States. The motivation, resources, discipline and conscious focus required to conceive, construct and maintain such a complex set of technologies may be beyond capitalism, as it is now for extra-terrestrial missions[11] – just as it may be beyond capitalism to construct and maintain a renewed and sustainable spaceship Earth.

Anthropocene

The blue marble 'spaceship' is a static image. Yet the Earth comprises dynamic processes and systems. The notion of an 'anthropocene' geological epoch is attributed to the Nobel atmosphere chemist Paul Crutzen and the biologist Eugene Stoermer, who reflected on the inadequacy of the post-glacial term 'Holocene' in signifying the increasing human impact on the planet. Both thought of the term 'anthropocene' independently and published a joint paper with that title in 2000. In that brief essay they acknowledge the work of Lyall and others in establishing the concept of geological epochs, but saw a need for a better name for the current era, which they propose began at the time of the Industrial Revolution. Writing in 2000, they also point out that the epoch may last for a very long time:

> It seems to us more than appropriate to emphasise the central role of mankind in geology and ecology by proposing to use the term "anthropocene" for the current geological epoch. The impacts of current human activities will continue over long periods...because of the anthropogenic emissions of CO_2, climate may depart significantly from natural behaviour over the next 50,000 years.

The newsletter in which the article was first published also contains several articles on 'Earth systems'[12] as it was around this time that ecological concerns were developing into larger and more dynamic concepts. 'Earth systems', which encompass the dynamics of the entire planet, as well as its energy and matter relationships with astronomical phenomena, was pioneered by the early computer modelling of Donella Meadows and colleagues in the early 1970s.

The philosopher Clive Hamilton (2014) recently became quite heated on the difference between the static, restricted notion of 'ecology' and the dynamic, vast notion of 'Earth systems' approaches. His blog is provocatively titled *ecologists-butt-out-you-are-not-entitled-to-redefine-the-anthropocene*:

> changes in landscapes, forest clearing, extinction of megafauna, "ecosystem engineering" and so on...[are] entirely irrelevant to the Anthropocene, unless it can be demonstrated that they changed the functioning of the whole Earth system in a detectable way...Ecological thinking focuses on ecosystems delimited by their spatial boundaries...[This] has been transcended by Earth system science with a deeper conception of the Earth as a total entity, stretching from the core of the planet to the moon and in an unceasing state of flux driven by natural cycles great and small, a flux in which humans in the Anthropocene have recently become the dominant process.

Crutzen and Stoermer's original landmark paper does not emphasise these conceptual differences. Nevertheless, Hamilton's point is that 'Earth systems' appear to vastly extend the notion of 'ecology'. As such, the approach offers hope for a better understanding of sustainability, if not for its political saleability.

Brundtland

More immediate than systems dynamics and conceptions of the Earth, human desires for equity over both time and place are politically serious. The *UN World Commission on the Environment and Development* was set up in 1983 to address wide concerns about the impact of development on the environment. The resulting landmark 1987 Brundtland report, *Our Common Future,* is about 'sustainable development' rather than the more general 'sustainability', but the two topics overlap. Named for the woman former Labour prime minister of Norway who chaired the enquiry, this report famously defined 'sustainable development' as development that "meets the needs of the present without compromising the ability of future generations to meet their own needs". This definition is often popularly confused with 'sustainability', whereas it is equally concerned with the politics of development, in particular Third World development and the alleviation of poverty. The issue of how to facilitate economic growth to reduce poverty without environmental damage remains important, but was regarded as a major conundrum during the era.

The report links technology, economic growth, equity and political will to the environmental limits on human impact:

> The concept of sustainable development does imply limits – not absolute limits but limitations imposed by the present state of technology and social organisation on environmental resources and by the ability of the biosphere to absorb the effects of human activities. But technology and social organisation can be both managed and improved to make way for a new era of economic growth…Painful choices have to be made…in the final analysis, sustainable development must rest on political will.

In his book, *Principles of Sustainability*, Simon Dresner (2008) observes that Brundtland's famous definition of sustainable development is both simple and vague, qualities that are both its strength and weakness. Adams' paper for the IUCN however, is more affirmative:

> *it cleverly captured two fundamental issues, the problem of the environmental degradation that so commonly accompanies economic growth, and yet the need for such growth to alleviate poverty.*

Jeremy Seabrook (2002), however, believes the term 'sustainable development' is an oxymoron:[13]

> Yet it was believed that the solution to the great clash between ecology and economy had been discovered in the 1980s: this was the idea of "sustainable development", triumphantly enshrined in the Rio declaration. Intra-generational equity would be balanced with inter-generational justice to ensure that we do not take more from the Earth than we give back to it. The excitement generated by this formula concealed the possibility that it might be a contradiction in terms: when unlimited desire is unleashed in a world of limited resources, something has to yield. The "fruits" of industrialism turn out to be strange hybrids – perhaps, ultimately, inedible.

Sharachchandra Lélé (1991) says the term needs to be more precise. John Dryzek (1997) believes that the term has meaning, but that Brundtland more asserted rather than argued that the environment and social justice did not have to confront normal material growth, and since then the term 'sustainable development' has become increasingly resigned to conventional concepts of economic growth.

Ultimately, Brundtland is a political document that reflects the undercurrents of the time. It manages to appear both balanced and progressive and it remains an icon in sustainability research, although three decades later it still attracts differing interpretations. According to one respondent to a recent debate on the meaning of sustainability in Australia, the essential aim of the Brundtland report was to stabilise global population (Lewis 2015). Yet the report's own statement of its goals is much wider. Brundtland looked at eight key issues, of which population was a part, but a relatively small one. Energy, industry, food security, human settlement and international economic relations were also areas included for analysis.

Adams (2006), however, points out that there is a measurement issue, because "there is no agreed way of defining the extent to which sustainability is being achieved in any policy programme". Brundtland's compatriots, Asheim and Kjell (1993, p. 3), writing several years after Brundtland's report, look at intergenerational sustainability using mathematical formulations that attempt to operationalise the concept:

> What kind of rules must our generation follow in order to manage the resource base in such a way that it constitutes a first part of a sustainable development? The problem of finding such rules can only — if at all — be resolved through an analysis of the long-term global production possibilities.

They also introduce the notion of risk: "The crucial question is whether the risk of decreasing future quality of life is acceptable" (ibid.).

As far as the sustainability discourse is concerned, the Brundtland report indicates an awareness of contemporary connotations of key words, especially 'environment' and 'development'.

> The environment does not exist as a sphere separate from human actions, ambitions, and needs, and attempts to defend it in isolation from human concerns have given the very word "environment" a connotation of naivety in some political circles. The word "development" has also been narrowed by some into a very limited focus, along the lines of "what poor nations should do to become richer", and thus again is automatically dismissed by many in the international arena as being a concern of specialists.
>
> (Brundtland 1987, foreword)

In the time since the report, there have been changes in terminologies and different emphases on different words, possibly attempts to escape the sort of connotations Brundtland identifies. For example, an emphasis on 'the greenhouse effect' became 'global warming' then more recently, 'climate change', which underlines its negative effects. The 'environment', something separate from humanity, has tended to be replaced by the 'biosphere', which emphasises its global relationship to life, or 'ecosystem', which underlines its characteristic as an inter-related system.

From a critical neo-Marxist perspective, Arturo Escobar argues that the discourse of 'sustainable development' only reshuffles the elements of neo-liberal development theories: "basic needs, population, resources, technology, institutional cooperation, food security, and industrialism". And overarching it is that modernity can only be achieved through the development expert – "the wise white man from the West".[14] Whether sustainable or not, it is argued that the very concept of development reinforces racial and cultural stereotyping. Similarly, Hilary Hove (2004, p. 49) suggests that sustainable development fails to transcend its ethnocentric flaws. She argues that it does not account for how the West "contributes to the inferiority and subordination of poorer parts of the world". Escobar also points out that the bond with sustainability essentially just means that the 'wise white man from the West' is as likely to be an environmental scientist as an economist:

> The Western scientist continues to speak for the Earth. God forbid that the Peruvian peasant, an African nomad, or a rubber tapper of the Amazon would have something to say in this regard.
>
> (Escobar 1995, p. 194)

Critics from the political left emphasise that the sustainable development concept does not address the issue of unsustainable consumption of the West, which must be a central concern. Rather, it both assumes and reinforces a globalised homogeneity, in which social diversity is 'disciplined' according to the dictates of the interests of capital. Jude Fernando (2003, p. 6), for example, asserts that achieving the goals of sustainable development – addressing both socio-economic inequality and environmental degradation – means that it "must be liberated from the ideology and institutional parameters of capitalism". But while much of such criticism is compelling, it leads towards unlikely social and political outcomes. As there appears to be little coherence in leftist political movements towards a sustainable development that is separate from global capitalism, it is doubtful that there will be significant change in that direction, despite some localised efforts in South America. This study therefore attempts a more pragmatic approach in which it is assumed that a form of capitalism will continue, but also that it may be channelled more responsibly.

Earth summits

After Brundtland, the UN 1992 Rio *Earth Summit* encouraged ways of merging development with environmental protection, producing the 300-page *Agenda 21*, and the *Rio Declaration on Environment and Development,* as well as opening conventions for signature on biodiversity, climate change and desertification. Three follow-up Earth Summits were held in 1997 (*Rio+5*), 2002 (*+10*) and 2012 (*+20*) to revisit the commitments made under Agenda 21.

Lack of progress on implementing the principles of the Earth summits is often ascribed to North-South attitudes to economic development and environmental impact. Agenda 21, however, launches into an enthusiasm for trade liberalisation, a policy area still scarred from battles between the environment and the (neoliberal) economy:

> An open, equitable, secure, non-discriminatory and predictable multilateral trading system that is consistent with the goals of sustainable development and leads to the optimal distribution of global production in accordance with comparative advantage is of benefit to all trading partners.
>
> (UNCED 1992, s.2.5)

This may help explain slow progress. Contrary to such fervour, the World Trade Organisation (WTO) and regional trade agreements that have evolved since the late 1980s have tended to reduce rather than extend environmental protections in the name of economic efficiency.

Trade

One example of these battles is the WTO tuna-dolphin cases of 1990 and 2008. The outcome of these disputes was that US import restrictions on tuna caught with fishing methods that incidentally killed dolphin were not upheld (Oxley 2001; Miles 2012). In effect, the WTO asserted that its purpose was purely the pursuit of trade liberalisation. Environmental matters were separate, for other forums. Since the stalling of progress with trade liberalisation at the Doha round, however, emphasis on multi-lateral instruments has slowed in favour of increasing numbers of bi-lateral and limited multi-lateral free trade agreements.

But trade can be antipathetic to sustainability more generally, and further, trade volumes are affected by technology. The classic Smith-Ricardo theory of comparative advantage shows that it is mutually beneficial for countries to trade goods that they can produce relatively efficiently compared with other goods in the same country, with countries that have different relativities. But transport costs can alter this picture. Paul Krugman (2010) used the same example as Ricardo to show that if transport and related costs exceed the production advantage, there is no economic advantage in trade. He speculates that changes in trade costs relative to production costs affect trade volumes. If trade costs fall, as they did in the late nineteenth century due to steam-powered shipping and the inter-ocean canals, trade increases. Conversely, if trade (especially transport) costs rise or if production costs fall faster than transport costs (as with the electrification of factories between the wars), then trade decreases. Containerisation, larger ships, jumbo aircraft and progressive tariff removal in the late twentieth century meant that trade increased because transport was effectively cheaper (ibid., paras 6–8).

The significance of this for sustainability is that more trade implies more emissions associated with the transport of goods. Therefore, reduced trade is beneficial to sustainability, at least while transport is based on emissions-producing technologies. Krugman's insight is especially significant in the present era of complex international supply chains that draw heavily on long-distance transport to assemble and bring the final product to market. The impact of trade is substantial. If it were a country, international shipping would be the ninth largest emitter of carbon dioxide – more than Canada, the United Kingdom, Brazil or Mexico. International aviation would be 16th – more than Italy, Australia or France (EDGAR 2014).

Components

Trade can be detrimental to sustainability due to its emissions, but sustainability is often conceived of as other than just environmental impact. The IUCN's Adams (2006, p. 2) has one of three diagrams describing the concept of sustainability as the centre of a triple intersection of economic, social and environmental concerns. A related image is at Figure 2.1, which suggests that sustainability is at the centre of the socially and environmentally bearable, the socially and economically equitable and the environmentally and economically viable.

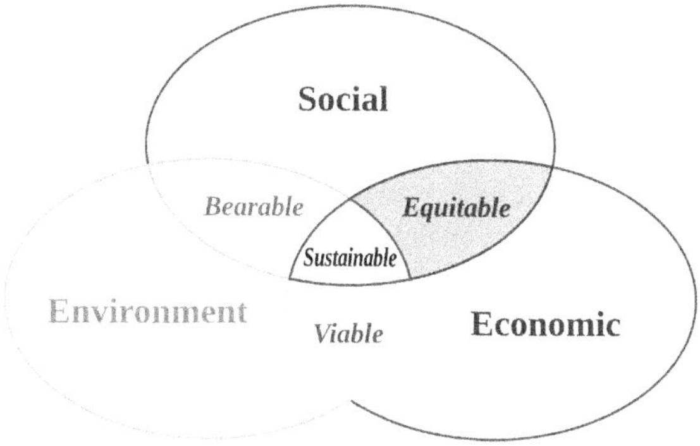

Figure 2.1 Sustainability: bearable, equitable and viable.
Source: Wikipedia. Sustainable Development. https://en.wikipedia.org/wiki/File:Sustainable Development.svg.

These three areas are routinely used in business to describe what is termed the 'triple bottom line', whereby an organisation aims to be economically and socially, as well as environmentally sustainable. While arguably the environment is the most important since the other two both depend on it, these differing perspectives are based on different values making an interdisciplinary approach to sustainability more challenging (Ostrom 2009; Gale et al. 2015).

Environmental sustainability can itself be split into three parts, all of which relate to natural capital.[15] In 1990 the pioneer ecological economist, Herman Daly (1990) proposed that (i) for renewable resources, the rate of harvest should not exceed the rate of regeneration (sustainable yield); (ii) for non-renewable resources, depletion should not exceed the creation of renewable substitutes (sustainable depletion); and (iii) for pollution, the rate of waste generation should not exceed the capacity of the environment to assimilate it (sustainable assimilation). Thus, environmental sustainability is the rate of renewable yield, non-renewable depletion and waste assimilation that can be continued indefinitely.

While the issue of renewable substitutes for non-renewable resources is potentially the most challenging – including substitutes for rare earths used in electrical storage – there is the issue of quality as against quantity in all three of Daly's dimensions. For example, a sustainable harvest might still occur in volume, but the quality of the harvest may decline. This is arguably true of meat production in some countries where animals destined for slaughter are fed chemicals to speed growth, but quality suffers as a result. Where fewer crops replace more diverse plantings, production may be sustainable, but biodiversity suffers. Also in many cases substitutes are less suitable than the original resource – plastics instead of metal for example in many applications. Further, the determination of when waste assimilation is complete is problematic.

Governance

Furthering such aims implies goal-setting and regulation. There has been at best mixed success with global regulation for sustainability. While the 1992 Rio Earth Summit proclaimed lofty principles, many of them still await implementation and enforcement. More recently there has been the 2004 *Rotterdam Convention* on informed consent for trade in toxic substances, and the *Stockholm Convention* of 2004 limiting persistent organic pollutants. The UN Secretary-General's High level panel on global sustainability (UNHPGS 2012) made several recommendations on governance, including ensuring a stronger interface between policy-making and science (R.44) and the creation of the SDGs at R.48. But many aspects such as the protection of biodiversity, measures to prevent desertification and to control climate change have yet made "regrettably few" advances (Pope Francis 2015, s.168–169). What is needed, he says, are ways to effect systems of governance for the whole range of the "global commons" (ibid., s.174).

John Dryzek (2014, p. 12) remarks that what counts in governance are similar expectations rather than rules and regulation:

> Institutions are in large measure discursive constructions: they work because of a convergence of expectations and understandings, not just formal rules. So, for example, market liberal globalisation is so powerful in large measure because it permeates the understandings of actors in the political economy. Policy deviations from its orthodoxy are punished not just by impersonal market forces, but because people in key positions in financial and economic institutions believe those deviations will have negative economic consequences.

Iconic imagery is also important in the force of ideas. The idea of an 'ozone hole' in the negotiations leading up to the *Montreal Protocol* contributed to its adoption and apparent success (ibid.). However, the metrics of sustainability regulation often present greater difficulty. Adams (2006, p. 4) says that at the program level sustainability is less measured than 'genuflected', less achievable than aspirational.

More positively, however, the Sustainable Development Goals (SDGs) for 2016–2030 combine indicators of progress with some of the major elements of sustainability. For example, target 6.3, as well as minimising water pollution, involves "halving the proportion of untreated wastewater and substantially increasing recycling and safe reuse globally" by 2030. The proportion of untreated wastewater will therefore be measured and progress monitored. This promises to be more than an empty gesture and should also inform program measurement at the national level.

Paradox and pessimism

Simon Dresner's influential treatise on sustainability, re-written at the time of the GFC and the subsequent global recession, makes the case for "an attempt to make our civilisation more sustainable than it presently is" despite the impossibility of drawing up "a detailed blueprint of a sustainable society or even a route to get to it". His historical analysis emphasises economic growth and the displacement of traditional religion by triumphant Western consumerism. He compares the failure of communism, due to both the impossibility of central planning of a complex modern economy (Hayek) and transformative communication technologies that destroy tradition and stability (Giddens) with socialism, which although "discredited" entails valuable concerns with equality and fairness. These concerns he says, now need to be associated with concerns for "nature, risk, growth and technology" (Dresner 2008, p. 179).

Dresner's take on neoliberal economics with its emphasis on growth is that it is a reflection of Smithian rationalist influence over the past 200 years, which has predominated the alternative 'respect for nature' Enlightenment tradition of Rousseau. He points out that the neoliberal "seductive message" that markets alone will make government planning unnecessary (ibid., p. 175) and the problem of "accumulation without end" assumed by Fukuyama (1992) and criticised by Giddens (1994), is clearly absurd (ibid., p. 160; Higgs 2015).

While he uses the terms 'liberal democratic capitalism' and 'liberal democracy' interchangeably, Dresner (pp. 144–153) believes that the market is central. "We can try to be sustainable or we can pursue the free market, but we can't do both." This is rather different in Gore (2013), who prefers a system of "sustainable capitalism", where the market is still paramount, but it must factor in environmental impact.

The notion of *paradox* is sprinkled throughout Dresner's book. One paradox of sustainability is that its pursuit depends on a global community linked through the Internet and other technologies, yet these technologies at the same time tend to hamper needed prediction due to the rapid social change they engender. Post-modernists believe in a diversity of moral values and are therefore unable to argue against fundamentalist values that would deny moral diversity, including those that might be needed to save the planet. And the paradox of utilitarianism is that people are expected to accept less for the sake of those who are already more fortunate. Elsewhere in this book is found the Jevons paradox whereby more fuel efficient machines result in more fuel used. There is also the paradox of biodiversity – that the more species we render extinct, the more 'new' species we discover.

Dresner also points out the irony that sustainability arose on the political agenda during the 1980s, just as the ideologies that might support it were being abandoned for neoliberalism. Similarly, Garnaut (2014b) points out that it took half a century for neoliberal economics to shrug off the taint of the Great Depression and achieve the sort of global dominance it enjoyed until the GFC in 2008. Wesley Widmaier (2015) argues that since the 1987 crash all neoliberal economics has given us is a series of "bubbles and bailouts" based on demand created by asset price increases rather than the wage increases of the Keynesian era. In this economic period where growth has been listless[16] and volatility rules, sustainability struggles for the attention it demands.

Dresner identifies a mood of pessimism about the future, which contrasts with the attitude of the Victorians. The very slogan 'sustainable development' is but an attempt to sound optimistic about a doubtful future, he says (2008, p. 19). This fits with Arthur Clarke's earlier observation that "this is the first age that has paid much attention to the future, which is ironic since we may not have one". And if the results of a recent four-nation survey are indicative beyond the Anglosphere, pessimism is the mood of our era. More than half the 2000 people surveyed rate the risk of our civilisation ending within 100 years as at least 50 percent; almost a quarter believe there is the same risk of actual human extinction within the same timeframe (Randle and Eckersley 2015).

Principles

Dresner collaborated with Ekins and Dahlström (2008 pp. 68–69), who outlined four principles that affect what method is best for evaluating sustainable development policies, both at point of contemplation and after implementation:

> The environmental *sustainability* principle concerns the need to maintain critical environmental functions for future generations. These functions include climate stability, the ozone layer and biodiversity, ensuring environmental quality and avoiding pollution. Clearly we are not doing well against this one.

The *precautionary* principle is about avoiding irreversibility. The loss of particular environmental functions may cost dearly to remedy, or may not be recoverable at all. Therefore because of risk uncertainty, actions that could result in very large costs should not be embarked on, even if there is little chance of their occurrence and high chance of benefits. This principle tends to dampen the zeal of geo-engineering enthusiasts. Exactly where the principle is invoked, however, is a political matter to be decided – or ignored – in each case.

The *appropriateness* principle relates to the ethics of trade-offs. 'Willingness to pay' needs to be consistent with 'ability to pay' – what poor people are willing to pay for environmental benefits is unlikely to be what they think something is actually worth, because they cannot afford to pay much at all. Therefore 'willingness to pay' approaches to environmental resource allocation are often inappropriate because the priorities they indicate will automatically favour the better off.

The principle of *justice* also applies. While the point above demonstrates marginal valuations may be in conflict with justice, the example of social costs and benefits in the IPCC's Second Assessment Report was of a different order to mere moral inconsistency. The authors say the report was "met with outrage" because lives in rich countries were valued 15 times greater than lives in poor countries. Further, "no distinction was made between those responsible for anthropogenic climate change and those likely to suffer its consequences".

However, while these principles apply to sustainable development policies, those policies are applied in a world driven by considerations of the 'dismal science': economics.

Ecological economics

The words 'economy' and 'ecology' are from the same root. While in English the former derives from sixteenth century usage, the latter was a German invention of the nineteenth century (Kurtz 2006). While economy was about management of the household's financial resources, ecology was about management of the household's physical resources (Ellerbrock 1998). The origin of the two terms is now apposite as we have reached the limiting household size – the extent of the planet Earth.

Eriksson and Andersson (2010) in outlining the elements of ecological economics, articulate the 'trilemma' of global social justice, prosperity and ecological sustainability. While the difficulty of resolving any two is high, resolving all three is much harder. They point to several recent reports (including the World Bank and the ILO) that attempt to address two of the three, with little or no mention of the third. Ecological economics arose from the need to address all three of these meta-issues that we now face on this planet.

Building a bridge between ecological and the still dominant neoclassical[17] economics, however, is difficult. Ecological and neoclassical economists do not necessarily share the same concepts, if one large study is typical. In an international workshop on economic sustainability held in Berlin in 2003, participants became involved in a heated discussion about whether economic growth is desirable for sustainable development. Ecological economists were concerned about the physical aspects of economic growth – material resources, emissions and waste. Neoclassical economists rather had a monetary concept of growth measurable by GDP and which would not necessarily increase environmental harm. While both groups ultimately agreed that decoupling economic activity from harmful environmental effects is a desirable goal, opinions differed on how to achieve it (Illge and Schwarze 2006, p. 3).

Nevertheless, there was agreement that sustainability means preserving development opportunities for future generations. It was also agreed that economic growth is not the only way to solve the conflict between efficiency and equitable distribution, that sustainability is an important field of economic research and that sustainability research must rise above the boundaries between disciplines.

Ecological economists, however, were alone in believing that an integrated understanding of economic, ecological and social welfare is essential to sustainability. Only they believe that human-made capital rarely substitutes for natural capital, that the economy depends on nature, and that the value of an intact environment cannot be measured in money. They alone believe that changing societal values is important to achieving sustainability rather than through a self-interested view of human nature. Sustainability economics must deal with the question of how to make decisions in an intergenerational context.

Neoclassical environmental economists, on the other hand, do not believe that inter-generational justice presupposes intra-generational justice. They are alone in believing that sustainability does not require restrictions on material consumption, that it may be achievable within the present economic system by setting the 'right' prices. They uniquely believe that international specialisation leads to more long-lasting wealth worldwide, and that economics and science should be objective and 'value-free'. Ecological economists tend to have a post-modern view of science as necessarily value-laden.

The former World Bank chief economist and Nobel laureate, Joseph Stiglitz (2014a, p. 6), steers a course between the two camps in terms of practical policy. While at present there are considerable subsidies involved in the production and consumption of polluting substances such as fossil fuels in many countries, Stiglitz says that it is "better to tax bad things than good things, because that way you get a double benefit" – revenue from the tax as well as less harmful activity. Hence the most critical thing for sustainability is to put a price on pollutants such as carbon – a carbon tax. Removal of subsidies on carbon production would also improve government revenue while at the same time reduce environmental harm and increase efficiency.

Others (notably Costanza et al. 2014), have grasped the nettle of putting a monetary value on nature. While they emphasise that doing so is not privatising or commodifying ecosystem services, and that the value estimates are more *use (or non-use)* rather than *exchange* values, it is obvious from their paper that there has been marked criticism of the concept. Original estimates in an earlier article in *Nature* (Costanza et al. 1997) showed that the value of ecosystem services exceeded global GDP at the time. By contrast, Jesse Ausubel (2014) argues that "we must make nature worthless" so that it can be owned by those wishing to conserve it and rendered unnecessary in the pursuit of an economy based on common, rather than rare, substances such as silicon. Humanity would thrive in dense communities that do not rely on chlorophyll, leaving larger areas to return to nature, as has happened already in the US North East, for example.

Decoupling

'Decoupling' refers to economic growth without environmental harm. The concept is similar to the idea of 'dematerialisation', by which economic growth is increasingly based on non-material outputs such as services, entertainment and software rather than on physical goods. There is a distinction between absolute and relative decoupling. In relative decoupling the growth rate of the environmental parameter is less than the economic parameter, but is still positive. Whereas absolute decoupling involves a zero or negative growth rate of the environmental parameter as the economy grows (UNEP 2010, p. 18).

At least in respect of carbon pollution of the atmosphere, decoupling has not yet been achieved:

> Globally, economic and population growth continue to be the most important drivers of increases in CO_2 emissions from fossil fuel combustion. ... Between 2000 and 2010, both drivers outpaced emission reductions from improvements in energy intensity. ... Increased use of coal relative to other energy sources has reversed the long-standing trend of gradual decarbonisation of the world's energy supply.
>
> (IPCC 2014d, p. 9)

Dietz and O'Neill (2013, p. 117) propose that our "obsession with GDP" can promote economic growth that is harmful. They say that our ecological overshoot is at least partly due to our narrow group of economic indicators, especially GDP. Ward et al. (2016) agree:

> We have shown that there is little evidence that GDP growth can be decoupled in the long-term (i.e. it is not sustainable) ...it is ultimately necessary for nations and the world to transition to a steady or declining GDP scenario.

Capitalism

Dennis Meadows of the Massachusetts Institute of Technology (MIT), one of the original authors of *The Limits to Growth* (1972), now believes that sustainability cannot be achieved within the present financial system:

> A fish will never create fire while immersed in water.[18] We will never create sustainability while immersed in the present financial system. There is no tax, or interest rate, or disclosure requirement that can overcome the many ways the current money system blocks sustainability. … I now understand, as proven clearly in this text, that the prevailing financial system is incompatible with sustainability in five ways:
>
> - it causes boom and bust cycles in the economy
> - it produces short-term thinking
> - it requires unending growth
> - it concentrates wealth
> - it destroys social capital.
>
> (Meadows in Lietaer et al. 2012, p. 6)

Seabrook (2002) is in accord. For him, 'sustainability' has been distorted to mean the economic rather than the environmental kind:

> 'Sustainable' now means what the market, not the Earth, can bear; what originally meant adjusting the industrial technosphere so that it should not destroy the planet has now come to indicate the regenerative power of the economy, no matter how it may degrade the "environment". Sustainable is what the rich and powerful can get away with.

Much earlier, Karl Polanyi (2001 [1944], p. 136) had written of the natural world that the "market economy, if left to evolve according to its own laws would create great and permanent evils". These would result because "food supplies … the climate … the denudation of forests…erosions, and dustbowls" do not respond to the supply-and-demand mechanism of the market (ibid., p. 193).

And yet, reflecting on Meadows' five incompatibilities, none of them is necessarily irreconcilable with a sustainable world. Certainly they are all discordant social issues, but they do not directly conflict with sustainability. Boom and bust economic cycles are undesirable socially and politically, but they do not necessarily have adverse environmental effects. Short-term thought is less revered than long-term thinking, but grand masterplans are no guarantee of success, as Napoleon and Hitler exemplify. Whereas the art of 'muddling through' has serious underpinning in public policy (Lindblom 1959 and 1979; Forester 1984). A series of short-term incremental policy steps are often the only practical way to proceed.

Unending growth, or as Higgs (2015) puts it "endless growth on a finite planet", seems the most cogent of these incompatibilities. Yet, ultimately the growth involved is – as our Berlin neoclassical economists assert – ephemeral. It is economic. It is money. It is a concept of exchange value. It is a number on a screen, not necessarily the consumption of materials or energy. A Rembrandt has an exchange value millions of times its physical content. A software program may be extremely valuable but is immaterial. And increasingly economies depend on growth in services rather than goods. At least in the developed world, most people have more than enough 'stuff' and economic growth is decoupling from physical possessions. It seems possible that not only can growth be de-materialised, but the innovation that it enables can lead to sustainability rather than to degradation. According to Nicholas Stern in *The Economics of Climate Change* (2006), for example:

> growth is part of the solution to climate change. Most growth is the result of innovation – the development of new products, new techniques and new ways of doing things that are an improvement on what went before. The next wave looks like it will be dominated by digital technology, robotics, biotech, lighter materials and renewable energy…To say that we have to stop growing – that we have to go backwards – I think is factually wrong, and also politically unlikely to be successful…We absolutely can have growth and protection of the climate at the same time, and in doing so we will construct a much better form of economic activity and growth in terms of clean air, less-congested cities and so on.

Concentration of wealth is not only inequitable and socially undesirable; it can also have indirect adverse environmental effects – through both excessive consumption and the insulation of the wealthy from the consequences of unsustainable practices. Yet this is not necessarily disastrous. Governments can still make laws that assuage inequities while keeping reasonable incentives. Information is increasingly available to anyone. New platforms can democratise the media and political activism is a counterweight to the power of wealth.

Lastly, the destruction of 'social capital', or the promotion of selfish behaviours, is documented. Certainly it would reduce societal cohesion. And it would weigh against environmental well-being, which would be ignored in pursuit of individual gain. But to say that the financial system destroys social capital is absurd. The central study on the issue ascribed the decline of US social capital to television, not the financial system (Putnam 1995). Also, while the financial system has been around for hundreds of years, social capital has apparently yet to be destroyed. Finance is not intrinsically evil – although money may well be at its root.

As to Seabrook's view that the economically powerful distort 'the environment' to mean 'the economy', there are means to counter such forces – the glare of social media on environmental issues for one. Polanyi's warning of unrestrained markets implies that markets must be restrained for the good of both natural and human sustainability.

Still others have more aggressive views, but along similar lines to Meadows. Joel Kovel, ecosocialist academic and former US Green party presidential candidate for example, believes that "capital is the efficient cause of the [ecological] crisis". He is concerned about "the general acceptance of capitalism as having a kind of divine right to organise society, and the coordinated refusal to face up to its essential ecodestructivity" (Kovel 2007, p. 164). This is an observation that resonates, especially since Fukuyama's 'end of history' neoliberal capitalist triumphalism at the end of the Cold War, although traditional socialist economic models have been less than exemplary in this regard as well.

Two visions

The US biologist John Vucetich observes ruefully that in the modern world we can have human prosperity in the absence of once-abundant life. "There are so few black-footed ferrets on the planet that we've already proved we don't need them," he says. "You can list hundreds of species we can get along fine without". Vucetich sees the fight over green modernism as a clash between two visions of sustainability: one that exploits nature as much as we like without infringing on future exploitation, and one that exploits nature as little as necessary to lead a meaningful life. "It's hard to imagine those two world views would lead to the same place," he says. "I think they would lead to wildly different worlds" (quoted in Keim 2014, paras 48–49).

However, human prosperity does not necessarily mean the elimination of other forms of life. The return of some areas to the wild ('rewilding') is happening in parts of the United States and Europe, for example (Monbiot 2014b). And when such habitats are physically linked and their scale increased, nature becomes more resilient, more sustainable.

Further, Dame Ellen MacArthur (2015), who has real-life experience of sustainability at sea, believes that it is not economic growth or capitalism that is the problem, rather it is the linear nature of the economy that extracts, uses and degrades in the pursuit of material wealth. She believes that a circular economy like a life system would enable sustainability – one in which people buy transport services, not cars and buy lighting services, not lights. This is a world where we 'use things' rather than 'using them up' and where organic wastes are recycled as fertiliser. Some of this may have been informed by Hawken, Lovins and Lovins (2015, p. 8) who argue that the abundant natural capital and scarce labour of the industrial revolution is giving way to the converse – scarce natural capital and abundant labour. Nature can no longer be regarded as free to exploit. Hence economic and production systems must be redesigned to be vastly more efficient and to produce no waste at all.

Nightfall

Isaac Asimov's *Nightfall* (1941) imagines a world that orbits several suns, producing continuous daylight throughout the planet, except for a period of nightfall that occurs but once every thousand years. The people of the planet do not expect this new phenomenon and are driven insane by darkness and the vista of the heavens that opens infinitely above them. The civilisation they build collapses, as fires are widespread in desperate attempts to produce light. Ultimately, a new civilisation emerges from the ashes of the old, until the darkness again returns after a further millennium. The plot tension centres on archaeological discoveries of several previous civilisations separated by layers of ash. Can this evidence be interpreted correctly, or will the next imminent darkness overwhelm the discovery of the truth and the means to deal with it? Arguably this scenario is not far from the current predicament of humanity on Earth. We already have strong evidence that our civilisation is unsustainable and we do know largely how to deal with it. The issue is rather how to deal with the political and economic consequences of such actions.

Roadmap

The underlying theme of Al Gore's book *The Future* (2013) is finding a way out of the fossil-fuel climate crisis, which has still wider application for sustainability. Gore outlines six 'drivers' that offer hope for a sustainable future. In Dryzek's discourse framework (Figure 2.1), this is a sustainability that is both imaginative and reformist.

Gore's sounding the alarm about global warming,[19] as well as his efforts to secure the 1997 Kyoto protocols when US vice president, entitle him to some regard in matters of sustainability. Not everyone endorses his views, however. From the left, Joel Kovel (2007, p. 166) believes that "history will be kind to Gore", but says it is wrong to set the logic of change within the dominant (capitalist) system. Nevertheless, Gore's drivers form a plausible roadmap:

The first driver is the integrated global web of corporations, finance and markets that now dominates liberal democracies. The second is the emerging "global mind" – humanity linked through the Internet and other digital communications that enables alternative political movements, new business forms that dispense with the middleman, and linked sensors, databases and intelligent machines. Gore's third factor is the shift in political and economic power from the global North to the global South. His fourth driver is "outgrowth", whereby the global population increases more in the South, especially Africa. While this puts added burdens on resources, it should also lead to the benefits of urbanisation and population restraint through the education of girls, the empowerment of women, the availability of contraception and the reduction of death rates.

His fifth driver is genetic and biological research, especially neuroscience. Combined with progress in cloud computing and artificial intelligence (AI), such technologies have yet to impact on public understanding of their potential. Due to interconnected sensors and other data gathering facilities, the cloud now contains vast amounts of data that are available for 'mining' – to create new models and ways of approaching intricate problems. Gore's final driver is the technological breakthroughs in renewable energy, materials and resource re-use. The main obstructions are the corruption of democracies through money from established industries and 'science deniers' who may fight the disruptive technologies of the future.

Consistent with this view of cloud computing and AI, the originators of the term 'anthropocene' believe that sustainability must be based on information technologies:

> To develop a world-wide accepted strategy leading to sustainability of ecosystems against human induced stresses will be one of the great future tasks of mankind, requiring intensive research efforts and wise application of the knowledge thus acquired in the noösphere, better known as the knowledge or information society.
>
> (Crutzen and Stoermer 2000, p. 18)

However, Adams (2006, p. 14) speaks of different factors:

> The elements needed for the future are easily stated, although very challenging to work through. They include imagination, vision, passion and emotion. The issue of emotion is probably central to success. Existing approaches to sustainability have depended heavily on natural science (from which the concept came), and economics. 'Dismal science' in all forms remains essential to charting a course to the future, but it is not enough to drive changes needed. The world is not run by technocrats (even economists), but politicians and the citizens they represent or govern. In the past sustainability has engaged the mind, but the future demands an engagement with the hearts as well.

Others, such as the historian Lynn White writing in 1967, see the origins of our sustainability crisis in the destructive combination of Western science, Christianity and technology, whereby not only do we see ourselves as rightly dominant over nature but have the means to assert it. His solution is based on the values of St Francis, who saw all creatures as equal, since echoed in Pope Francis' *Laudato si': on care for our common home* (2015). There is little doubt that sustainability must require a change of values. Yet the half century since White has not been encouraging. For example, former Google engineer, Anthony Levandowski's announced mission to develop a "Godhead based on artificial intelligence" to worship and help develop society (Sacasas 2017) is bizarre in its combined religion and technology; values are diminished in an idolatry of power over nature. Humanity will best focus where there is the most leverage. And in this Gore's drivers offer more practical hope in the shorter term – which may be all that we have.

Conclusion

The factors of sustainability have both origins and consequences. The IPAT relationship has changed since its conception, but still provides a framework on which to consider the causal elements of sustainability. Daly's 1990s considerations of the three natural capital aspects of sustainability (yield, depletion and assimilation) balance that structure on the effects side.

Much of the debate about sustainability is between biologists, who more easily see the environmental effects of our species, and those economists who have faith in the market as the ultimate arbiter and who therefore dispute the intervention required. A third group are the visionaries who view sustainability from wider perspectives, whether science fiction writers, astronomers or inventors. But while science, economics and imagination might inform the concept of sustainability, its achievement requires focus where there is significant leverage. Population policies, even if politically acceptable, take a long time to effect and still longer to show results. Although many favour its redistribution, few politicians are in favour of less affluence. But while our embrace with industrial technology has produced this crisis, our partner changes with bewildering speed, perhaps enough to baffle the conundrum of sustainability itself.

In the end, however, sustainability is a matter of politics that involves perceptions about equality – which is probably why the landmark *Brundtland* report on sustainable development was headed by an accomplished politician. In line with Dryzek (1997), there are several discourses that challenge the dominance of industrialism in different ways. The task of politics is still to establish a narrative that finds wide support for the actions needed to bring sustainability about – before the long night falls.

Notes

1 Both Anne and Paul Ehrlich were the original authors, although the publisher listed only Paul.

2 Although the notion of 'sustainable yield' emerged with scientific forestry in Germany during the eighteenth century (Worster 1993, p. 145).

3 According to its lead author, Mark Malloch Brown of the UNDP, goal seven was a last-minute inclusion due to a chance meeting in the corridor he had with "the beaming head of the UNEP. Brown says "a terrible swearword crossed my mind when I realised we'd forgotten an environmental goal" (Tran 2012).
4 Three such universal orders have been proposed by the Israeli historian, Yuval Harari: monetary, imperial and religious.
5 While New Zealand was settled by Europeans a little later – in 1814 – the Maori ended its autonomy at the end of the Polynesian expansion several hundred years earlier.
6 Although Lovelock acknowledges that the idea of a living Earth system originated with the 'father of geology', James Hutton, in a 1785 lecture (cited in Young 1991, p. 122).
7 John Holdren originally studied aeronautics and physics, and was Professor of Environmental Policy at the Kennedy School of Government at Harvard.
8 For example, the counter-cultural *First Whole Earth Catalogue* was published in 1968. Its front cover showed an image of the Earth taken from space, which reputedly its editor had campaigned for NASA to release.
9 Such as a slice of New York cheesecake.
10 His immortality results from his photonic, rather than carbonic, essence.
11 The private Space-X project is acknowledged, but it has yet to put a person into space, which was achieved by the USSR in 1961. Also Space-X depends on NASA contracts that support its operations.
12 Such as *Earth System Models of Intermediate Complexity* (Clausen et al. 2000) and *Full-Form Earth System Models: Coupled Carbon-Climate Interaction Experiment (the "Flying Leap")*, Fung et al. 2000).
13 Seabrook's observation is consistent with Michael Redclift's (2002 [1987]) argument that the term sustainable development is "founded on contradiction". Others, notably Herman Daly, have been attributed with similar observations (Redclift 2006, p. 66).
14 See Ronny Röwert 2011, About Arturo Escobar: Encountering Development, p. 2.
15 The term 'natural capital' remains controversial due to its perceived diminution of nature as simply a resource for humanity.
16 For example, "longer-term potential growth rates remain subdued across the globe compared with past decades, especially in advanced economies" (IMF 2017b, p. xii).
17 'Neoclassical' economics assumes rational actors within markets that balance supply and demand. Much of its approach is used in the political ideology of neoliberalism.
18 This is an odd metaphor: fish do not create fire, whether immersed in water or not.
19 As in the film *An Inconvenient Truth* (2006).

References

Adams, W 2006, *The Future of Sustainability: Re-Thinking Environment and Development in the Twenty-First Century*, International Union for the Conservation of Nature (IUCN), Gland, Switzerland.

Asheim, G and Kjell, A 1993, Sustainability When Resource Management Has Stochastic Consequences, Discussion Paper no. 86, March, Central Bureau of Statistics, Oslo, Norway.

Asimov, I 1971 [1941], 'Nightfall' in *Nightfall One*, Granada, Frogmore, UK.

Ausubel, J 2014, 'We Must Make Nature Worthless', Conference paper, Resources for the Future, November, published in *Real Clear Science*, 19 September 2015.
Berman, R, Piller, M and Taylor, J, 1995–2001, *Star Trek Voyager*, Paramount Network Television series, Los Angeles, US.
Boulding, K 1956, 'General Systems Theory – the Skeleton of Science', *Management Science*, vol. 2, pp. 197–208.
Boulding, K 1966, 'The Economics of the Coming Spaceship Earth' in Jarrett, H (ed.), *Environmental Quality in a Growing Economy*, Johns Hopkins University Press, Baltimore, US, pp. 3–14.
Brondizio, E, O'Brien, K, Bai, X, Biermann, F, Steffen, W, Berkhout, F, Cudennec, C, Lemosh, M, Wolf, A, Palma-Oliveira, J, and Chen, C-T 2016, 'Re-Conceptualizing the Anthropocene: A Call for Collaboration', *Global Environmental Change*, vol. 39, July, pp. 318–327.
Brundtland, G 1987, *Report of the World Commission on Environment and Development: Our Common Future*, United Nations, Geneva, Switzerland.
Buckminster Fuller, R 1969, *Operating Manual for Spaceship Earth*, Design Science Lab.
Carson, R 1952 [1950], *The Sea Around Us*, Staples Press, London, UK.
Carson, R 2002 [1962], *Silent Spring*, Houghton Mifflin Harcourt, New York, US.
Choi, Y 2015, 'Intermediary Propositions for Green Growth with Sustainable Governance', *Sustainability*, vol. 7, no. 11, pp. 14785–14801. doi:10.3390/su71114785.
Commoner, B 1972, *The Closing Circle,* Knopf, New York, US.
Costanza, R, d'Arge, R, de Groot, R, Farber, S, Grasso, M, Hannon, B, Limburg, K, Naeem, S, O'Neill, R, Paruelo, J, Raskin, R, Sutton, P and van den Belt, M 1997, 'The Value of the World's Ecosystem Services and Natural Capital', *Nature*, vol. 387, 15 May, pp. 253–260.
Costanza, R, de Groot, R, Sutton P, van der Ploeg, S, Anderson, S, Kubiszewski, I, Farber S, and Turner, R 2014, 'Changes in the Global Value of Ecosystem Services', *Global Environmental Change*, vol. 26, pp. 152–158.
Crutzen, P and Stoermer, E 2000, 'The "Anthropocene"', *Global Change Newsletter*, no. 41, pp. 17–18.
Daly, H 1990, 'Toward Some Operational Principles of Sustainable Development', *Ecological Economics*, vol. 2, pp. 1–6.
Dietz, R and O'Neill, D 2013, *Enough Is Enough: Building a Sustainable Economy in a World of Finite Resources*, Earthscan (Routledge), Oxford, UK.
Dietz, T and Rosa, E 1994, 'Rethinking the Environmental Impacts of Population, Affluence and Technology', *Human Ecology Review*, vol. 1, no. 2, pp. 277–300.
Dresner, S 2008, *The Principles of Sustainability*, second edition, Earthscan, London, UK.
Dryzek, J 1997, *The Politics of the Earth: Environmental Discourses*, Oxford University Press, New York, US.
Dryzek, J 2014, 'Institutions for the Anthropocene: Governance in a Changing Earth System', *British Journal of Political Science*, vol. 46, no. 4, pp.1–20.
EDGAR: EU Emissions Database for Global Atmospheric Research 2014, European Environment Agency, Copenhagen, Denmark.
Ehrlich, P 1968, *The Population Bomb*, Ballantine Books, New York, US.
Ehrlich, P 2009, *Late Night Live*, Australian Broadcasting Corporation: ABC Radio National, 19 November.
Ehrlich, P and Holdren, J 1971, 'Impact of Population Growth', *Science*, vol. 171, pp. 1212–1217.

Ekins, P, Dresner, S and Dahlström, K 2008, 'The Four-Capital Method of Sustainable Development Evaluation', *European Environment*, vol. 18, pp. 63–80.

Ellerbrock, M 1998, *Roots of Economics, Ecology and Ecumenism: Foundations of the Land-Grant Household*, Department of Agricultural and Applied Economics, Virginia Tech, Virginia, US.

Eriksson R and Andersson, J 2010, *Elements of Ecological Economics*, Routledge, London, UK.

Escobar, A 1995, *Encountering Development: The Making and Unmaking of the Third World*, Princeton University Press, Princeton, US.

Fernando, J 2003, 'The Power of Unsustainable Development: What Is To Be Done?', *The Annals of the American Academy of Political and Social Science*, vol. 590, no. 1, November, pp. 6–34.

Forester, J 1984, 'Bounded Rationality and the Politics of Muddling Through', *Public Administration Review*, vol. 44, no. 1, pp. 23–31.

Friedman, T 2014, 'Obama on Obama on Climate', *New York Times*, 7 June.

Fukuyama, F 1992, *The End of History and the Last Man*, Hamish Hamilton, London, UK.

Gale, F, Davison, A, Wood, G, Williams, S and Towle, N 2015, 'Four Impediments to Embedding Education for Sustainability in Higher Education', *Australian Journal of Environmental Education*, vol. 31, no. 2, December, pp. 248–263.

Garnaut, R 2014b, 'Australia's Economic and Social Future after the Resources Boom', public lecture, University of Tasmania, Hobart, Australia, 5 July.

Giddens, A 1994, *Beyond Left and Right*, Polity Press, Cambridge, UK.

Gore, A 2013, *The Future*, Random House, London, UK.

Hamilton, C 2014, *Ecologists Butt Out: You Are Not Entitled to Redefine the Anthropocene*, blog. http://clivehamilton.com/ecologists-butt-out-you-are-not-entitled-to-redefine-the-anthropocene/#sthash.pM8NkgcS.dpuf.

Harari, Y 2014, Sapiens: A Brief History of Humankind, Vintage, London, UK.

Hawken, P, Lovins, A and Lovins, L 2015 [1999], *Natural Capitalism*, Rocky Mountain Institute, Snowmass, US.

Higgs, K 2015, *Collision Course: The Impossibility of Endless Growth on a Finite Planet*, MIT Press, Cambridge, US.

Hove, H 2004, 'Critiquing Sustainable Development: A Meaningful Way of Mediating the Development Impasse?', *Undercurrent*, vol. I, no. 1, pp. 48–54.

Hoyle, F 1950, *The Nature of the Universe*, 1957 edition, Bail, Blackwell and Mott, Oxford, UK.

Hynes, H Patricia 1993, *Taking Population Out of the Equation: Reformulating I=PAT*, Institute on Women and Technology, Cambridge, US.

Illge, L and Schwarze, R 2006, *A Matter of Opinion: How Ecological and Neoclassical Environmental Economists Think about Sustainability and Economics*, DIW-Diskussionspapiere, No. 619, German Institute for Economic Research (DIW), Berlin, Germany.

IPCC: Intergovernmental Panel on Climate Change 2014d: *Synthesis Report*. Contribution of Working Groups I, II and III to the Fifth Assessment Report of the Intergovernmental Panel on Climate Change [Core Writing Team, R.K. Pachauri and L.A. Meyer (eds.)], IPCC, Geneva, Switzerland.

IUCN: International Union for Conservation of Nature and Natural Resources 1980, *World Conservation Strategy: Living Resource Conservation for Sustainable Development*, United Nations Environment Programme (UNEP), World Wildlife Fund (WWF),

Food and Agriculture Organization of the United Nations (FAO) and the United Nations Educational, Scientific and Cultural Organization (UNESCO).

Kaya, Y and Yokobori, K (eds) 1997, *Environment, Energy, and Economy: Strategies for Sustainability*, United Nations University Press, Tokyo, Japan.

Keim, B 2014, 'Earth Is Not a Garden', *Aeon*, 18 September.

Kolbert, E 2011, 'Enter the Anthropocene – Age of Man', National Geographic, March.

Kovel, J 2007, *The Enemy of Nature: The End of Capitalism or the End of the World?* Zed Books, London, UK.

Krugman, P 1995, 'Growing World Trade: Causes and Consequences', *Brookings Papers on Economic Activity*, vol. 1, pp. 327–377.

Krugman, P 2010, 'A Globalisation Puzzle', *New York Times* op ed, 21 February.

Kurtz, M 2006, *Economy and Ecology*, Open University, Milton Keynes, UK, 30 June.

Lélé, S 1991, 'Sustainable Development"A Critical Review', *World Development*, vol. 19, no. 6, pp. 607–621.

Leopold, A 1977 [1949], 'The Land Ethic' from *A Sand County Almanac*, Oxford University Press, UK in *Morality and Social Policy*, pp. 865–875.

Lewis, D 2015, Online response to interview with Haydn Washington, *Late Night Live*, ABC RN: Australian Broadcasting Corporation, Radio National, 17 March.

Lietaer, B, Arnsperger, C, Goerner, S and Brunnhuber, S 2012, *Money and Sustainability: The Missing Link*, a report from the Club of Rome, EU chapter to Finance Watch and the World Business Academy, Triarchy Press, Brussels, Belgium.

Lindblom, C 1959, 'The Science of 'Muddling Through", *Public Administration Review*, vol. 19, pp. 79–88.

Lindblom, C 1979, 'Still Muddling, Not Yet Through', *Public Administration Review*, vol. 39, no. 6, November–December, pp. 517–526.

Lovelock, J 2000 [1979], *Gaia: A New Look at Life on Earth*, Oxford University Press, Oxford, UK.

Lovelock, J 2014, *James Lovelock Talks to David Freeman – A Rough Ride to the Future,* video.

MacArthur, E 2015, *The Surprising Thing I Learned Sailing Solo Around the World*, TED talk. www.ted.com/talks/dame_ellen_macarthur_the_surprising_thing_i_learned_sailing_solo_around_the_world.

McLuhan, M and Logan, R 2015 [1976], 'The Future of the Library, Circa 1976', *Island*, no. 140, pp. 14–28.

Meadows, D H, Meadows D L, Randers, J and Behrens, W 1972, *The Limits to Growth: A Report for the Club of Rome's Project on the Predicament of Mankind*, Universe Books, New York, US.

Miles, T 2012, 'WTO Rules against U.S. "Dolphin Safe" Tuna', *Reuters*, 16 May.

Monbiot, G 2014b, 'How Wolves Change Rivers', *Sustainable Human*, YouTube.

Ostrom, E 2009, 'A General Framework for Analyzing Sustainability of Social-Ecological Systems', *Science,* 24 July, vol. 325, no. 5939, pp. 19–422.

Oxley, A 2001, *WTO and the Environment*, Asia-Pacific Economic Cooperation, Singapore.

Polanyi, K 2001 [1944], The Great Transformation, Beacon Press, Boston, US.

Pope Francis 2015, *Laudato si'*, Libreria Editrice Vaticana, 24 May.

Putnam, R 1995, 'Bowling Alone: America's Declining Social Capital', *Journal of Democracy*, vol. 6, pp. 65–78.

Randle, M and Eckersley, R 2015, 'Many Fear the Worst for Humanity, So How Do We Avoid Surrendering to an Apocalyptic Fate?', The Conversation, 12 October. https://theconversation.com.

Rifkin, J 2009, *The Empathetic Civilisation: The Race to Global Consciousness in a World in Crisis*, Polity Press, Cambridge, UK.

Sacasas, M 2017, *Idols of Silicon and Data*, blog, 1 October. http://thefrailestthing.com/2017/10/01/idols-of-silicon-and-data/.

Sachs, J 2008, *Commonwealth: Economics for a Crowded Planet*, Penguin, London, UK.

Seabrook, J 2002, 'Sustainable Development Is a Hoax: We Cannot Have It All', The Guardian, 5 March.

Sharp, P 2014, 'Meeting Global Challenges: Discovery and Innovation through Convergence', *Science*, vol. 346, no. 6216, 19 December, pp. 1468–1471.

Snow, C 1959, *The Two Cultures: The Rede Lecture*, Cambridge University Press, Cambridge, UK.

Sorrell, S 2010, 'Energy, Economic Growth and Environmental Sustainability: Five Propositions', *Sustainability*, vol. 2, pp. 1784–1809.

Steffen, A 2009, *Bright Green, Light Green, Dark Green, Gray: The New Environmental Spectrum*, World Changing website, 27 February. https://web.archive.org/web/20160112194947/http://www.worldchanging.com/archives/009499.html.

Stern, N 2006, *The Economics of Climate Change: The Stern Review*, Cambridge University Press, Cambridge, UK.

Stiglitz, J 2014a, Reforming Taxation to Promote Growth and Equity, White Paper, Roosevelt Institute, 28 May.

Tickell, C 2013, 'The Human Future', Public lecture, Julie Anne Wrigley Institute of Sustainability, Arizona State University.

UNCED: United Nations Conference on Environment & Development 1992, *Agenda 21*, Rio de Janeiro, Brazil, 3–14 June.

UNEP: United Nations Environment Program 2010, ABC of SCP: Clarifying Concepts on Sustainable Consumption and Production, 100 Watt, Paris, France.

UNHPGS: United Nations Secretary-General's High-level Panel on Global Sustainability 2012, *Resilient People, Resilient Planet: A Future Worth Choosing*, United Nations, New York, US.

Ward, J, Sutton, P, Werner, A, Costanza, R, Mohr, S, and Simmons, C 2016, 'Is Decoupling GDP Growth from Environmental Impact Possible?', *PLoS One*, vol. 11, no. 10, e0164733. doi: 10.1371/journal.pone.0164733.

White, L 1974, 'The Historical Roots of Our Ecologic Crisis', in *Ecology and Religion in History*, Harper and Row, New York, US.

Widmaier, W 2015, 'It's the End of the World as We Know It ... Again', *The Conversation*, 26 August.

Worster, D 1993, *The Wealth of Nature*, Oxford University Press, New York, US.

Young, J 1991, *Sustaining the Earth*, New South Wales University Press, Marrickville, Australia.

3 Impact

> Between the idea
> And the reality
> Between the motion
> And the act
> Falls the Shadow
> –T. S. Eliot, *The Hollow Men*

This chapter is about how to establish the current and future impact of our species, *Homo sapiens*, on this planet, especially on our own sustainability. In line with the 'Earth systems' concept, ultimately we must approach the question from a planetary perspective, rather than any narrower viewpoint.

There is no other world to which we can escape should human impact on the planet be terminal. Yet the wider perspective of astrobiology is instructive. Adam Frank and Woodruff Sullivan (2014, p. 37) writing in *Anthropocene* have used such an approach to shed light on the chances that humanity may become sustainable in terms of our impact on the planet. In astrobiology, sustainability is a subset of habitability: a planet may be habitable, but it is only sustainable if a human-like civilisation can develop over a long time-scale.

Drake equation

Such is the scale of its field, astronomy is not known for precision. Members of other disciplines have said that astronomers are usually content with an accuracy "within a couple of orders of magnitude" or hundreds of times the estimate. With that reservation in mind, a key to Frank and Sullivan's analysis is the classic Drake equation[1] that was originally developed to predict the chances of other civilisations with radio technology existing within the known Universe. The Drake equation takes the rate of star formation, the fraction of stars with planets, of planets that are habitable, of habitable planets likely to contain life, of those with intelligent life and of those with radio technology to arrive at the number of civilisations that may exist (ibid., p. 33).

Like the IPAT relationship, the Drake equation is not strictly mathematical. For example, contrary to the equation, it is the *net* rate of star formation, rather than the gross rate that determines the number of stars currently in the Galaxy. Stars form, but stars also die. And stars with planets have been forming in the galaxy long before the birth of our own sun (Campante et al. 2015). A modification to the equation results in an estimate for the mean lifetime of *species with energy-intensive technology* (SWEIT), which frustratingly the article fails to develop. Nonetheless, it does provide an estimate of the commonality of our plight:

> Thus even with the odds of evolving a SWEIT on a given habitable planet being one in one million billion, at least 1000 species will still have passed through the transition humanity faces today within our local region of the cosmos. (Frank and Sullivan 2014, pp. 34–35)

Dynamic modelling

The 'transition' Frank and Sullivan refer to is the possibility of imminent collapse of our civilisation, or worse, the demise of the human species due to our own impact on our habitat. Therefore, the analysis compares systems dynamics across two and three dimensions. The two-dimensional model plots the SWEIT population (N) against energy consumption per capita (Ec) over time. It shows "a stable dynamical system [that] experiences oscillations with decreasing amplitude until a steady state is achieved" (Figure 3.1).

In this model, energy per capita increases to high levels then declines, then oscillates until stability is reached at a moderately high level. Population at first increases to high levels, then declines as energy per capita still increases. After reaching a low level, population then varies until stability is reached at a moderate level.

The three-dimensional model, however, includes an estimate of feedback on planetary systems or 'forcing' (F) that result from energy use, such as atmospheric carbon dioxide increase. It finds current circumstances — high population of the energy-intensive species (N), high energy consumption per capita (Ec) and resultant high feedback on the planetary system (F) – to be unstable, with rapid volatility and deterioration after only a short time. However, *low* population and high energy consumption per capita is ultimately stable (ibid., p. 36) – that is, 'sustainable' (Figure 3.2):

This implies our present energy dilemma may not be unique to our species. Rather, it could be "an issue for petrol-heads across the universe", because having ready access to a highly concentrated stock of fossil fuel creates a dependency that is hard to break. Perhaps we have not been contacted by other advanced civilisations because of this common 'great filter': alien civilisations

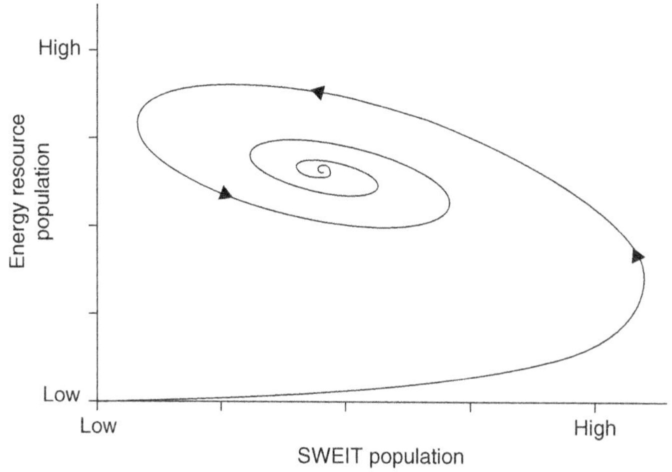

Figure 3.1 Species with energy-intensive technology: population and energy per capita over time.

Source: Frank and Sullivan 2014.

may have wrecked their planet or run out of fossil fuels before they could make the transition to renewables (Le Page 2014, p. 39).[2]

UK astrobiologist Lewis Dartnell (2015) argues that while difficult, it may be possible to construct an advanced civilisation without fossil fuels by combining hydroelectricity and renewable forest practices. This includes charcoal production for metal smelting, electric cars and public transport. Such an industrial revolution would require "very favourable natural environments" such as Canada or Scandinavia whose rivers enable hydroelectric power and where there are large forests that can be harvested sustainably for thermal energy. This, however, also implies a much lower human population and reduced consumption.

This view is not far from Paul Ehrlich's 1960s contention that population is the key factor in the welfare of our species and the planet. Nor is it far distant from *The Limits to Growth* (Meadows et al. 1972), which pioneered computer modelling to estimate the effects of continuing exponential growth.

Lately, computer modelling of impact was featured by complex systems geophysicist Brad Werner, who famously delivered a talk at a large meeting of the American Geophysical Union (AGU) in 2012 with the provocative title *Is the Earth F***ed? – Dynamical Futility of Global Environmental Management and Possibilities for Sustainability via Direct Action Activism.* Although a transcript of the talk may not exist, according to reports (Kintisch 2012), it drew on the author's

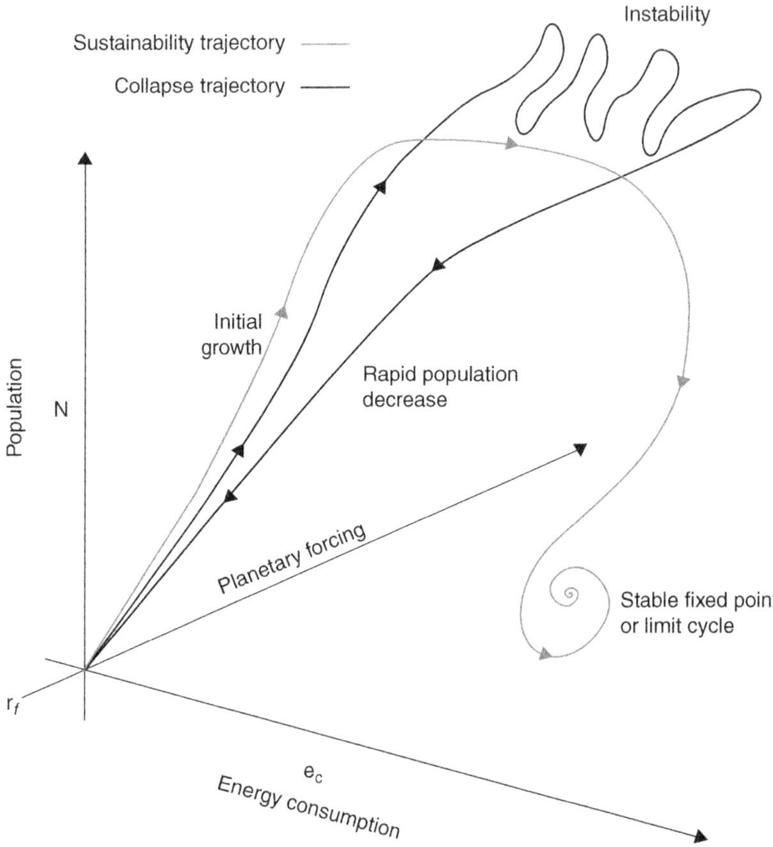

Figure 3.2 Species with energy-intensive technology: population, energy per capita and planetary forcing.

Source: Frank and Sullivan 2014. Copyright © 2014 Elsevier Ltd. All rights reserved. Reprinted with permission.

computer models of humanity's relationship with environmental systems. According to the abstract – and in contradiction to Hinchcliffe's assertion[3] – its conclusion is that:

> the dynamics of the global coupled human-environmental system within the dominant culture precludes management for stable, sustainable pathways and promotes instability.
>
> (Werner 2012)

Yet as the talk's subtitle implies, Werner proposes that sustainability may be possible through "direct action activism", rather than global environmental management.

However, public perceptions of impending catastrophe tend to focus on the impacts of climate change (droughts, floods, hurricanes, rising sea levels), as well as resource depletion, ecosystem degradation, human pandemics, economic collapse, nuclear and biological conflict and "runaway technological change" (Randle and Eckersley 2015, p. 3). Arguably all of these potential catastrophes contain technology as their central factor.

Confusion

Even a landmark paper such as *The Anthropocene: From Global Change to Planetary Stewardship* (Steffen et al. 2011) can illustrate that measuring impact is not entirely straightforward. To show the acceleration of environmental impact due to change in population, affluence and technology since the 1950s, the paper reprints a diagram from an Elizabeth Kolbert article in the National Geographic of the same year. The diagram purports to show the product of the three factors at 1900, 1950 and 2011 (Figure 3.3), in a vivid illustration of how they form a volume of impact. Accepting the units used in the paper (population, GDP in US dollars and patents), the impact of 1950 is about 10.8 times that of 1900, but the impact of 2011 is an estimated 134 times that of 1950 – or 1442 times[4] that of 1900!

Yet arguably there are two shortcomings in this representation. First, unless affluence (A) is per person, rather than total, population (P) is effectively multiplied by itself, which considerably inflates the result. Second, technology (T) is represented by *patents*, which also tend to increase with population. Patents better represent *innovations*, which now often reduce rather than exacerbate environmental impact (Chertow 2000, p. 18). Further, the scope of what is patentable has drastically expanded over the past three decades (Mercurio 2014, p. 4).

But even accepting that 'patents' is a valid cipher for T (technology), when total affluence is reduced to affluence *per person* and patents are reduced to patents *per person*, the resultant impact in 2011 is 95 times times that of 1900. This is still a large figure, but considerably less than that implied by the illustration. The 'per person' approach is more consistent with the Holdren-Ehrlich formulation that proposed "environmental degradation = population × consumption per person × damage per unit of consumption" (Holdren and Ehrlich 1974, p. 282). Professor Steffen disagrees with this assessment,[5] saying that much of the increase in GDP in that time was not due to population increase because it largely occurred in OECD countries with low population growth rates, and

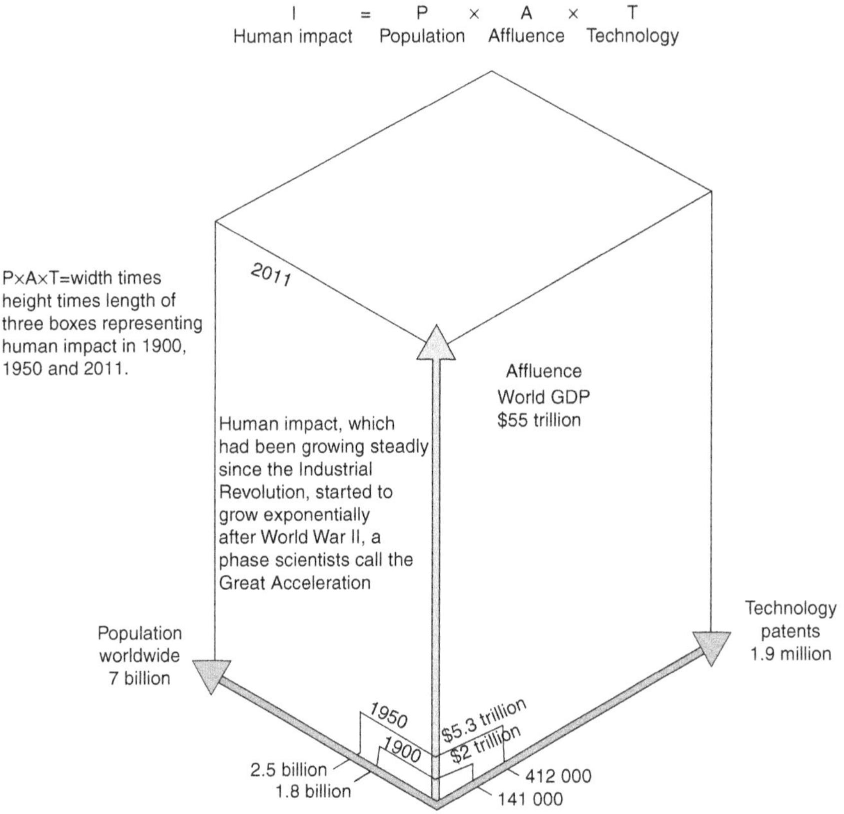

Figure 3.3 Global impact of population, affluence and technology 1900, 1950 and 2011. Source: Kolbert 2011. © Copyright Bryan Christie Design, reproduced with permission.

that the diagram was only for illustrative purposes, not meant to be mathematically rigorous. Nevertheless, while acknowledging these points, the illustration may be interpreted in different ways.

Relative impact

In terms of the impact of humanity compared with other species, the US marine biologists Charles Fowler and Larry Hobbs have shown just how much of an outlier species we humans are. Their Royal Society paper, *Is humanity sustainable?* (2003), measures the impact of humans compared with various other species in biomass consumption, CO_2 production, geographical range and population size. They compare humanity with seabirds, marine mammals, fish and terrestrial mammals. The differences in impact are vast. For example:

> [the impact of] the human population is over two orders of magnitude [hundreds of times] greater than the upper 95 percent confidence limit of populations for non-human mammalian species of similar body size, and over four orders of magnitude [36,760 times] greater than the mean.
>
> (Fowler and Hobbs 2003, p. 2580)

Their conclusion, that the human species is abnormal and that atypical elements of human ecology "are among the primary factors" contributing to the environmental problems the world now faces, is hardly surprising. What is surprising is the enormous scale of that abnormality. The features of that abnormality are equally important, discussed later in the *extinctions* section.

Extinctions

Apart from relative consumption and waste production, our impact on biodiversity is profound. Carson's 1962 *Silent Spring* evoked a world devoid of animal life due to synthetic pesticides. The spectre of what has been termed 'biosimplification' still haunts our planet. In 2014 Peter Fisher wrote of a future eerie silence:

> we could be alone except for pets, chickens, livestock, and an unknown suite of microbes and freeloaders such as mice and cockroaches. For a sneak preview of this "biosimplification", look no further than the swathes of European countryside where there has been a crash in bird populations – no songs, no glimpses of plumage, just an eerie silence – as a result of the wholesale ripping up of hedgerows, draining of wetlands and ploughing over of meadows robbing farmland birds of their homes and sustenance in order to boost farming production. That would leave us living in a drab, crummy landscape where surviving native plants cower in small niches away from the weeds; zoos exhibit a lost fauna; and biophilia is reduced to watching carp.
>
> (Fisher 2014)

In the seas, climate change is a major driver of biodiversity loss as oceans warm and marine species follow thermal niches (Molinos et al. 2015, p. 1). Overall, threats to biodiversity and whole species on both land and sea are due to over-hunting, over-fishing and and over-farming, or in blunt technological terms "guns, nets and bulldozers" (Maxwell et al. 2016, p. 143).

All this is in contrast to the Enlightment's Thomas Jefferson, who was of the firm view that nature never let any species become extinct (Gerbi 2010, p. 257). Then, North America was sway to 40 million bison and as well as five billion prairie dogs, both now reduced to pathetic remnants[6] of that earlier "infinitude" (Worster 1993, p. 4). At least, what was once 40 million white-tailed deer, reduced to about half a million by 1900, have since rebounded to be about 15 million in the same habitat (Smith 1991, p. 5).

The concept of 'great extinctions' was popularised by Elizabeth Kolbert in *The Sixth Extinction: An Unnatural History* (2014). The fossil record shows that there have been five brief eras in which many forms of life on the planet became extinct; the extinction of the dinosaurs and the rise of mammals about 60 million years ago were the most recent – until now. This sixth great extinction is driven by humanity and industrial technology, and is extremely rapid, about "a thousand times the rate of background extinctions" (Pimm et al. 2014).

Whilst extinctions occur due to chance natural causes, only humans drive other life forms to extinction with an organised intense and ruthless brutality. The passenger pigeon, once the most common bird in North America that flew in flocks of millions, finally was made extinct over a century ago. According to the Encyclopedia Smithsonian (2014), this species once constituted 25 to 40 percent of the total bird population of the United States. There were estimated to be up to five billion of them when Europeans reached America. That extinction was by direct human agency:

> The professionals and amateurs together outflocked their quarry with brute force. They shot the pigeons and trapped them with nets, torched their roosts, and asphyxiated them with burning sulfur. They attacked the birds with rakes, pitchforks, and potatoes. They poisoned them with whiskey-soaked corn.
>
> (Yeoman 2014)

Using industrialised methods that exploited the new technologies of the railway and the telegraph, the birds were slaughtered more intensely even as their numbers dramatically declined, until the final one, named 'Martha', that withered away the last years of her life in the Cincinnati zoo until her death in 1914.

Extinction of the Tasmanian tiger (thylacine) was even swifter and more recent. A top predator that had co-existed with an isolated indigenous human population for forty thousand years, it was wiped out in the space of little more than a century by the arrival of Europeans, who saw it as a threat to farming. European settlement began in Tasmania in 1803. The very last of the species, called 'Benjamin', died in a Hobart zoo in 1936 (Shreeve 2013).

Many large fauna extinctions at our hand have occurred on islands, as in Tasmania. The human wave that populated the Pacific over the last few thousand years meant the end of much animal life, not only many bird species in Hawaii, but a species of eagle and all of the genus moa, including the giant Moa[7] in New Zealand during the twelfth century, when it became the last land on Earth to be permanently occupied by humans. In the Indian Ocean, the extinction of the dodo in Madagascar in the seventeenth century is infamous. On an island there is no refuge once humans assert it as their territory.

The 2014 report from the World Wildlife Fund, *Living Planet: Species and spaces, people and places*, indicates that between 1970 and 2010 more than *half* of all vertebrate species (fish, amphibians, reptiles, birds and mammals) have been made extinct. Already, humans together with our domesticated animals form 97 percent of all vertebrates on Earth by weight. 'Wild' vertebrates form only three percent (Smil 2011, cited in Hamilton 2014, p. 1). According to the report (WWF 2014, p. 20), the main factors in these extinctions are over-exploitation (37 percent) habitat degradation (31 percent) and habitat loss (13 percent). Other factors are climate change (7 percent), invasive species or genes (5 percent), pollution (4 percent) and disease (2 percent). The first three are all products of increased consumption, or increased population, or both.

This stark picture echoes the concept of the 'Eremozoic era', found in biologist EO Wilson's book, *The Creation: An Appeal to Save Life on Earth* (2006). The destruction of fauna that began when human ancestors began eating meat is now at cataclysmic proportions. Megafauna on each continent began to disappear as soon as humans arrived, such that its destruction marks the spread of humanity through the planet. In Australia the animals included the giant kangaroo and the diprotodon (Flannery 1994; Roberts and Jacobs 2008). The Eremozoic era will be a consequence of the sixth great extinction, when the Earth will lack nearly all vertebrate life due to the impact of human activities. This permanent loss will probably reach final Mesozoic proportions by the end of this century, or much sooner according to Ceballosa, Ehrlich and Dirzob (2017), who have deemed it a "biological annihilation", when humanity enters the Eremozoic — the *Age of Loneliness*.

Is this destruction intentional? Steven Freeland (2015) points out that the environment is often a silent victim of warfare. Like rape once was, it is regarded as an inevitable consequence of mass conflict. Dutch ecologist, Sanne van der Hout, asks a related question in her 2014 article "Is de natuur partner of jachtgrond?" ("Is Nature a Partner or a Hunting Ground?"). This devastation of species and habitat looks more and more like an intentional organised conflict rather than just the unfortunate result of overconsumption or a by-product of human warfare. According to legal geographer Bronwyn Lay's paper *Ecocide as War* (2014), ecological destruction may be "a deliberate power strategy and attack against nature, not as collateral damage, but nature as centre stage both as participant and victim in warfare". She remarks that "the mass destruction of forests, rivers and soil pollution increasingly resemble battlefields: policed by privatised 'armies' masked as corporate security" (ibid., p. 4). Certainly this rings true for the manner in which native forests are treated in for example, Tasmania and New Zealand. In these places entire hillsides are laid waste to monochrome grey mud and fibre like the battlegrounds of Flanders, torched by firebomb then poisoned with chemicals to kill any wildlife that may return to nibble a new-sprouting alien seed.

If ecocide were an international crime, then many now unmonitored areas around the world would come under legal scrutiny as crime sites rather than the consequence of 'environmental degradation'. If 'nature' thus becomes a legal subject, as 'humanity' became after World War II with invocation of the crime of genocide, then recognition of ecocide would change the current paradigm from potential regulation to "resultant rights and protections" (ibid.). In fact, an international crime of ecocide was proposed to be included as one of five *Offences Against the Peace and Security of Mankind* in a 1996 UN draft that also listed *Genocide, War Crimes, Crimes Against Humanity* and *Crimes of Aggression* that since became International Crimes Against Peace under the jurisdiction of the International Criminal Court (Higgins 2012, p. 10).

While extinctions are global, populations can be decimated locally. Atlantic cod survive in some seas and have been a food staple of Europe for hundreds of years. The Grand Banks, off Newfoundland, were regarded as a limitless resource that were exploited to sustain a growing human population. As Figure 3.4 shows, the amount of cod landed from this area gradually increased until the 1960s when catches trebled to a peak of 800,000 tons a year. The catch then suddenly collapsed in two stages, until 1992 when there were virtually no cod left and the fishing grounds were closed. Again it was industrialisation of the industry that enabled the huge increase in catch during the 1960s and its subsequent exhaustion. Combined with poor regulation, resulting in increased competition, the fishery was doomed (Kurlansky 1998).

The Grand Banks story is an example of the industrialised approach to fishing since World War II, using diesel engines, sonar, vast nylon nets, and satellite imagery to locate and harvest increasing amounts of fish. Since the war, "fishing was transformed from a local endeavor into a global one" using much larger ocean-going ships and the "permanent occupation of marine ecosystems, instead of the local raids practiced by previous generations" (Greenberg and Worm 2015). This impact is such that despite increased use of active fishing technologies such as trawling and purse seining, globally fish caught *per fisher* halved from more than five to less than three metric tons a year between 1970 and 2012 (World Bank 2017b, p. 14).

In contrast with the Grand Banks, however, the Atlantic cod fishery asserted by Iceland during the three 'Cod Wars' with the United Kingdom in the 1950s–1970s has remained sustainable and is certified as such under Food and Agriculture Organisation (FAO) guidelines (IRF 2014). Iceland attained exclusive rights to manage the seas 370 kilometres from its coast through international arbitration that followed confrontations between the British navy and Icelandic trawlers. Sustainability was achieved because the fishery is one of

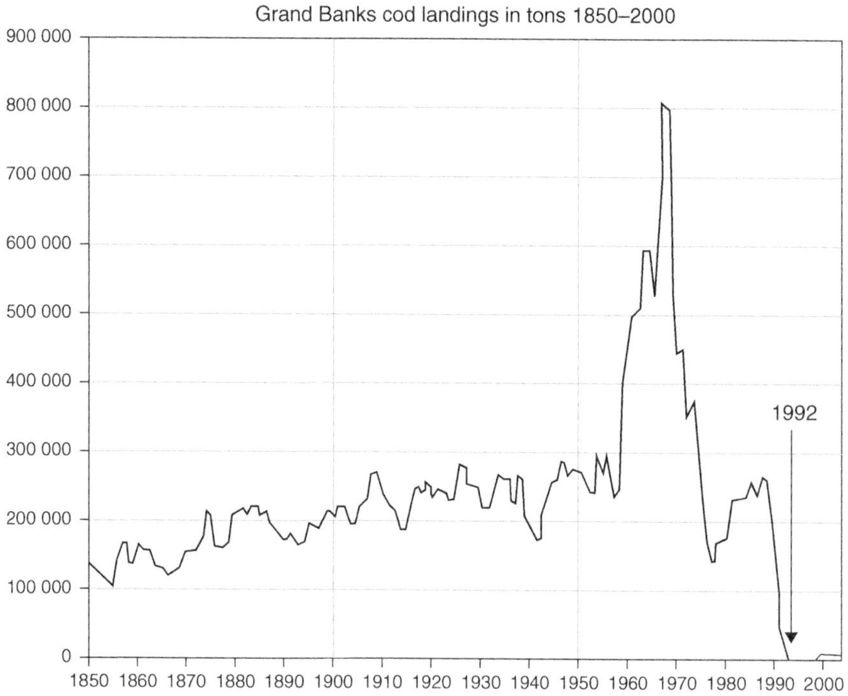

Figure 3.4 Collapse of Atlantic cod fisheries off Newfoundland.
Source: Millennium Ecosystem Assessment 2005, Synthesis Report, p. 58.

Iceland's few resources, resulting in a strong focus on its regulation, which is supported by small tight-knit population who understand its importance. There is also evidence that stocks of Atlantic Cod are again becoming viable in the North Sea following the implementation of fishing bans in that area (Smith 2015). Atlantic cod are not at risk of extinction of the entire species, but the remnant population off the Grand Banks remains closed to further fishing.

These examples illustrate how new technologies as well as the pressure of a growing human population were involved in increasing environmental impact during the last two centuries. However, Chertow (2000) shows how the impact of 'T' in the IPAT has increasingly become regarded as benign or more favourable in reducing impact since the identity was formulated fifty years ago. At first, technologies were used to clean up the residues of other production technologies "at the end of the pipe". Later, technologies were developed that were less harmful in their composition. Also some technologies are like Janus. For example, the same electronic technologies that are used to track schools of fish are now also used to track fish poachers, an increasingly important endeavour as fish stocks decline and marine reserves are created in response (Greenburg and Worm 2015).

Yet even if newer processes are less harmful or even beneficial, there is a conundrum facing environmental impact policy due to the relationship between population and technology. Does an increased population call for improved technology that can reduce environmental impact, or does an improved technology allow an increased human carrying capacity? While there has been a productivity revolution in both energy and resources since the IPAT identity was formulated, new technologies allow for more of us, but eventually too many of us may be overwhelming: "Does technology merely delay the inevitability of environmental destruction, or is better technology our best horse in the race toward sustainability?" (Chertow 2000, p. 18).

The rate of extinction may be compared with the natural rate of speciation, the evolutionary creation of new species. Comparisons of the two estimated rates are jaw-dropping. Robert May, writing in *Science* attempts to quantify the possible number of species on Earth. Through various extrapolations he produces numbers up to 50 million, although he "will not trust any estimate" without further research. Uncertainties are due to the myriad very small species with body sizes less than a millimetre, which have received "relatively little attention from taxonomists" (May 1988, pp. 1447–1448). Ehrlich and Pringle (2008, pp. 330–331) estimate there is of the order of tens of millions of species on Earth, and that estimates of extinction rates "are similarly imprecise". They cite May et al. (1995) saying that while only a little more than a thousand extinctions have been certified in the past four centuries, "this is a small fraction of the true number" of extinctions because our knowledge of biodiversity is "pitiful". Further, "current extinction rates vastly exceed background ones, perhaps by two to three orders of magnitude", or hundreds to thousands of times.

But while he is unable to establish the precise number of species, in a coda to the article, May (1988, p. 1448) remarks that rates of extinction have roughly matched rates of speciation for most of the period of life on Earth. Assuming that half of all existing species evolved within the last 100 million years and that half of them will become extinct within the current century, then contemporary extinction rates are around a million times faster than contemporary rates of speciation.

More recent estimates of the number of species appear diminished by comparison. The biologist and 'ecopragmatist', Stewart Brand, for example, in *Rethinking Extinctions* (2015) mentions "the 1.5 million species so far discovered, and most of the estimated 4 million or so species yet to be discovered". However, because he is writing in the context of the IUCN red list of endangered species, the apparent inconsistency between his estimates and those of May, Ehrlich and Pringle is plausible: May, Ehrlich and Pringle discuss *all* species, including those at the micro-level, whereas Brand discusses larger organisms (mammals, birds, reptiles, amphibians, fish, insects, molluscs and plants), especially those that have been studied at least to some degree.

According to the IUCN the estimated number of (larger) species as at 2007 was 1,388,137 in eight groups. Of those, 15,790 are threatened, or just over one percent of those studied. Dell'Amore (2013), not unreasonably given the IUCN figures, mentions that 20,000 species are now facing extinction. And irrespective of extinctions, the Living Planet Index (LPI) shows a 58 percent drop in overall numbers of vertebrates between 1970 and 2012, or two percent per year (WWF 2016, p. 6).

The key assertion of Brand's article though, is that biodiversity is continuing to *increase*, rather than dramatically plummet. In support of this he includes a graph (Figure 3.5) based on data produced by University of Chicago palaeontologist J. J. Sepkoski Jr that shows the five past great extinctions as substantial, but ultimately mere blips in a long-term trend increase in biodiversity based on the fossil record of *genera*, the level above species:

This result appears counter-intuitive and seems to contradict the findings of May, Fisher, Ehrlich and Pringle, Kolbert and Dell'Amore. There are reasons for this, however. First, increasing biodiversity as suggested by the graph still allows for declining populations. Whether there are a billion passenger pigeons or one Martha, there is still one species in existence. Now there are far fewer cod than

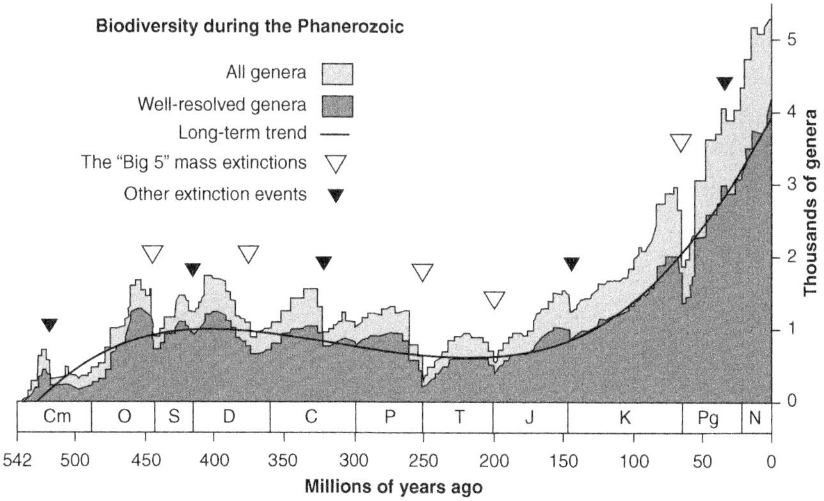

Figure 3.5 Biodiversity: number of genera over time.

Source: Schuttenhelm 2017, based on Rohde and Muller 2005, from data in Sepkoski 2002.

before, but the number of the species remains the same: one. Palaeobiologist Professor Tim Wootton of the University of Chicago adds that the graph may be distorted as fossil preservation is better in more recent material and there are wider sources of information in later periods ("the pull of the recent").[8]

But most tellingly, Professor David Jablonski, also of the University of Chicago and former associate of Sepkoski, points out that the data in the figure "are binned into million year or larger increments" and that the curve is at genus rather than species level, so that anthropogenic extinctions of the last 12,000 years would be undetectable at the scale it encompasses.[9] So it seems that Brand has misinterpreted the graph. There is no doubt that over the past *millions* of years, biodiversity has continued to increase. But over the latest few *thousand* years it has almost certainly declined – precipitously – at human hand.

Pollution

Contamination from pollutants has been with us for a long time. French and US geologists found evidence of lead pollution in ancient times – considerably before the advent of Industrial Revolution:

> Analysis of the Greenland ice core covering the period from 3000 to 500 years ago—the Greek, Roman, Medieval and Renaissance times—shows that lead is present at concentrations four times as great as natural values from about 2500 to 1700 years ago (500 B.C. to 300 A.D.). These results show that Greek and Roman lead and silver mining and smelting activities polluted the middle troposphere of the Northern Hemisphere on a hemispheric scale two millennia ago, long before the Industrial Revolution. Cumulative lead fallout to the Greenland ice sheet during these eight centuries was as high as 15 percent of that caused by the massive use of lead alkyl additives in gasoline since the 1930s.
>
> (Hong et al. 1994)

Apart from the longevity of the phenomenon, the operative word here is 'cumulative'. It is the incapacity of the biosphere to absorb them that essentially defines pollutants. And arguably, it was the exposition of cumulative effects by two women that defined modern environmentalism – the American Rachel Carson's 1962 book *Silent Spring* on pesticide accumulation, and the so-called 'Japanese Carson', Ariyoshi Sawako's 1974 novel *Fukugō Osen* ('*Compound Contamination*' or '*Cumulative Pollution*') that took Japan by storm. Ariyoshi's book sparked intense public debate about environmental toxins, and led to more than sixty articles in the major newspaper *Asahi Shimbun* about cumulative pollution during the subsequent year (Hartley 2014).

The impact of *Fukugō Osen* was magnified by awareness of the outbreak of the lethal "Minamata disease' from the late 1950s to the 1970s. The disease, which had profound neurological effects similar to cerebral palsy, was found to be due to concentrations of heavy metals in shellfish in Minamata Bay and its surroundings on the West coast of Kyushu, the result of industrial pollution. While tens of thousands of people were affected and more than 1500 died from the toxins, Japanese authorities and the business concerned moved at glacial pace to rectify the source of the problem.

Before World War II, "smoke, sewage, and soot were the main environmental concerns" as they had been since the Middle Ages (Chertow 2000), but it was the revolutionary thinker Barry Commoner, who observed that there had been a remarkable change since then:

> most US pollution problems are of relatively recent origin. The postwar period, 1945–46, is a convenient benchmark, for a number of pollutants—man-made radioisotopes, detergents, plastics, synthetic pesticides, and herbicides—are due to the emergence, after the war, of new productive technologies.
>
> (Commoner 1972, p. 345)

These pollutants were not only being produced at vastly increasing rates, but they were artificial – synthetic – alien to the structure of the biosphere and therefore difficult to break down and absorb. Hence they tended to accumulate in ever-increasing concentrations. However, pollution is a global issue. Former World Bank economist and US Treasury Secretary under Clinton, Larry Summers, infamously once asked in an email to Bank staff if pollution should be shifted to poor countries:

> shouldn't the World Bank be encouraging more migration of dirty industries to the LDCs [less developed countries]?...The economic logic behind dumping a load of toxic waste in the lowest wage country is impeccable, and we should face up to that...Under-populated countries in Africa are vastly under-polluted; their air quality is probably vastly inefficiently low compared to Los Angeles or Mexico City.
>
> (Summers 1992, p. 66)

While it has been argued that the email was sardonic (Johnson, Pecquet and Taylor 2007, p. 398), the idea has failed to find overt support. But as a matter of reality, rather than Africa, much polluting industry has shifted to Asia. The ship breakers of Bangladesh are renowned for their lack of protection from pollutants such as asbestos, for example. In China, power generation and associated manufacturing has displaced much of the industry of the West, but at the cost of considerable pollution.

Contamination also takes other forms, such as pesticides that may cause the decline of bees, which are essential to the pollination of many crops (Saunders 2016). Derived from nicotine as their name suggests, *neonicotinoids* are widely used including for 95 percent of the US corn crop. Part of their appeal is that they are simple to use. The water-soluble pesticide is absorbed by the growing plant. But when bees feed on the nectar of flowering crops, they are exposed to the pesticide. Bees may even prefer treated crops due to the effects of their nicotine content[10] (Philpott 2016). 'Neonics' as they are often called are also sprayed, which risks direct exposure for bees and other insects. According to a US Environmental Protection Agency,

> for all crops and application methods where on-field exposure is expected, values exceeded risk levels of concern [LOCs]. ... For all use patterns where residue data were available, LOCs were exceeded.
>
> (USEPA 2016, p. 18)

And even in the deepest ocean, artificial persistent organic pollutants (POPs)[11] have been found to be accumulating in marine amphipods (tiny crustacea), at depths over 7,000 metres (Qiu 2017). Nowhere is safe from chemical contamination.

Air

According to the World Health Organisation (WHO), air pollution causes more deaths worldwide than AIDS, diabetes and road injuries combined, causing "one in eight of total global deaths" or more than seven million people each year, making it the single largest environmental health risk on Earth (WHO 2014b). The impacts of air pollution fall mainly on the poor and disadvantaged. People living near major roads or large industrial sites are particularly affected, as are many people in rural areas. Nearly 84 percent of all people live with ambient concentrations of fine particulates greater than WHO guidelines,[12] and the number of people exposed increased by ten percent between 1990 and 2010 (World Bank 2015b, p. 63). Pollutants such as sulphur dioxide, nitrous oxides and particles resulting from energy generation, industry and motor vehicles have serious health and environmental consequences, according to the American Meteorological Society (AMS) (2015).

Air pollution has been a public issue for centuries, but effective action has been limited to mainly rich countries since World War II. In 1306 King Edward I of England proclaimed a ban on the use of sea coal[13] in London due to the smoke it caused. In the United States, clean air legislation was enacted in several cities in the 1880s. But during the late 1940s serious smog incidents raised political concern to new levels, resulting in the 1955 US Air Pollution Control Act, the first of several clean air laws that are still in effect. In 1949 the USSR enacted an air pollution law and established a department to monitor emissions. In 1952 dense smog in London UK lasted for four days causing about four thousand deaths. Four years later the United Kingdom passed the Clean Air Act and air quality began to improve. During the 1970s lead-based products began to be banned in the West, the catalytic converter was introduced to cut motor vehicle emissions, and chlorofluorocarbon (CFC) refrigerants that destroy the ozone layer were banned. During the 1980s radioactive radon gas was found to cause thousands of cancer deaths each year, and thousands were killed or injured in Bhopal, India, due to isocyanate gas leak from Union Carbide plant in 1984. In 1985 the hole in the ozone layer over Antarctica was discovered and attributed to CFCs. Between 1985 and 1995 leaded petrol (gasoline) was phased out in OECD countries. In 1986 the United States and Canada recognised acid rain as a "transboundary" issue and industry clean air programs were increased. In 1989 the UN *Montreal Protocol* phased out CFCs in favour of less harmful substitutes such as hydrofluoroolefins (HFOs), that rapidly degrade in the lower atmosphere (Rae 2016). In 1990 the US *Pollution Prevention Act* passed, dealing with pollution at source (American Meteorological Society 2015; UK Met Office 2018).

In 2016, India announced plans for 56.5 percent of its electricity capacity to be from non-fossil-fuelled sources by 2026–27 (CEA 2016, p. xxv). In 2017, China announced cancellation of plans to build an additional 103 coal-fired power plants, representing more than the entire coal-fired capacity of Germany (Forsythe 2017).

Carbon

Carbon pollution is largely unnoticeable because its main vehicle, the gas carbon dioxide (CO_2), is both colourless and odourless, as well as safe to breathe providing there is sufficient oxygen present. Its indirect effects due to increasing concentrations are nevertheless highly significant, both in its impact on the atmosphere as the major greenhouse gas and on the oceans as the major acidifier.

During the nineteenth century it was known that atmospheric carbon dioxide was increasing and that it could cause global warming through the greenhouse effect. In 1895 the Swedish chemist Svante Arrheinius quantified the impact showing that a doubling of its proportion in the atmosphere would result in a global temperature increase of between five and six degrees Celsius (Arrhenius 1896, p. 266). However, he did not specifically link the CO_2 increase with the burning of fossil fuels.

Nevertheless, over the next century the link and its magnitude became clear. After winning the Pulitzer Prize for her book *The Sixth Extinction*, Elizabeth

Kolbert, was interviewed by the *New York Times*. She spoke of how fast we are changing the world:

> When we use fossil fuels, we are reversing geological history by taking organisms that were buried millions of years ago and pumping their carbon back into the atmosphere at a very fast rate...Humans have sped up the rate by which we change the world, while the rate at which evolution adapts is much slower. There's a mismatch between what we can do and what nature can sustain.
>
> (Dreifus 2014)

Pumping ancient stored carbon back into the atmosphere at millions of times the rate it was fossilised is of course the source of global warming now experienced throughout the biosphere. The issue is how to de-carbonise the economy before the impact of resultant global warming brings irreversible change including accelerating sea level rise, climate disruption, habitat destruction and even more extinctions. The main sources of global carbon pollution are coal and petroleum equally, both about a third of the total of almost 10 billion tonnes per year, an amount rapidly increasing. Gas accounts for about 18 percent and cement production at about five percent of global carbon pollution (Boden, Marland and Andres 2015).

Cement is the key ingredient of concrete, the ubiquitous building material of modernity, by volume second only to water in amount consumed world-wide (WBCSD and IEA 2009, p. 2). While the rate of carbon emissions per tonne of cement has fallen since 2000, without mitigation total carbon emissions from this industry will still rise by mid-century, such is the likely increase in production (ibid.).

Still, the energy sector has by far the greatest emissions. Figure 3.6 shows the relative contribution to global carbon and other greenhouse gas pollution by sector. Not only does energy production emit the most greenhouse gases by far, but emissions are growing the most in that sector. Much of this energy production is of course grid electricity from thermal power plants that are fired by coal and gas. Where grids do not exist or are unreliable, much electricity is produced from diesel generators that use petroleum-based fuel.

The IPCC takes a broader view of energy sector emissions by assessing the total greenhouse gas emissions contributed by various sources of electricity generation over the life cycle of the technology. Thus, for example, a concrete hydro-electric dam includes emissions incurred during the extraction, transport and conversion of materials used in its construction, as well as its operation. The resulting table (Table 3.1) shows that coal-fired electricity generation is orders of magnitude more harmful than, say, solar, wind or hydro power as far as greenhouse gas emissions are concerned.

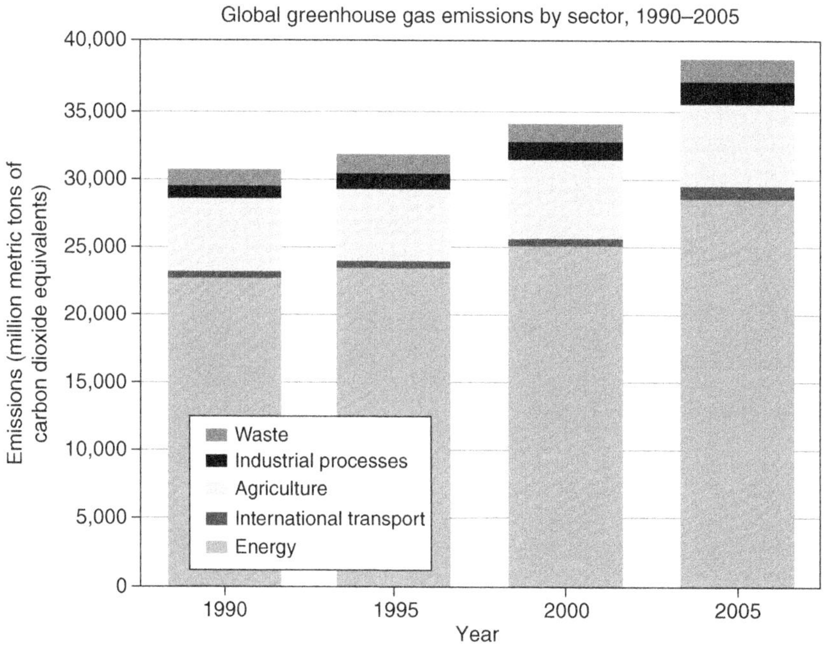

Data Source: World Resources Institute, 2009. Climate Analysis Indicators Tool (CAIT). Version 6.0. Accessed January 2009. httpc//cait.wri.org.

For more infomation, visit U.S. EPA's "Climate Change Indicators in the United States" at www.epa.gov/climatechange/science/indicators.

Figure 3.6 Greenhouse gas emissions by sector.

Source: World Resources Institute (WRI) 2009, reprinted with kind permission of WRI.

Even 'pre-commercial' fossil fuel 'carbon capture and storage' technologies are far more polluting over their life cycle than all renewable and nuclear technologies.

Deforestation

Globally, net deforestation is less than might be expected. Generally speaking, deforestation is continuing in developing countries, while reforestation is continuing in developed countries. As a result, tropical forests are in decline whereas temperate forests are regenerating. As far as tropical forests are concerned, one widely cited World Bank study, based on the results of 140 analyses, indicates ambiguity about how factors such as population, income and security of tenure affect tropical deforestation. According to the study, economic liberalisation

Table 3.1 Median carbon dioxide emissions by source of electricity generation: lifecycle CO_2 equivalent per kilowatt-hour (gCO_2eq/kWh)

Source	*Median emissions (gCO_2eq/kWh)*
Existing commercial	
Pulverised coal	820
Biomass, co-firing with coal	740
Combined cycle gas	490
Dedicated biomass	230
Solar photo-voltaic: utility scale	48
Solar photo-voltaic: rooftop	41
Geothermal	38
Concentrated solar	27
Hydro	24
Offshore wind	12
Nuclear	12
Onshore wind	11
Pre-commercial	
Carbon capture and storage – pulverised coal	220
Carbon capture and storage – coal – integrated gasifier combined cycle	200
Carbon capture and storage – gas – combined cycle	170
Carbon capture and storage – coal – oxyfuel	160
Ocean – tidal and wave	17

Source: IPCC 2014c, p. 1335, with permission.

reforms may increase pressure on tropical forests (Angelsen and Kaimowitz 1999). More recently, there is evidence that the rate of tropical deforestation can be reduced using accurate satellite imaging combined with cadastral reform (Chomitz 2015).

But forests are more than just trees. To the extent that natural forests are replaced with monocultural plantations, there is an inevitable loss of ecosystem and biodiversity. The UN Forum on Forests (UNFF), which aims to promote sustainable forest management and prevent forest degradation was only established as a permanent body at the turn of the millennium (UNFF 2004). As the overall net rate of forest loss is unclear[14] and depends on definitional factors such as percentage of crown cover, it is not possible to claim that the planet's forests are at present sustainable overall, and especially in respect of the biodiversity they embrace.

Risks

To summarise the risks now faced, the thirty-year old Brundtland report identified the stand out risks of a high energy future. They are climate change, urban air pollution and acidification of the environment – all produced from the combustion of fossil fuels. There are also the risks of accident, waste disposal and proliferation – all associated with nuclear energy (Brundtland 1987, pp. 121–122).

Another major problem arises from the growing scarcity of fuelwood in developing countries. Brundtland indicated that if trends continued, by the year 2000 around 2.4 billion people would be living in areas where fuelwood is extremely scarce. In fact, a more recent study points to fuelwood dependency in parts of Kenya as well as in Bangladesh and Ethiopia at around 95 percent in rural areas, and that between 84 and 94 percent of people experienced fuelwood shortage. Much of this shortage appears due to agriculture replacing forest (Stacey et al. 2013, p. 4140), which due to its unsustainability creates demand for alternative cheap energy in the developing world.

Metrics

Whilst impacts such as the extinction of particular species can be blatant, there is less clarity about other impacts, or overall impact for that matter. There are four variables in the IPAT identity. Two of them are fairly clear: 'population' simply means the number of humans alive at any one time. 'Affluence' in the equation means the amount of goods and services consumed, on average, by each person in a year. Affluence is usually measured by GDP per person, and is therefore usually expressed in US dollars per person.

However, a persistent thread over the decades has been how the 'T' in I=PAT has changed. And if the meaning of 'T' has changed, then so must 'I', because the equation says that impact is proportional to technology, as well as to population and affluence. Early on, 'T' meant *pollution*, then *emissions*, then *waste*. It has also meant *damage per unit of production*, or *energy per unit GDP*, or '*the residual of all that impacts the environment that is not A or P*', or *patents* as in the Kolbert-Steffen example above, and so on. Because the nature of human impact on the environment critically depends on the meaning of T in the IPAT equation, an important question is, what is the best metric for T?

An appropriate metric for technology might follow from the notion of impact as 'ecological footprint', – the area of ecosystem required to sustain an existing human population (Wackernagel 1994; Wackernagel and Rees 1996). If the I in I=PAT is impact, then the equation says impact is proportional to T (technology). Footprint is commonly measured in hectares. Recasting I=PAT, we find that T=I/PA. Therefore,

T = I: (footprint (hectares)) / P: (population) x A: (GDP dollars / population).

The *populations* cancel out, leaving *T = footprint (hectares)/GDP (dollars)*. Thus T is measured in hectares of impact per dollar of consumption. This is an annual flow, given that GDP is income or consumption over one year. The original I=PAT formulation then becomes:

$$I:\left(\textit{footprint (hectares)}\right) =$$
$$P:\frac{\left(\textit{population}\right)}{1}\times A;\frac{\left(\textit{GDP (dollars)}\right)}{\left(\textit{population}\right)}\times$$
$$T:\frac{\left(\textit{footprint (hectares)}\right)}{\left(\textit{GDP (dollars)}\right)},$$

– which is self-evidently true, since the populations cancel out as do the GDPs, leaving *footprint (hectares)* = *footprint (hectares)*.

Therefore, what is required for a smaller footprint (I) is a technology (T) that minimises footprint independent of consumption (A). Technologies that enable more intensive agriculture are examples of this sort. There is also support for such a possibility in the *'environmentalist's paradox'*. Not only is human well-being now increasing as ecosystems are degraded, but technology can be used to enhance ecosystems rather than just replace them or mitigate their damage (Raudsepp-Hearne et al. 2010, p. 586).

Alternatively, because we are a *species with energy intensive technology* (SWEIT) that distinguishes us from all other species on this planet, and which importantly has caused much of our present conundrum, technology and therefore impact may be measured in terms of energy intensity. The Earth is essentially a solar-powered life-support system. As well as direct radiation, solar services include weather, photosynthesis (food), fossil fuels and natural waste disposal. The rate of energy usage per person fulfils the metric of energy intensity and is measurable in units such as kilowatts per person. Following this logic, the dimension of impact (I) is power, or energy per unit time.

Assuming that we are dealing in rates of change rather than accrued stocks, a footprint using this metric would measure the fraction of the Earth's life support services consumed per unit time. Because the Earth's life-support services are solar powered, it is reasonable that footprint can be measured in units of solar power, or the rate of incoming energy. Elements of the I=PAT equation would be based on: e = energy, t = time, and c = consumption:

I is Impact, with dimensions e/t,
P is population, which is simply a number, and
A is affluence with dimensions c/t/person.

Recasting I=PAT results in T=I/PA as before. Therefore, the dimensions of technology (T) are:

(energy/time) / (population x ((consumption/time) / person))). The 'times' cancel and so do the 'populations'. Therefore, T = energy / consumption. This is essentially the affordability of energy. A valid unit would be kilowatt-hours per dollar. The original I=PAT formulation then becomes:

$$I\left(\textit{footprint : energy / time, or power}\right) =$$
$$P:\frac{\left(\textit{population}\right)}{1}\times A;\frac{\left(GDP\right)}{\left(\textit{population}\right)}\times T:\frac{\left(\textit{energy / time}\right)}{\left(GDP\right)},$$

Thus the more affordable energy becomes through technology, the greater the impact of a given human population. This makes sense in the real world. For example, where petrol (gasoline) is cheap as in the United States, the environmental impact of inefficient 'gas guzzler' cars is greater than in say France, where fuel is more than double US prices and cars are smaller and more efficient, and rail transport is widely used. Similarly, where electricity is cheap, more is used. There are three main impacts: there are resource impacts for the stored energies used, there are pollution impacts due to the energy conversions involved and there are direct impacts on habitat resulting from processes and machinery using the energy that technologies make available.

And yet … there is still another way of looking through the IPAT lens. Cheap energy is not necessarily destructive of the environment. It depends on both its source and how it is used. If it pollutes or it is used to destroy habitat, then impact is greater. But if it is non-polluting and it minimises impact through closed systems, then it can benefit both humanity and the biosphere.

Minimising

A different approach – that of factors – shows promise in policy terms, and does relate to resource productivity (Chertow 2000, p. 24). The Factor 10 Club wants to reduce resource use to the extent that resource productivity must increase by a factor of ten over the 30–50 years from the mid-1990s, based on better technologies and methods.

Vaclav Smil in *Harvesting the Biosphere: What We Have Taken from Nature* writes of the "metabolic imperatives" that still drive us and which have already transformed the planet. The harvesting of plants for food remains "the quintessential activity of modern civilisation" (Smil 2013, p. viii). He later summarises the conditions necessary to "minimise human claims on the biosphere's productivity", remarking that this is not difficult to do.

First on his list is stabilising the world population at less than nine billion. This is then aided by using "best agronomic practices" for food production including crop rotation rather than monoculture, limiting per capita food requirements to healthy levels and through much greater attention to post-harvest food loss and household food waste. Smil says we should develop cereal permacultures, and grain staples that can fix nitrogen. Timber harvesting should be limited to the long-term capacities of forest ecosystems and its demand restricted by banning disposable packaging. As well as demand reduction through different technologies:

> the greatest savings of woody phytomass could result from a universal adoption of efficient rural wood stoves, such as those that have been widely diffused in China (Smil 2004); by whole tree utilisation and expanded production of engineered timber (Williamson 2001); by even higher rates of paper recycling (McKinney 1994); and by further shift from paper-based records to purely electronic files. Looking further ahead, expansion of crop and wood harvests may not require conversion of substantially larger undisturbed areas to cropping or to wood plantations thanks to new high-yielding transgenic plants.
>
> (Smil 2011, p. 630)

Indicators and boundaries

In order to provide a clearer view of human impact from the beginning of the Industrial Revolution, a series of global-scale indicators has been developed. Figure 3.7 shows 12 indicators, four concern atmospheric gas concentrations, two are climate effects, three concern ocean and coastal ecosystems, two are about land use, and one is biodiversity measured by extinctions. Shrimp production as a proxy for 'coastal zone alteration' (h) might be questioned, but shrimp farming, much increased since 1950, clearly affects low-lying coastal zones (Hernândez-Cornejo and Ruiz-Luna 2000). Overall, the shape of each graph indicates strong tendencies to unsustainable impact against all of the indicators.

What these indicators do not show, however, is how close to unsustainable limits the Earth is at already. The concept of planetary boundaries – impact limits on key features of the earth system beyond which sustainability is jeopardised – was proposed by Steffen, Rockström and Costanza (2011), who argue for boundaries in ten major "Earth-system processes". The concept helps clarify the dynamics of human impact and which earth systems are most affected. Notably, it is far broader than the issue of climate change. A revised assessment of the concept showed at least three earth systems processes, *genetic diversity*, and the *phosphorous* and *nitrogen cycles* are already well beyond safe operating space at high risk of abrupt change. Two more are in the zone of uncertainty or increasing risk: *climate change* and *land system change*. Three, *novel*

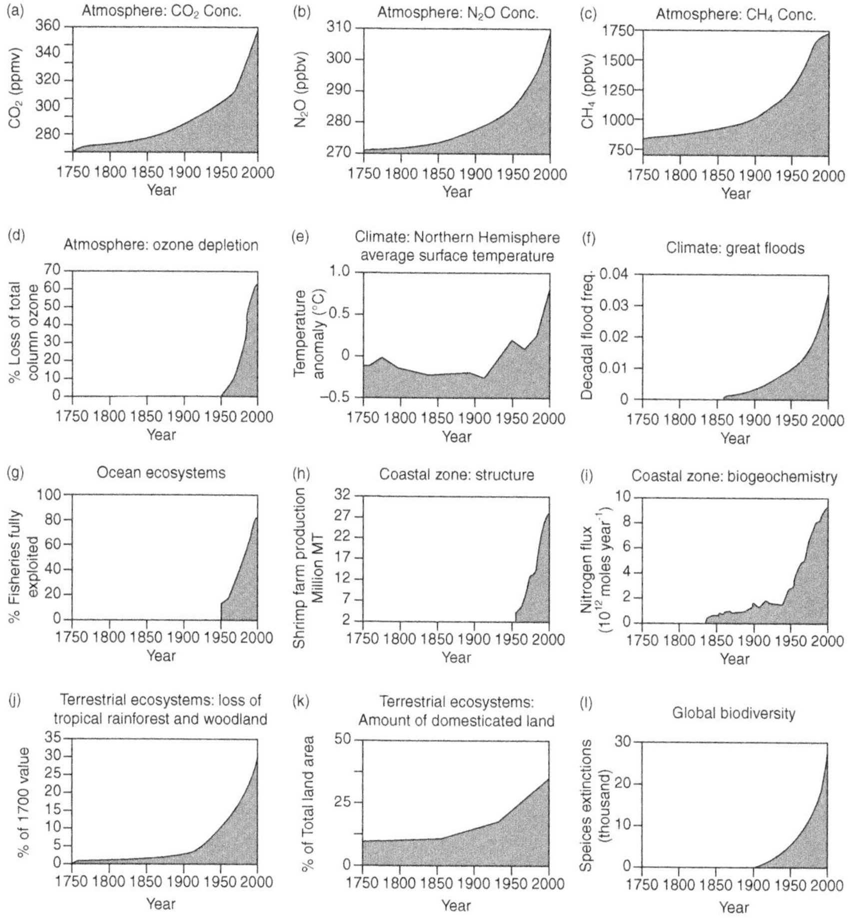

Figure 3.7 Global-scale environmental indicators of human impact.

Source: Steffen et al. 2011. Reprinted by permission from Springer Nature Switzerland AG, copyright 2011.

entities (or 'chemical pollution'), *atmospheric aerosol loading* and *functional diversity*, part of *biosphere integrity* have not yet been determined. Only *freshwater use*, *ocean acidification* and *ozone depletion* are yet (surprisingly) at relatively safe levels (Steffen et al. 2015).

Intriguingly, the authors return obliquely to the IPAT identity in their final comments:

> the planetary boundaries approach doesn't say anything explicit about resource use, affluence, or human population size. These are part of the

> trade-offs that allow humanity to continue to pursue increased well-being…Because the planetary boundaries approach says nothing about the distribution of affluence and technologies among the human population, a "fortress world," in which there are huge differences in the distribution of wealth, and a much more egalitarian world, with more equitable socio-economic systems, could equally well satisfy the boundary conditions… remaining within the planetary boundaries is a necessary—but not sufficient—condition for a bright future for humanity.
>
> (Steffen, Rockström and Costanza 2011)

It is nevertheless implied in all this that human impact upon the biosphere is unevenly distributed. Particulate emissions from factories, power stations and arterial roads are concentrated in and around cities. Compound contamination from heavy metals most affects rivers and especially estuarine bays. The Grand Banks remain devoid of cod, while in Icelandic waters cod still thrive. Animals and vegetation adapted for high altitudes and low temperatures have nowhere to go as glaciers melt; their low-altitude cousins may still move to higher ground. Continents allow for some migration; but islands become killing grounds from which there is no escape. Coastal mangroves are destroyed; rocky cliffs remain. The people of Kiribati prepare for inundation; the Sudanese, Ethiopians and Kenyans see desert encroaching.

Conclusion

Human impact on the biosphere is already unsustainable in relation to several earth systems, including the rapid decline of many species. Much impact results from the use of cheap energy. We are the only species on the planet that artificially harnesses intensive energy and human impact on this planet is orders of magnitude greater than any other species. Compared with other SWEITs that may have existed or yet exist within the galaxy the plight we face may not be unique, but the continuation of our civilisation is precarious and clearly interwoven with our energy practice.

Notes

1 Attributed to the US astrophysicist Frank Drake (1962).
2 This is one possible answer to the 'Fermi paradox': if other civilisations are common throughout the universe, why haven't they contacted us?
3 Hinchcliffe asserted that the answer to any interrogative title will always be 'No', which is confounded by Peon's riposte: 'Is Hinchcliffe's rule true?', which demonstrates that the assertion is false – but only if it is true (Peon 1995).
4 1900: 1.8 billion X 2 trillion X 141,000, whereas 2011: 7 billion X 55 trillion X 1.9 million. Ratio: 3.89 X 27.50 X 13.48 = 1442 times that of 1900.

5 Personal email of 26 September 2014.
6 Bison numbers were down to 1091 by 1889, and have rebounded to about 500,000 according to Defenders of Wildlife. Prairie dog numbers have been reduced by 98 percent since 1900 according to the National Geographic 2015.
7 The word 'Moa' throughout the Pacific now has a reduced meaning: 'chicken'.
8 Wootton, personal email of 25 April 2015.
9 Jablonski, personal email 25 April 2015.
10 The effects might be described as a 'mild buzz'!
11 These include polychlorinated biphenyls used in plastics and antifoulants as well as polybrominated diphenyl ethers, used as flame retardants, which are either banned or being phased out.
12 Fine particulates are less than 2.5 microns diameter; WHO guidelines allow up to ten micrograms per cubic meter.
13 'Sea coal' was mined from easily accessible deposits on coastal cliffs, especially along the Yorkshire coast.
14 The net yearly rate of forest lost was 0.18 percent in the 1990s, falling to 0.08 percent between 2010 and 2015. However, this apparent improvement included a reduced definition of forest crown cover (Lang 2015).

References

American Meteorological Society 2015, *History of the Clean Air Act*, website.

Angelsen, A and Kaimowitz, D 1999, 'Causes of Deforestation: Lessons from Economic Models', *The World Bank Research Observer*, vol. 14, no. 1 February, pp. 73–98.

Ariyoshi, S 1977 [1974], *Fukugō osen sono go*, Ushio Shuppansha, Tōkyō, Japan.

Arrhenius, S 1896, 'On the Influence of Carbonic Acid in the Air upon the Temperature of the Ground', *London, Edinburgh and Dublin Philosophical Magazine and Journal of Science*, 5th series, vol. 41, pp. 237–276.

Boden, T, Marland, G and Andres, R 2015, '*Global, Regional, and National CO_2 Emissions*', Carbon Dioxide Information Analysis Center, Oak Ridge National Laboratory, United States Department of Energy, Oak Ridge, US.

Brand, S 2015, 'Rethinking extinction', *Aeon*, 21 April. https://aeon.co/essays/we-are-not-edging-up-to-a-mass-extinction.

Brundtland, G 1987, *Report of the World Commission on Environment and Development: Our Common Future*, United Nations, Geneva, Switzerland.

Campante, T, Barclay, T, Swift, J, Huber, D, Adibekyan, V, Cochran, W, Burke, C, Isaacson, H, Quintana, E, Davies, G, Silva Aguirre, V, Ragozzine, D, Riddle, R, Baranec, C, Basu, S, Chaplin, W, Christensen-Dalsgaard, J, Metcalfe, T, Bedding, T, Handberg, R, Stello, D, Brewer, J, Hekker, S, Karoff, C, Kolbl, R, Law, N, Lundkvist, M, Miglio, A, Rowe, J, Santos, N, Van Laerhoven, C, Arentoft, T, Elsworth, Y, Fischer, D, Kawaler, S, Kjeldsen, H, Lund, M, Marcy, G, Sousa, S, Sozzetti, A, and White, T 2015, 'An Ancient Extrasolar System with Five Sub-Earth-Size Planets', *Astrophysical Journal*, vol. 79, no. 2. Viewed 29 January 2015.

Carson, R 2002 [1962], *Silent Spring*, Houghton Mifflin Harcourt, New York, US.

CEA: Central Electricity Authority 2016, Draft National Electricity Plan, Volume 1 Generation, Government of India, Ministry of Power, December.

Ceballosa, G, Ehrlich, P, and Dirzob 2017, 'Biological annihilation via the ongoing sixth mass extinction signalled by vertebrate population losses and declines', *Proceedings of the National Academy of Sciences of the United States of America*, online edition, 10 July. doi: 10.1073/pnas.1704949114.

Chertow, M 2000, 'The IPAT Equation and Its Variants: Changing views of Technology and its Environmental Impact', *Journal of Industrial Ecology*, vol. 4, no. 4, Yale-MIT, New Haven, US.

Chomitz, K 2015, *Can Tradeable Forest Conservation Obligations Enhance Economic, Biodiversity, and Carbon Outcomes?* Seminar, Institute for the Study of Social Change, University of Tasmania, Australia, 15 October.

Commoner, B 1972, *The Closing Circle,* Knopf, New York, US.

Dartnell, L 2015, 'Out of the Ashes', *Aeon*, 13 April.

Dell'Amore, C 2013, '20,000 Species Are Near Extinction: Is it Time to Rethink How We Decide Which to Save?', *National Geographic*, Tampa, US, 16 December.

Dreifus, C 2014, 'Chasing the Biggest Story on Earth: 'The Sixth Extinction' Looks at Human Impact on the Environment', *New York Times*, 10 February.

Ehrlich, P and Pringle, R 2008, 'Where Does Biodiversity Go from Here?', *In the Light of Evolution, Volume II: Biodiversity and Extinction*, Avise, J, Hubbell, S and Ayala, S (eds), National Research Council, National Academies Press, Washington, DC, US, pp. 329–346.

Fisher, P 2014, 'The 'Pre-Holocene' Climate Is Returning – and It Won't Be Fun', *The Conversation*, 24 July.

Flannery, T 1994, *The Future Eaters: An Ecological History of the Australasian Lands and People*, Grove Press, Melbourne, Australia.

Fowler, C and Hobbes, R 2003, 'Is Humanity Sustainable?', *Proceedings of the Royal Society*, vol. 270, pp. 2579–2583.

Forsythe, M 2017, 'China Cancels 103 Coal Plants, Mindful of Smog and Wasted Capacity', *New York Times*, 18 January.

Frank, A and Sullivan, W 2014, 'Sustainability and the Astrobiological Perspective: Framing Human Futures in a Planetary Context", *Anthropocene*, vol. 5, pp. 32–41.

Freeland, S 2015 'Crimes against the Environment: The Silent Victim of Warfare', *The Conversation*, 9 November.

Gerbi, A 2010, *The Dispute of the New World*, University of Pittsburgh Press, Pittsburgh, US.

Greenberg, P and Worm, B 2015, 'When Humans Declared War on Fish', *New York Times*, Sunday Review, 8 May.

Hamilton, C 2014, 'Ecologists Butt Out: You Are Not Entitled to Redefine the Anthropocene', *Anthropocene Review*, vol. 5, p. 5.

Hartley, B 2014, *Unnatural Futures*, seminar notes, University of Tasmania, Hobart, Australia, August.

Hernândez-Cornejo, R and Ruiz-Luna, A 2000 'Development of Shrimp Farming in the Coastal Zone of Southern Sinaloa (Mexico): Operating Characteristics, Environmental Issues, and Perspectives', *Ocean and Coastal Management*, vol. 43, no. 7, pp. 597–607.

Higgins, P 2012, 'Seeding Intrinsic Values: How a Law of Ecocide Will Shift Our Consciousness', *Cadmus*, vol. 1, no. 5, October.

Hong, S, Candelone, J-P, Patterson, C and Boutron, C 1994, 'Greenland Ice Evidence of Hemispheric Lead Pollution Two Millennia Ago by Greek and Roman Civilisations', *Science*, vol. 265, no. 5180, 23 September, pp. 1841–1843.

IPCC: Intergovernmental Panel on Climate Change 2014c, *Climate Change 2014,* annex III, pp. 1329–1356 (Schlömer S, T. Bruckner, L. Fulton, E. Hertwich, A. McKinnon, D. Perczyk, J. Roy, R. Schaeffer, R. Sims, P. Smith, and R. Wiser, 2014: Annex

III: 'Technology-Specific Cost and Performance Parameters' in '*Climate Change 2014: Mitigation of Climate Change*', Contribution of Working Group III to the Fifth Assessment Report of the Intergovernmental Panel on Climate Change [Edenhofer, O., R. Pichs-Madruga, Y. Sokona, E. Farahani, S. Kadner, K. Seyboth, A. Adler, I. Baum, S. Brunner, P. Eickemeier, B. Kriemann, J. Savolainen, S. Schlömer, C. von Stechow, T. Zwickel and J.C. Minx (eds)]), Cambridge University Press, Cambridge, UK, and New York, US.

IRF: Iceland Responsible Fisheries, 2014, *Iceland Cod Fisheries Certified.*

Johnson, J, Pecquet, G and Taylor, L 2007, 'Potential Gains from Trade in Dirty Industries: Revisiting Lawrence Summers' Memo', *Cato Journal*, vol. 27, no. 3, pp. 397–410.

Kolbert, E 2011, 'Enter the Anthropocene – Age of Man', *National Geographic*, March.

Kolbert, E 2014, *The Sixth Extinction: An Unnatural History*, Bloomsbury, London, UK.

Kurlansky, M 1998, *Cod: A Biography of the Fish That Changed the World*, Penguin Books, New York, US.

Lang, C 2015, 'Deforestation Denial: FAO Claims that Forest Loss Has Halved, While Global Forest Watch Raises the Alarm about "Dramatic Forest Loss"', REDD-Monitor, 9 September.

Lay, B 2014, *Ecocide as War: Military Ecology*, paper for Understanding Ecology through the Humanities: An Environmental Humanities Symposium, Working Group of Environmental Humanities through SAGUF, Zurich, Switzerland, 16–21 June.

Le Page, M 2014, 'World without Fossil Fuels', *New Scientist*, no. 2991, 18 October, pp. 34–39.

Marland, G, Boden, T and Andres, R 2007, 'Global, Regional, and National CO_2 Emissions' in *Trends: A Compendium of Data on Global Change*, Carbon Dioxide Information Analysis Center, Oak Ridge National Laboratory, United States Department of Energy, Oak Ridge, US.

Maxwell, S, Fuller, R, Brooks, T and Watson, J 2016, 'Biodiversity: The Ravages of Guns, Nets and Bulldozers', *Nature*, vol. 536, 11 August, pp. 143–145.

May, R 1988, 'How Many Species Are There on Earth?', *Science*, New Series, vol. 241, no. 4872, 16 September, pp. 1441–1449.

McKinney, R (ed.) 1994, *Technology of Paper Recycling*, Springer-Verlag, New York, US.

Philpott, T 2016, 'The EPA Finally Admitted that the World's Most Popular Pesticide Kills Bees—20 Years Too Late', Mother Jones, 7 January.

Pimm, S, Jenkins, C, Abell, R, Brooks, T, Gittleman, J, Joppa, L, Raven, P, Roberts, C and Sexton, J 2014, 'The Biodiversity of Species and Their Rates of Extinction, Distribution, and Protection', *Science*, vol. 344, no. 6187, 30 May, 1246752. doi: 10.1126/science.1246752.

Qiu, J 2017 [2016], 'Man-Made Pollutants Found in Earth's Deepest Ocean Trenches', Nature, February. doi: 10.1038/nature.2016.20118.

Raudsepp-Hearne, C, Peterson, G, Tengö, M, Bennett, E, Holland, T, Benessaiah, K, MacDonald, G and Pfeifer, L 2010, 'Untangling the Environmentalist's Paradox: Why is Human Well-Being Increasing as Ecosystem Services Degrade?', *BioScience*, vol. 60, no. 8, pp. 576–589.

Roberts, R and Jacobs, Z 2008, 'The Lost Giants of Tasmania', Australasian Science, vol. 29, October, pp. 14–17.

Rohde, R, and Muller, R 2005, 'Cycles in Fossil Diversity', *Nature*, vol. 434, March, pp. 208–210.

Saunders, H 1992, 'The Khazzoom-Brookes Postulate and Neoclassical Growth', *The Energy Journal*, vol. 13, no. 4, pp. 131–148.

Schuttenhelm, R 2017, 'Climate Change & Holocene-Anthropocene Mass Extinction 2: Biodiversity graph shows Garden of Eden is Now', *Bits of Science* (online), 6 June.
Sepkoski, J 2002, *A Compendium of Fossil Marine Animal Genera, Bulletins of American Paleontology*, Vol. 363, Ithaca, US.
Shreeve, J 2013, 'Species Revival: Should We Bring Back Extinct Animals?', National Geographic News, 5 March.
Smil, V 2004, *China's Past, China's Future*, RoutledgeCurzon, London, UK.
Smil, V 2011, 'Harvesting the Biosphere: The Human Impact', *Population and Development Review*, vol. 37, no. 4, 13 December, pp. 613–636.
Smil, V 2013, *Harvesting the Biosphere: What We Have Taken from Nature*, MIT Press, Cambridge, US.
Smith, Winston P 1991, 'Odocoileus virginianus', *Mammalian Species*, no. 388, 6 November, pp. 1–13.
Stacey, W, Maraga, J and Koech, M 2013, 'Factors Influencing Fuelwood Scarcity among Rural Households in Lurambi District, Kakamega County, Kenya', *International Journal of Current Research*, vol. 5, no. 12, December, pp. 4138–4141.
Steffen, W, Persson, A, Deutsch, L, Zalasiewicz, J, Williams, M, Richardson, K, Crumley, C, Crutzen, P, Folke, C, Gordon, L, Molina, M, Ramanathan, V, Rockström, J, Scheffer, M, Schellnhuber, H, and Svedin, U 2011, 'The Anthropocene: From Global Change to Planetary Stewardship', *Ambio*, vol. 40, no. 7, pp. 739–761.
Steffen, W, Richardson, K, Rockström, J, Cornell, S, Fetzer, I, Bennett, E, Biggs, R, Carpenter, S, de Vries, W, de Wit, C, Folke, C, Gerten, D, Heinke, J, Mace, G, Persson, L, Ramanathan, V, Reyers, B and Sörlin, S 2015, 'Planetary Boundaries: Guiding Human Development on a Changing Planet', *Science*, vol. 347, no. 6223, 13 February, 1259855. doi: 10.1126/science.1259855.
Steffen, W, Rockström, J, and Costanza, R 2011, 'How Defining Planetary Boundaries Can Transform Our Approach to Growth', *Solutions*, vol. 2, no. 3, pp. 59–65.
Summers, L 1992, 'Let Them Eat Pollution', *The Economist*, vol. 322, no. 7745, 2 August, p. 66.
UK Met Office 2018, *The Great Smog of 1952*, case study, website.
UNFF: The United Nations Forum on Forests 2004, Fact Sheet 1.
USEPA: United States Environmental Protection Agency, Office of Chemical Safety and Pollution Prevention 2016, *Preliminary Pollinator Assessment to Support the Registration Review of Imidacloprid*, Memorandum, 4 January, Washington, DC, US.
Van der Hout, S 2014, 'The Homeotechnological Turn: Sloterdijk's Response to the Ecological Crisis', *Environmental Values*, vol. 23, pp. 423–442.
Wackernagel, M 1994, Ecological Footprint and Appropriated Carrying Capacity: A Tool for Planning Toward Sustainability, PhD thesis, OCLC 41839429, University of British Columbia, Vancouver, Canada.
Wackernagel, M and Rees, W 1996, *Our Ecological Footprint: Reducing Human Impact on the Earth*, New Society Publishers, Gabriola Island, British Colombia, Canada.
WWBCSD and IEA 2009, *Cement Technology Roadmap: Carbon Emissions Reductions up to 2050*, World Business Council for Sustainable Development Geneva, Switzerland and International Energy Agency, Paris, France.
WHO: World Health Organisation 2014b, 7 *Million Premature Deaths Annually Linked to Air Pollution*, news release, Geneva, Switzerland, 25 March. Viewed 20 April 2015.
Williamson, T 2001, *APA Engineered Wood Handbook*, McGraw-Hill, New York, US.
Wilson, E 2006, *The Creation: An Appeal to Save Life on Earth*, Norton, New York, US.

Wilson, E 2013, 'Beware the Age of Loneliness', *The Economist: The World in 2014*, p. 143.
World Bank 2017b, *The Sunken Billions Revisited: Progress and Challenges in Global Marine Fisheries*, World Bank Environment and Sustainable Development Series, World Bank, Washington, DC, US. doi: 10.1596/978-1-4648-0919-4.
Worster, D 1993, *The Wealth of Nature*, Oxford University Press, New York, US.
WRI: World Resources Institute 2009, *Greenhouse Gas Emissions by Sector*.
WWF: World Wildlife Fund 2014, *Living Planet Report: Species and Spaces, People and Places*, WWF, Gland, Switzerland.
WWF: World Wildlife Fund 2016, *Living Planet Report: Summary*, WWF, Gland, Switzerland.

4 Population

Leaving rural France and arriving in Hong Kong for a rugby match, I wondered if I were in the future. Under an angry soot-stained sky in this swarming city, the buildings like close-packed daggers thrust around the harbour, there was graffiti in a dark alley. *As density grows, so does our loneliness,* it said in block-lettered English. In the crowded streets it felt true. Faces were the inscrutable of cliché, often weary-looking, and each apparently alone with their thoughts. But do people packed together equal loneliness any more than life with less human mass? And why in English in a Cantonese city? Is it that language is a source of isolation where density is greatest because difference becomes more obvious? Before the Industrial Revolution, the divergent dialects of separate rural groups mattered little because they seldom met. But isolation is made apparent when one cannot understand the language of a neighbour who may live only a thin wall alongside one's apartment.

This chapter evaluates the relationship of population to sustainability, how population policy has evolved over time and space, and how it may apply to a present and future world. It argues that while the rate of population change is critical, and non-coercive measures such as migration, education and contraception are more effective in checking growth in our numbers, a viable population policy will result in increasing density this century such is the result of momentum.

Half a century ago the 'population explosion' had the same central role in public discourse that 'global warming' does now (Friedman 2013). The Ehrlichs' 1968 *Population Bomb* was a runaway best seller. Population then was considered the central factor leading to poverty, unsustainability and catastrophe.[1] Policies designed to curb the growth in our numbers were applied over much of the developing world – by China with its one-child policy, by India with its sterilisation program, and by the United States as part of its foreign aid agenda.

Friedman's brief contemporary calculation of population density designed to test the belief that overpopulation caused poverty had a counter-intuitive result: the five most densely populated countries were all relatively affluent. They were Belgium and the Netherlands, both already rich and developed as well as the rapidly developing 'Asian tigers' of Taiwan, South Korea and Singapore. If Hong Kong were an independent country it would have been included as it had about ten times the population density of Singapore (ibid., para. 20).

Limits

As far as the wider impact of population is concerned, including poverty, this particular variable has a long history of contemplation and warning of dire consequence, beginning more than two millennia earlier and especially including the views of Adam Smith, Thomas Malthus and David Ricardo in the late eighteenth and early nineteenth centuries.

Not all are doomsayers, though. For example, one of the original authors of the chilling *The Limits to Growth* (Meadows et al. 1972), Jørgen Randers (2013), since concluded, based on demographic trends caused primarily by urbanisation and the availability of contraception, that the human world population will peak at less than nine billion by 2040 and then begin to decline. Europe, Japan, Russia and China have already begun to do so. Africa has already reduced its fertility rate from seven to five per woman during the past 50 years. UK science author, Fred Pearce agrees, having written about the 'coming population crash', and particularly links the rise of feminism and education as factors in declining birth rates (Pearce 2010; Eby 2010). Randers further advocates that it is both desirable and possible to reduce the human population to around four billion by 2100 given a focused global public policy. The central factor in this optimism is that while population still continues to grow, the rate of increase is slowing and may arrive at below the rate needed just to maintain our numbers, as Figure 4.1 illustrates.[2]

A zero population growth rate after 2050 does appear possible, but the achievement of any significant population reduction after that time seems unlikely. Since Randers' population control optimism, more recent UN estimates project that there could be more than 11 billion of us by the end of this century (Figure 4.2). Despite the observed decline in fertility rates, much of the increase will be from Africa (McKenna 2017), such is the momentum from earlier increases in our numbers.

Making accurate predictions of total human population over the long-term is particularly difficult because of the cumulative effect of variables affecting birth and death rates. For those specialised in disciplines other than demography, it is especially so. For example, the Cambridge astronomer, Fred Hoyle, in a 1964 discussion on the desirability of populating the galaxy, indicated that

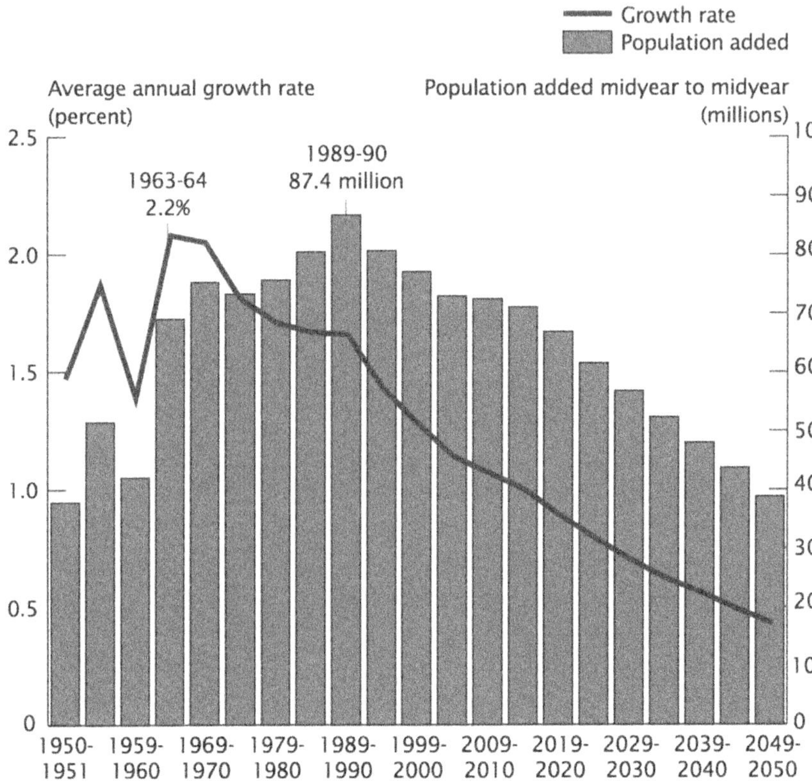

Figure 4.1 World population growth rates 1950–2050.
Source: US Census Bureau, Global Population Profile 2002, p. 15.

"some 10^{10} humans … will be on Earth by the year A.D. 2000" (Hoyle 1964, p. 42) – ten billion people rather than the little more than six billion it actually was at the end of the millennium. The difference le between his prediction and reality is remarkable. It may have been due to to his discipline as an astronomer which, as mentioned above under *Impact*, is reputedly unconcerned by imprecision. It may have been due to the widespread concerns of the population alarmists at the time; or perhaps simply due to a flawed assumption that the very high population growth rate at the time would continue.

Fears

Fears of both over and under-population have in the past gained sway over relatively short periods of time. In writing about the connections between early

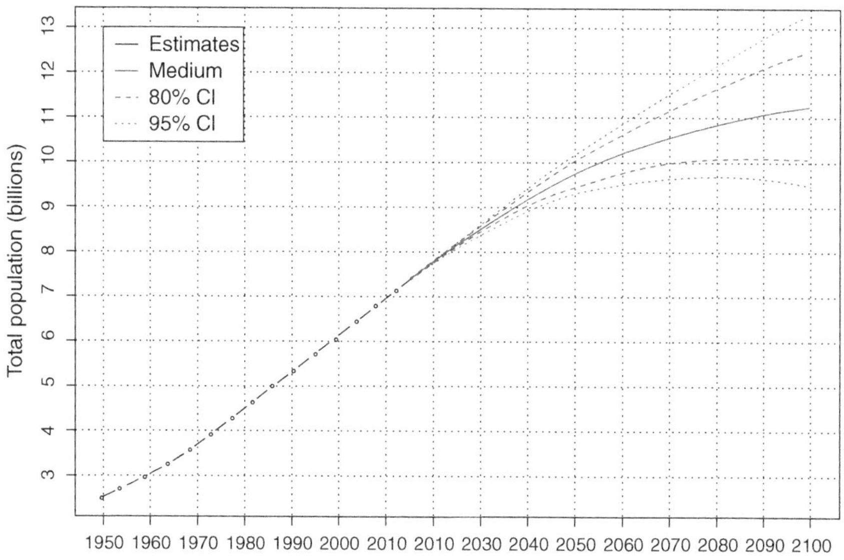

Figure 4.2 World population projections 1950–2100.
Source: United Nations, Department of Economic and Social Affairs, Population Division 2015, *World Population Prospects: The 2015 Revision, Key Findings and Advance Tables,* Working Paper No. ESA/P/WP.241, p. 2.

capitalism and slavery, Eric Williams (1944, p. 16) observes that in relation to Britain, "fear of overpopulation at the beginning of the seventeenth century gave way to a fear of underpopulation in the middle of the same century".

This fear was based on the mercantilist economic argument that in order to compete with other countries it was best to reduce costs by paying low wages. And the way to ensure low wages was to have a large population. The mercantilist economist and later governor of the British East India Company, Sir Josiah Child, in commenting on the colonisation of America as an exception, still advanced the then majority opinion that "whatever tends to the depopulating of a kingdom tends to the impoverishment of it" (Child 1751 [1668], p. 134) and further:

> Most nations in the civilised parts of the world, are more or less rich or poor, proportionally to the plenty or paucity of their people, and not to the sterility or fruitfulness of their lands.
>
> (Ibid.)

While this argument – that greater wealth arises from cheap labour – contains obvious contradiction,[3] it retained popular currency in that time. A century later, however, Adam Smith disagreed. Writing at about the period of the American

Revolution and the beginnings of industrialisation, he drew attention to populous China and the impoverishment of its citizens. He referred to China as an example of an advanced nation that had changed little since the time of Marco Polo, where its economic stagnation produced a general poverty that "far surpasses that of the most beggarly nations of Europe" (Smith 1952 [1776], pp. 31–32). Having noticed that every species multiplies according to its means of subsistence and not beyond it, Smith favoured economic growth rather than absolute numbers of people as the path to national wealth, a view that again holds sway, so far, in the current century.

Malthus

At the edge of the nineteenth century and a generation after Smith, the cleric and Cambridge political economist Thomas Malthus (1766–1834), himself one of eight children, found popularity, influence and controversy with his best-seller, *Essay on the Principle of Population* (1798) that went through six editions. This is the work to which much of the modern concern about overpopulation can be traced. The essence of his 'principle' is that food supply limits population growth because it can only increase arithmetically, whereas population can increase geometrically. Thus Malthus linked overpopulation to poverty, encouraging harsh measures aimed at making the poor more 'responsible' and less likely to reproduce themselves. In fact, Malthus' views were developed as another Victorian 'iron law', which implied that poverty, war, abstinence and birth control were needed to hold the population in check lest food supplies ran out.

Malthus' views were provocative to say the least. In the second edition of his *Essay*, he included the following before it was removed in later editions due to the antagonism it created:

> "A man who is born into a world already possessed, if he cannot get subsistence from his parents on whom he has a just demand, and if the society do not want his labour, has no claim of right to the smallest portion of food, and, in fact, has no business to be where he is. At nature's mighty feast there is no vacant cover for him. She tells him to be gone, and will quickly execute her own orders, if he does not work upon the compassion of some of her guests. If these guests get up and make room for him, other intruders immediately appear demanding the same favour.
>
> (Malthus 1803, p. 531)

According to Garrett Hardin (1998, p. 182), pointedly writing during the bicentennial of Malthus' essay, while those words almost guaranteed controversy, they also "walked right past the central problem of population" – the denial of "limits in the supply of terrestrial resources".

In a Malthusian spirit, the UK poor law was amended in 1834 to preclude any public money or assistance to the able-bodied. Malthus' friend, David Ricardo (1772–1823), had urged such a measure, writing in 1817 that it was in the best interests of the poor:

> It is a truth which admits not a doubt that the comforts and well-being of the poor cannot be permanently secured without some regard on their part, or some effort on the part of the legislature, to regulate the increase of their numbers, and to render less frequent among them early and improvident marriages. The operation of the system of poor laws[4] has been directly contrary to this. They have rendered restraint superfluous, and have invited imprudence, by offering it a portion of the wages of prudence and industry.
>
> (Ricardo 1911 [1817], p. 61)

But while Ricardo and Victorian biologists such as Wallace and Darwin were impressed by Malthusian ideas in relation to the "survival of the fittest", other political economists were appalled. Socialists of the era especially tended to believe that poverty was rather a result of maldistribution or economic inequality, reputedly leading the French anarcho-socialist Pierre-Joseph Proudhon (1809–1865) to declare that "*there is only one man too many on Earth, and that is Mr. Malthus!*"

In direct refutation of Malthus, Friedrich Engels, for example, wrote the following tirade:

> Malthus...maintains that ... the inherent tendency of the population to multiply in excess of the available means of subsistence is the root of all misery and all vice. ... The implications of this line of thought are that since it is precisely the poor who are the surplus, nothing should be done for them except to make their dying of starvation as easy as possible ... it is after all better to establish a state institution for the painless killing of the children of the poor. ... Charity is to be considered a crime, since it supports the augmentation of the surplus population. Indeed, it will be very advantageous to declare poverty a crime and to turn poor-houses into prisons, as has already happened in England.
>
> (Engels 1844)

Engels was unconcerned about human population pressure because:

> The productive power at mankind's disposal is immeasurable. The productivity of the soil can be increased ad infinitum by the application of capital, labour and science ... day by day science increasingly makes the forces of nature subject to man.
>
> (Ibid.)

Others were not so sure. The American economist, Henry Carey (1793–1879), although optimistic about the benefits of progress and economic growth, was nevertheless "more than half agreed with Malthus" on the procreative power of humanity (Galbraith 1976 [1958], p. 40). Carey wrote that the time may come "when there will not even be standing room" (quoted in Gray 1931, p. 256). At least, despite the power of compound increase, we are not yet at that point.

A corollary to Malthus written by Ricardo further demonstrates the interplay between population, affluence and technology. Ricardo contended that the level of income at which the Malthusian population would be stable depended on the tastes of the poor. In a remarkable insight, he argued that the more luxurious the tastes of the masses, the more likely population would stabilise, as the relative cost of an additional child was an incentive to hold down the birth rate. Hence he concluded that the "friends of mankind" should wish that the workers have more luxurious tastes (cited in Friedman 2013). Before the industrial revolution, less attractive "friends of mankind" held the population in check. These included the scourges of "war, violence, disorder, harvest failure, collapsed infrastructures, [and] bad sanitation" (Clark 2007, p. 5).

It was only after 1800 in the West that population growth escaped from its long flat trajectory – before held back by such scourges, especially the food supply. With accelerating production technologies, including in agriculture, affluence increased and human population began to exceed its Smith-Malthusian boundaries.

In the East, Japan now leads an apparent transition to a much lower population. Its numbers peaked at about 128 million in 2008 and is in subsequent rapid decline, currently losing about 250,000 a year and the rate of loss is increasing:

> … the Japanese population gradually grew from the beginning of the Meiji period, increased by 70 million in the 140 years up to 2008, and is about to decrease by 70 million in the 90 years leading up to the year 2100.
>
> (Masuda 2015)

This means that the Japanese population is now decreasing faster than its earlier explosive growth during the twentieth century. Even if Japan does increase its birthrate significantly in line with national targets, its population will not increase because there are now too few women of child-bearing age. Geographically, the effect of population decline is evident in the concentration of people in Tokyo and the depopulation of the countryside, such that Tokyo's numbers will not begin to decline until after 2020, whereas half of all municipalities have faced severe population loss since 2010 (ibid.). While especially foreign commentators point to the obvious measure of increasing immigration to slow population loss (Panda 2014) the Japanese cultural attitude to blood purity and unique identity[5] prevents any implementation. Tokyo University's Hiroya Masuda, for example, fails to even mention it in his public policy lecture cited above. Ninety-eight percent of its population are ethnically Japanese. As population shrinks, the proportion of older – and more culturally conservative – people grows, making policy change increasingly difficult politically. At the same time the concentration of the population in Tokyo further depresses fertility – lengthy commute times, high expenses and extended work and study hours in the big city are not conducive to breeding.

However, especially if Japan is a miners' canary for population, it is the *rate* of decrease that is of policy concern rather than where Japan's numbers ultimately stabilise. Some very small countries (Norway, Sweden, Switzerland) seem to cope well on the human development index (HDI). Absolute population is largely irrelevant to economic success. But a teetering, runaway population plunge distorts the demographics and, at least over two or three generations, leads to a rapidly ageing nation. In Japan, the void of young people is being filled with robots.

Optimism

In a spirit of technological optimism similar to Engels, but with more respect for nature than to require its subjugation, Arthur Clarke foresaw a transformed future. This twenty-third century world, in which the "endless forests of the American Midwest…so much the Earth must have looked in ancient days" and which were recently all farmland, are restored. Wolves, buffalo and the grizzly bear roam free from the threat of humankind (Clarke 1975, p. 116). The twenty-first-century return of much New England farmland to the wild and the recent re-introduction of wolves into the world's first national park, Yellowstone, indicates this concept is feasible. This 'rewilding' has in a short time cascaded[6] positive environmental effects including the return of several species and even the change of the course of rivers (Monbiot 2014b). But Clarke's future Earth is based on a much reduced population of only a few hundred million, albeit with additional human colonies established on the moons of Saturn.

Randers had earlier predicted environmental, resource and human catastrophe this century due to the exponential growth of our species and its impacts. Early computer modelling helped explain such scenarios. However, the US-based libertarian Foundation for Economic Education believes that the risks of overpopulation are overstated. In *Overpopulation: The Perennial Myth* libertarian David Osterfeld (1993) writes that "the prospect of the Malthusian nightmare is growing steadily more remote." In support he quotes the second century Carthaginian, Tertullian (A.D. 160–c. 220), a Christian priest who lived among a world population of about 190 million:

> Our numbers are burdensome to the world, which can hardly support us... In very deed, pestilence, and famine, and wars, and earthquakes have to be regarded as a remedy for nations, as the means of pruning the luxuriance of the human race.
>
> (quoted in Osterfeld 1993)

During the 1960s and despite the predominance of Ehrlich's *Population Bomb*, Buckminster Fuller was typically optimistic about population growth:

> The population explosion is a myth. As we industrialise, down goes the annual birth rate. If we survive, by 1985, the whole world will be industrialised, and, as with the United States, and as with all Europe and Russia and Japan today, the birth rate will be dwindling and the bulge in population will be recognised as accounted for exclusively by those who are living longer.
>
> (Buckminster Fuller 1969, p. 43)

Other than Tertullian, early concerns were expressed about overpopulation by, for example, Confucius in the sixth century B.C., Plato and Aristotle in the fourth century B.C., as well as by the sixteenth-century Italian Botero. And today, Malthus' shadow still looms over contemporary science. An article, 'The Millennium Assessment', in *BioScience* (Powledge 2007a) that relates the increasing pressures on ecosystems attracted this criticism in the next issue:

> What I find lacking in the whole approach, not only in this article but in the profession as a whole, is the failure to openly recognise that none of it matters as long as we fail to correct the underlying cause of nearly all of our problems: overpopulation.
>
> (Bennett 2007, p. 101)

In his response, the article's author points out that in fact there was reference to overpopulation as a fundamental cause: "increasing consumption per person, multiplied by a growing human population, are the root causes of the increasing demand for ecosystem services". However, he laments that what is missing from most such discussions is a plan for dealing with overpopulation that the world will accept. "That's a terribly difficult cat to bell" (Powledge 2007b, p. 101).

Lebensraum

Population is one thing. Space to accommodate it is another. Alison Bashford, in an article on Karl Haushofer's geopolitics of the Pacific, draws attention to the notion of *Lebensraum,* infamously associated with Nazi policy in the second quarter of the twentieth century. The concept was promoted by Haushofer (1869–1946) through his writings around the turn of the century. The link with Nazism is quite direct: Haushofer visited his assistant and former student, Rudolf Hess, in a Munich prison in 1924. He also visited a friend of Hess in the same prison, one Adolf Hitler, who was then writing *Mein Kampf*, and confined as he was personally, may well have felt a particular attraction to the concept of 'living room'. Certainly the loss of the German Pacific colonies[7] after the Great War was used as a partial justification for Germany's quest for eastern expansion during the Nazi era (Bashford 2012, p. 121).

The concept is also associated with the ideas of Malthus and Darwin both. The German zoologist-anthropologist-geographer, Friedrich Ratzel (1844–1904), coined the term "Lebensraum" in the late nineteenth century (Smith 1980, p. 52). Ratzel was influenced in turn by the biologist Ernst Haeckel (1834–1919), who had earlier coined the term "ecology".[8] Both were familiar with Malthus and Darwin and were especially interested in the concept of life forms competing within a confined habitat. Ratzel observed that it was no accident that Malthus' essay on population derived from an island nation, where population pressures were often more apparent. Confined countries could either (positively) expand through emigration or colonisation, or lacking new territory, (negatively) put a low value on human life while remaining within the same boundaries. By this logic, the United Kingdom (as well as France and Germany) sought virtuous outlet through colony and empire. Japan, another populous island nation, sought Lebensraum in Manchuria and Micronesia, much of the latter formerly German. But in this latter respect, where there was no accessible new territory, Pacific islanders had been observed to practise ways of limiting population growth – cannibalism, sacrifice, infanticide and gender segregation on different islands, amongst others (Bashford 2012, p. 132).

Haushofer was closely associated with this thinking and worked with the military in Japan before the Great War (ibid., p. 121). Interestingly, Haushofer saw Australia as an antipodean Germany. Not only hemispherically opposite, but also as an island continent of few people rather than a confined nation of many (ibid., p. 134). He also saw Australia as a country that, with California, had implemented the Pacific "colour line" where Asian immigration was forbidden as a result of experience with the overly industrious Chinese diaspora during the mid-nineteenth century gold rushes. And at the 1919 Paris Peace Conference Australia had helped resist Japan's attempt to propose a clause about racial equality in immigration (ibid., p. 132). As an aside on population representation, when at the peace conference US President Wilson queried Australia's right to annex former German colonies in the Pacific on the basis that its five millions were opposing the wishes of twelve hundred millions, Prime Minister Billy Hughes retorted that he represented not five million living but rather 60,000 dead – those sacrificed during the war (DVA 2015).

The gleam of Lebensraum lingered on well beyond Haushofer and fascist dreams, however. Osterfeld in his 1993 article on the 'myth of overpopulation' mentioned above, contains a paragraph intriguingly sub-headed '*Living space*':

> But even if food and resources are becoming more abundant, certainly this can't be true for living space. After all, the world is a finite place and the more people in it, the less space there is for everyone. In a statistical sense this is true, of course. But it is also irrelevant. For example, if the entire population of the world were placed in the state of Alaska, every individual would receive nearly 3,500 square feet of space...Less than one-half of one percent of the world's ice-free land area is used for human settlements.
>
> (Osterfeld 1993, para. 18)

Despite this sceptical view the population issue is not so much about space for individuals. It is about space for one species alongside habitable space for all other life forms at the same time. To argue otherwise denies the more threatening issue of sustainability.

This mix of population policy, migration, racism and geo-politics illustrates how complex the issue of population is. While there is still much resettlement due to war and poverty, now in a different era, there is no more *Lebensraum*. The idea of nationalist expansion driven by population pressures, whether contiguous to a country's borders or in distant colonies, seems ridiculous. Short of an unlikely *Battlestar Galactica* forced migration through space, or an "impossible journey" to a distant exoplanet (Kerins 2017), there just is no more room, no more political will, or any more need. The planet is obviously limited. Larger human populations over this century must be dealt with, not by Polynesian techniques, but by increasing densities – through migration and urbanisation.

Urbanisation

The world passed a milestone in 2008. For the first time in history, more people now live in urban than in rural areas. And while urbanisation has "increased tenfold in the past century" (Crutzen and Stoermer 2000, p. 17), the rate is still accelerating. So-called 'migrant workers' from rural western China still flood the industrialised east, especially since Deng Xiaoping's market-based economic reforms of the late twentieth century. The productivity of agribusiness means that many fewer workers are required in agriculture beyond subsistence. South Africans can have a better existence in Capetown than on the Veld. Formerly rural Tunisians and Senegalese increasingly populate the suburbs of Paris and Marseilles.

The proportion of urbanised people is more than 54 percent (UNDESA 2014, p. 2). And contrary to popular perception, cities are often associated with environmental benefit: it is easier to provide environmental services where people are more concentrated and urban populations are more demanding of them, as the Beijing administration is now acutely aware and as London and Pittsburg were a few generations ago in relation to air pollution. Cities strive to provide clean water, sewage treatment and coordinated waste disposal. A concentrated population not only demands them, but makes such facilities more economically feasible.

However, the megacity centres of Lagos, Mexico City, São Paulo and Dhaka for example, are encircled by slums, in which there are no basic services such as safe drinking water and sanitation, where tenure is insecure, housing lacks durability and there is overcrowding (UN 2014a, p. 46). According to the UN, the proportion of the urban population who are slum dwellers in the developing world declined from 47 to 37 percent between 1990 and 2005, an apparently dramatic and welcomed improvement. But unfortunately, closer inspection of the *Millennium Development Goals report* shows that this apparent advance was mainly due to a change of definition:

> In 2005, only a proportion of households using pit latrines were considered slum households, whereas in 1990 and 2001 all households using pit latrines were counted as slum households. The change affects estimates mostly in those countries where the use of pit latrines is more widespread, as in Sub-Saharan Africa.
>
> (UN 2007, p. 26)

However, within a further seven years there appeared to be a more genuine improvement with less than 33 percent of the urban population who were slum dwellers in the developing world. Such was the increasing rate of urbanisation combined with a growing population in the developing world, however, that this represents more people in absolute numbers than in 1990 – 863 million in 2012, compared with 650 million in 1990 (UN 2014a, p. 46). However according to the same UN report, there is new hope in technology: one partial answer to slum improvement is to increase the typically minimal amount of urban land allocated to streets. More and wider streets means that service arteries can be extended more easily. Mapping and planning slum upgrades can be facilitated by geospatial information systems. As a result of this approach it is known that Dar Es Salaam allocates only four percent land to streets outside its city centre. The ideal is about 25 percent (ibid., p. 47).

But while increasing concentrations of human beings, slums or not, is continuing, what does this mean for sustainability? Certainly it appears that there are economic benefits, and possibly there are further advantages. According to former cosmologist and now New York University growth economist, Paul Romer (2015), "policy-induced changes in the urban share of the population could have big effects on GDP per capita and ... the quality of life for billions of people". If both higher incomes and quality of life are linked with sustainability in the sense of the theoretical environmental Kuznets curve (EKC) that says environmental impact eventually reduces with affluence, then sustainability benefits may be real. Although, as discussed in more detail in the chapter on affluence, waste and carbon emissions may be exceptions to this principle, urbanisation and sustainability are partners. People living closely together facilitate the sharing of ideas, which can contribute to both economic and environmental improvement. Higher population densities in cities implies lower birthrates and less population pressure on rural land. And it is surprising how much of the territory of Hong Kong is uninhabited.[9]

Migration

Migration affects population. Globally of course migration makes absolutely no difference to the number of people on the planet at any one time. However, it does make a difference *over* time. This is because people tend to migrate from poorer to richer areas, whether attracted by better standards of living, or displaced by poverty, conflict or disaster. The continuing urban drift within countries is also part of this, typified by the millions of migrant workers from rural western China lured to the East in recent decades.

Between countries there is net migration from the global South to the global North, including from Africa to Europe (especially from former colonies), from South Asia and the Middle East to both Europe and North America, from Latin America to the United States and Canada, from the Pacific and the Middle East to Australasia, and from central Africa to South Africa.

However, it is the effect of the host culture and rising affluence that tends to reduce fertility within immigrant groups, as high birth rates are associated with rural poverty, lower birth rates with urban affluence. While it may take generations for this reduction to play out, there are fewer incentives in wealthier circumstance to have large families, and more incentives to produce fewer children. Thus immigrant birth rates tend to match the lower birth rates of the host urban, wealthier culture over time. While longevity tends to increase in wealthier environments due to better hygiene, welfare and medical services, the overall effect is that the rate of population growth tends to fall as the reduction in birth rates more than offsets the increased longevity.

On this basis, then, sustainability might be enhanced by migration as ultimately it means relatively fewer people. However, this would only be true if the increased affluence and its associated consumption did not result in proportionately greater environmental impact. The state of much of eastern China does not support this possibility at present, especially with regard to emissions and chemical pollutants that have been widely observed in the international media. However further development along greener lines may yet do so as implied by the EKC.

Elephants

US comedian Doug Stanhope alludes to the "elephant in the room" misperception about overpopulation. In his 2010 monologue *Abortion Is Green*, Stanhope says that "the major problem is obviously overpopulation" but that the issue never arises. "Even Al Gore has to be pressed to admit it's even a problem" he says, but if as Gore says "it's gonna correct itself by 2050, I guess fucking will go out of style". Moreover, individual efforts to reduce environmental impact are trivial. Ironically, the "combined uteri" of the environmentally aware in the US "wreak more havoc to the environment than a thousand Dow Chemical Corporation accidents combined" (Stanhope 2010).

In *Sex and Destiny: The Politics of Human Fertility*, Germaine Greer anticipated this last assertion in 1985. For her it is the overconsumption of the West interdependent with the 'serfdom' of the East and South that has caused environmental havoc, not the numbers of Indians living in poverty (Greer 1985, pp. 409–412). Her book contains pointed criticism of the Ehrlich's trip to India, from which their book *The Population Bomb* was conceived. The Ehrlich's middle class American sensibilities were shocked by one experience whereby, in a temperature "well over 100°F" and in a taxi "hopping with fleas":

> The streets seemed alive with people. People eating, people washing, people sleeping. People visiting, arguing and screaming. People thrusting their hands through the taxi window begging. People defecating and urinating. People clinging to buses. People herding animals. People, people, people, people. As we moved slowly through the mob, hand horn squawking, the dust, noise, heat and cooking fires gave the scene a hellish aspect.
>
> (Ehrlich 1968, p. 1)

For Greer however, it is normally hot in the tropics, she has never encountered a flea-ridden taxi in India and where the Ehrlichs were taken was probably less populated than Manhattan. It was just that all the people were at ground level, rather than stacked many stories above. Greer does not know how many people the Earth can support. She says though that "it is quite probable" the world has been overpopulated for some time. Nevertheless, it is clear that it can support far more people at a low calorie (as in contemporary India) than a high calorie intake (as in the United States).

From a contemporary business perspective, Joon Yun is concerned about a different elephant – depopulation rather than overpopulation. He notes that falling fertility and a future global depopulation means that demand for natural resources "could wane" leading to deflation in commodity prices, land, housing and debt. While "much would depend on whether the central banks would choose to allow low-grade deflation or otherwise continue to target low-grade inflation in the face of declining aggregate demand" Yun (2012) asks "where would yield be found in a depopulating world facing the forces of deflation?" This, he says, is in aged care and health services — an area in which he has a particular interest as CEO of a company that services such industries.

Both these views illustrate the tensions that are inherent in population policy and sustainability. As Stanhope implies, the issue is an unspoken concern because it is so difficult to deal with. And as Joon Yun infers, the depopulation that is desirable from a sustainability perspective runs counter to an economic system built on the growth in demand that increasing numbers underpin.

Perspectives

There are therefore different perspectives on the population problem both over time and between disciplines, as well as according to self-interest. However, for this book they tend to resolve into two broad questions: (1) Is overpopulation a critical issue for sustainability? and (2) What should be done about it in public policy terms?

In assessing the first question, it has been observed (Dorling, Crutzen and Stoermer, Purdy, Steffen, Crutzen and McNeill) that the last 200 years have been the most extraordinary period of human history. In that time population has increased tenfold and carbon dioxide levels risen significantly, both associated with the Industrial Revolution including the rise of public health measures and the burning of fossil fuels.

Still, it is revealing to undertake some mathematical modelling of the increase in human population. One absorbing question is how many people were on the planet around 10,000 years ago at about the dawn of agriculture, when our species first began to break from the immediate growth limits of the environment. Starting with an approximation of the current world population (7 billion) and assuming a net increase of 0.1 percent per year on average, then there were 319,000 people on the planet at that time.[10]

But assuming an average growth rate of 0.2 percent per year, there would have been *no-one* on the planet then, and there would have been only 17 people on the planet 1000 years ago. Therefore, the former average growth rate of 0.1 percent is clearly more accurate. The current world population growth rate is now far greater – around 1.4 percent per year and estimated to decline to around 0.5 percent by 2050 (Worldometers 2014). Looking 200 years into the future, a growth rate of only the long-term average of 0.1 percent would result in a definitely unsustainable 51 billion people on the planet! Hence it is important, or rather critical, that zero or negative population growth is effected.

In 1994, Gretchen Daily and the Ehrlichs concluded that an optimum world population would be between 1.5 and two billion people to enable both a creative critical mass, as well as to maintain biodiversity: "homo sapiens should foster the continued existence of its only known living companions in the universe" (Daily, Ehrlich and Ehrlich 1994). Garrett Hardin (1998) believes that there is no question about the need for population control. The key issue is how it can be achieved with minimal pain:

> The problem is simply this: can the necessity of population control be reconciled with the apparent demands of individualism, as that complex concept has developed since John Locke? I conclude that there is a fatal contradiction between these two necessities; and that the survival of civilisation will require us to modify significantly the powers we now grant to individual 'rights.' This social revolution will be painful, but it cannot, I think, be successfully evaded.

Dietz and O'Neill (2013) also say that overpopulation is a critical issue and that it should be stabilised using 'compassionate, non-coercive' methods, which may be less than politically palatable as a slogan or policy if only because it eerily echoes George W. Bush's 2000 presidential platform of 'compassionate conservatism'. An extreme view of the desirability of reducing human population is the *Voluntary Human Extinction Movement*, which has an extensive website and links on the issue:

> Phasing out the human race by voluntarily ceasing to breed will allow Earth's biosphere to return to good health. Crowded conditions and resource shortages will improve as we become less dense.
>
> (Knight 2016)

Of course it was not always thus. Further back in time, and as discussed in the 'technology' chapter, Homo sapiens was once itself an endangered species. DNA and archaeological evidence suggests that around 150,000 years ago our ancestors came close to extinction. This was due to an Ice Age that affected Africa – leaving perhaps an absolute total of only hundreds of people clinging to an existence on the southern coast. Genetic studies indicate that everyone now living is descended from this tiny number of ancestors (Marean 2010). We have come from a smudge of mere hundreds of us then to the prospect of a swarm of 51 billion of us by the twenty-third century. Between these extremes lies sustainability.

Anthropocene

Contrary to the long-held views of Paul Ehrlich and others, Andreas Malm and Alf Hornborg (2014, p. 4) argue in the *Anthropocene Review* that population growth is not a significant causal factor in the establishment of the Anthropocene epoch, in which the impact of humanity is profound. They point out that between 1820 and 2010, carbon dioxide emissions increased by more than 600 times, while population increased by less than seven times, "indicating that another, far more powerful engine must have driven the fires". And recently, there has been a negative correlation between population and carbon emissions – since 1985 population tended to rise fastest where emissions grew slowest, and vice versa (Satterthwaite 2009). However, much of this relationship is due to the unique case of China, which has many people but low population growth and high emissions growth due to recent industrialisation.

A significant chunk of humanity is not party to the fossil economy at all. Hundreds of millions rely on non-fossil charcoal, firewood, or dung for all domestic purposes. Satterthwaite (2009, pp. 547–550) concluded that one-sixth of the human population "best not be included in allocations of responsibility for GHG emissions", because their contribution is close to zero. Moreover, two billion people, or nearly one-third of humanity, have no access to electricity, and so "the difference in modern energy consumption between a subsistence pastoralist in the Sahel[11] and an average Canadian may easily be larger than 1,000-fold" (Smil 2008, p. 259). Given these enormous variations over both time and space, human population seems too slender an abstraction to carry the burden of causality.

Some of the apparent issue with population and sustainability is to do with simple accounting. We tend to divide nature amongst how many of us there are: *per capita*. Natural resources might appear to shrink simply because there are more people for them to be divided amongst:

> the general trend is that population has been growing in most of the countries, exacerbating thereby the decline in natural capital [per capita] growth rates, as resources are accounted for among a larger number of people. The growth in population explains more than half of the changes in natural capital per capita in 13 out of 20 countries.
>
> (UNU-IHDP and UNEP 2012, p. 51)

Will Steffen (personal communication, 2014)[12] points out that most of the economic growth in the last century did not result from population growth. Countries with higher population growth rates had lower increases in total GDP. This is consistent with both Satterthwaite and Smil above: impact depends on how much each person consumes, not just how many people there are. Of course politicians tend to gloss over the difference between growth in national GDP and growth in GDP *per person*. Without the population denominator it is not possible to know whether average affluence has increased, yet most public announcements of GDP growth are naked of how much they may be the result of population increase.

Still, popular opinions about the culpability of population are passionate and diverse, and may have to be politically accommodated. Online responses to a review of Elizabeth Kolbert's 'The Sixth Extinction: An Unnatural History' published in The Guardian (McKie 2014) included comments like "depressing", accusations of "climate denier paid for troll", "you forgot about drastically reducing the human population", "It's not so much 'save the planet', but 'save the humans'", as well as the succinct: "less people = less extinctions".

Nevertheless, Fischetti (2014, p. 80), writing in the *Scientific American* has a sobering message about total population: that it will continue to grow beyond 2100 now seems more likely. While Dorling (2013) and others were predicting that population would peak at less than ten billion by mid-century due to declining fertility in all regions, more recent evidence suggests that fertility in Africa may not decline as much as indicated and that as a result, global population may continue to increase to 10.9 billion by the end of the century.

Public policies

Aldous Huxley's utopian island of *Pala* introduced a population policy following experience of famine by its founder and new population growth spurred by better health and nutrition. In seeking a way between the unpleasant Malthusian horns of famine and 'moral restraint', the islanders decided that contraception "should be like education – free, tax-supported and, though not compulsory, as nearly as possible universal" (Huxley 1979 [1962], p. 97).

In reality, the two largest countries by population, China and India, together account for more than 2.6 billion people, well over a third of the world total. Both have attempted to control population through public policies.

China's 'one-child policy' was introduced in 1979 in order to curb an explosive population growth that stemmed from the peace after the 1948 revolution. It was renounced only in 2015. The policy rewarded couples who had just one child with cash bonuses and better housing, and discouraged larger families with fines, forced abortions and official examinations for pregnancy by village family planning officers (Jian 2013). The one-child policy was resisted, especially in rural areas, where the predominance of agriculture dictates larger families; and within minority ethnic groups. In the first case two children were tolerated where the first was a girl, whereas in the second case the family was exempt. Nevertheless, the birth rate of 1.4 children per woman is now well below the rate of 2.1 needed to maintain the population level. As a result, China's population is ageing as the spectacular economic growth rate of the past three decades is slowing. The one-child policy has also resulted in a skewed ratio of men to women because boys are traditionally preferred to girls among Chinese families and girl fetuses tend to be aborted more often than boys. Hence many men cannot find a marriage partner in China and look to other countries for brides.

India's national family planning program began earlier – in 1952. It has since promoted contraceptive use, especially sterilisation, among married women. The 1992–93 National Family Health Survey (NFHS) found that almost 41 percent of married women used contraception. Of those, two thirds had been sterilised, plus a further nine percent of their husbands (Adlakha 1997, p. 3). After declaring a state of emergency in 1975, then Prime Minister Indira Gandhi led an aggressive campaign under which more than six million sterilisations, many forced, were performed. After violent protests, Mrs. Gandhi's party lost office at the 1977 elections and the number of sterilisations declined. However, since those times sterilisations have again been adopted as policy, more involving cash incentives. But risks remain high: one hurried mass sterilisation of 83 women by one surgeon in one day resulted in 12 deaths from infection in 2014. The women were paid about 600 rupees, or about ten US dollars each (Barry and Raj 2014). Sterilisation remains common in India because it is cost-effective. It is also common in poorer countries such as Indonesia, especially among men.

The Chinese government is more authoritarian than its democratic Indian counterpart and this may have contributed to its more effective population control, although poverty reduction and urbanisation were also factors. Population control has been achieved at great cost, however. Women have had to undergo public scrutiny and forced abortions. The demographic profile of the country has distorted both between age groups and gender. The Indian program included forced sterilisations and still results in death and suffering from inadequate medical procedures. Its effects fall mainly upon the poor. As a result, neither country has succeeded in using coercive policies to reduce population growth without consequent resentment.

The opposite approach was applied during the 1960s when the Romanian government banned both contraception and abortion in an attempt to *increase* its population, following considerable decline during the World War II. But while the birthrate initially tripled, it soon returned to lower levels as families were too poor to support the extra children. Illegal abortions became widespread. Maternal mortality tripled. Many children grew up in orphanages. The architect of the policy, Nicolae Ceausescu, was executed when his government fell in 1989. The new government swiftly repealed the bans on contraception and abortion (Meadows 2008, p. 114).

This is not to say that population control cannot be achieved through non-coercive and less direct means. In crowded Bangladesh the demand for contraceptives outstrips supply[13] (Streatfield and Kamal 2013), but most families use family planning methods. Population growth slowed from over three percent in the 1970s to 1.34 percent in 2011. Educational and awareness programs have helped drive these results (Bangladesh Government 2012). In Ghana, women without education have an average of 5.7 children, women with secondary education have 3.2 and those tertiary educated 1.5 (Carrington 2014). Italy has changed from high fertility to a country almost devoid of children in two generations with the lowest birthrates in 150 years (Stanton 2015). Japan and Russia now both have declining populations, as would much of the developed world but for immigration. The public policy task is to apply such policies appropriately so that popular resentment is avoided. Poverty reduction and women's access to education are both proven indirect means of reducing population pressures. Consistent with Huxley's *Pala* (1979), according to the UN the relatively simple measure of providing access to dependable contraception may be the single greatest contribution to population restraint that can be made (Islam 2015, s.8).

Evaluation

While the issue of human population growth appears less important now that the rate of growth is slowing, our mounting numbers over the present century are a significant threat to sustainability due to the resources required and the waste produced, in a world where we are already exceeding the capacity of the planet.

Recognising that it is very difficult to halt the momentum of population growth due to its course over generations, the key question is how to contain population increases without using coercive methods. The recent population decline in Japan and Russia and the natural decrease in much of Europe[14] have occurred without compulsion and indicate how public policy might be applied. Migration, industrialisation, urbanisation, education, gender equality and the availability of contraception are all agents of population restraint. To encourage these agents will be to brake our impact upon our planet where there are already too many of us for the Earth to support.

Otherwise, inevitably, much of the world will look increasingly like Hong Kong does today – but without the beauty of its harbour. More like Mexico City or Lagos. Like Karachi or Beijing or Chongqing. If our civilisation remains viable and irrespective of an increase in loneliness, greater density through urbanisation seems inevitable. Sustainability depends on it.

Notes

1 "In the 1970s and 1980s hundreds of millions of people will starve to death in spite of any crash programs embarked upon now" (Ehrlich 1968, p. xi).
2 The sharp decline in population growth in 1959–1960 was due to famine in China.
3 The wealth must not accrue to the labourers, who would otherwise not be cheap.

4 The *Speenhamland* decision of 1795 in England in effect guaranteed to labourers a minimum income based on the price of bread irrespective of their production (Polanyi 2001 [1944]).
5 *Nihonjinron* (日本人論) is a concept of Japanese national and cultural identity, based on ideas of the Japanese race as unique in an isolated archipelago with a unique climate and way of thinking.
6 That is from a top predator down to other species, to flora and to landforms.
7 The German Pacific colonies comprised New Guinea, the Northern Solomons, much of Micronesia and Samoa.
8 Coined around 1870, as Ökologie in German. It appeared in English in 1873 (White 1967, p. 1204).
9 Only 21.1 percent of Hong Kong's land area is built-up (Wang and Lau 2002, p. 113).
10 This is roughly consistent with Purdy (2015), who estimates that the global human population was between one and ten million until the start of the agricultural revolution.
11 Region South of the Sahara Desert in Africa.
12 Personal email of 26 September 2014.
13 Largely due to bureaucratic procedures and procurement logistics.
14 Population growth Europe tends to be from immigration rather than natural increase.

References

Adlakha, A 1997, *Population Trends: India*, International Brief, US Bureau of the Census, Department of Commerce Economics and Statistics Administration, IB/97-1, April.

Bangladesh Government 2012, *Bangladesh Population Policy*, Ministry of Health and Family Welfare, Government of the People's Republic of Bangladesh.

Barry, E and Raj, S 2014, 'Twelve Women Die after Botched Government Sterilizations in India', *New York Times*, 11 November. www.nytimes.com/2014/11/12/world/asia/botched-government-sterilizations-india.html.

Bashford, A 2012, 'Karl Haushofer's Geopolitics of the Pacific Ocean', in Fullagar, K (ed.), *The Atlantic World in the Antipodes: Effects and Transformations since the Eighteenth Century*, Cambridge Scholars, Newcastle upon Tyne, UK.

Bennett, J 2007, Letter, *BioScience*, vol. 57, no. 2, p. 101.

Buckminster Fuller, R 1969, *Operating Manual for Spaceship Earth*, Design Science Lab.

Carrington, D 2014, 'World Population to Hit 11bn in 2100 – With 70% Chance of Continuous Rise', *The Guardian*, 19 September.

Child, Sir J 1751 [1668], *A New Discourse of Trade*, Foulis, Glasgow, UK.

Clark, G 2007, *A Farewell to Alms: A Brief Economic History of the World*, Princeton University Press, Princeton, US.

Clarke, A 1975, *Imperial Earth*, Pan, London, UK.

Crutzen, P and Stoermer, E 2000, 'The 'Anthropocene', *Global Change Newsletter*, no. 41, pp. 17–18.

Daily, G, Ehrlich, A and Ehrlich, P 1994, 'Optimum Human Population Size', *Population and Environment: A Journal of Interdisciplinary Studies*, vol. 15, no. 6, July, pp. 469–475.

Dietz, R and O'Neill, D 2013, *Enough Is Enough: Building a Sustainable Economy in a World of Finite Resources*, Earthscan (Routledge), Oxford, UK.

Dorling, D 2013, *Population 10 Billion*, Constable, London, UK.

DVA: Australian Government Department of Veterans' Affairs 2015, *Australians on the Western Front*.

Eby, M 2010, "The Coming Population Crash: The Overpopulation Myth – How Feminism and Pop Culture Saved Earth from Getting Too Crowded – And Are Helping To Avert Planetary Catastrophe', Interview with Fred Pearce, *Salon*, 19 April.

Ehrlich, P 1968, *The Population Bomb*, Ballantine Books, New York, US.

Engels, F 1996 [1844], 'Outlines of a Critique of Political Economy', Deutsch-Französische Jahrbücher, Marxists.org.

Fischetti, M 2014, 'Up, Up and Away', *Scientific American*, vol. 311, no. 6, December, p. 80.

Friedman, D 2013, *The Puzzle of the Other Hockey Stick*, blog, 15 July.

Friedman, G 2015, 'Population Decline and the Great Economic Reversal', *Geopolitical Weekly*, 17 February.

Friedman, G 2016, '1929, 1945, 1991 … and 2008?', *Real Clear World*, 3 October.

Friedman, M 1970, 'The Social Responsibility of Business Is to Increase Its Profits', *The New York Times Magazine*, 13 September.

Friedman, M 1982 [1962], *Capitalism and Freedom*, University of Chicago Press, Chicago, US, and London, UK.

Galbraith, J 1976 [1958], *The Affluent Society*, third edition, New American Library, New York, US.

Gray, A 1931, *The Development of Economic Doctrine: An Introductory Survey*, Longman Green, London, UK.

Greer, G 1985, *Sex and Destiny: The Politics of Human Fertility*, Picador, London, UK.

Hardin, G 1998, 'The Feast of Malthus', *Social Contract Journal*, vol. 8, no. 3, pp. 181–187.

Hoyle, F 1964, *Of Men and Galaxies*, 1965 edition, Heinemann Educational Books, London, UK.

Huxley, A 1979 [1962], *Island*, Panther, London, UK.

Islam, S 2015, *Inequality and Environmental Sustainability*, Working Paper No. 145, United Nations Department of Economic and Social Affairs (DESA), August.

Jian, M 2013, 'China's Brutal One-Child Policy', *New York Times*, 21 May.

Kerins, E 2017, 'When the World Is Not Enough: How to Find Another Planet to Live on', *The Conversation*, 12 September.

Knight, L 2016, *The Voluntary Human Extinction Movement (VHEMT):* "May we live long and die out", website.

Malm, A and Hornborg, A 2014, 'The Geology of Mankind? A Critique of the Anthropocene Narrative', *The Anthropocene Review*, vol. 1, no. 1, 7 January, pp. 62–69.

Malthus, T 1803, *An Essay on the Principle of Population*, second edition, Johnson, London, UK.

Marean, C 2010, 'When the Sea Saved Humanity', *Scientific American*, vol. 303, no. 2, August,.

Masuda, H 2015, *Municipalities at Risk of Vanishing*, lecture, University of Tokyo and Institute for Studies in Happiness Economy and Society, Japan for Sustainability, 23 April.

McKenna, J 2017, '6 Numbers That Prove the Future Is African', *World Economic Forum*, 2 May.

McKie, R 2014, 'The Sixth Extinction: An Unnatural History by Elizabeth Kolbert – Review', *The Guardian*, 16 February.

Meadows, D H 2008, *Thinking in Systems: A Primer*, Wright, D (ed.), Sustainability Institute, Earthscan, London, UK.

Meadows, D H, Meadows D L, Randers, J and Behrens, W 1972, *The Limits to Growth: A Report for the Club of Rome's Project on the Predicament of Mankind*, Universe Books, New York, US.

Monbiot, G 2014b, 'How Wolves Change Rivers', *Sustainable Human*, Youtube.

Osterfeld, D 1993, 'Overpopulation: The Perennial Myth', *The Freeman*, September.

Panda A 2014, 'Japan's Demographic Crisis: Any Way Out?' *The Diplomat*, 26 March 2014.

Pearce, F 2010, *The Coming Population Crash and Our Planet's Surprising Future*, Beacon Press, Boston, US.

Polanyi, K 2001 [1944], *The Great Transformation*, Beacon Press, Boston, US.

Powledge, F 2007a, 'The Millennium Assessment', *BioScience*, vol. 56, pp. 880–886.

Powledge, F 2007b, response to letter, *BioScience*, vol. 57, no. 2, p. 101.

Purdy, J 2015, 'Imagining the Anthropocene', *Aeon*, 31 March.

Randers, J 2013, *We Won't Be Nine Billion: Jørgen Randers at TEDxMaastricht*, September, Internet video.

Ricardo, D 1911 [1817], *The Principles of Political Economy and Taxation*, John Murray, London, UK.

Romer, P 2015, *Urbanization Passes the Pritchett Test*, blog, 04 May.

Satterthwaite, D 2009, 'The Implications of Population Growth and Urbanization for Climate Change', *Environment and Urbanization*, vol. 21, pp. 545–567.

Smil, V 2013, *Harvesting the Biosphere: What We Have Taken from Nature*, MIT Press, Cambridge, US.

Smil, V 2008, *Energy in Nature and Society: General Energetics of Complex Systems*, MIT Press, Cambridge, US.

Smith, A 1952 [1776], *An Inquiry into the Nature and Causes of the Wealth of Nations*, University of Chicago/Encyclopedia Britannica, Chicago, US.

Smith, W 1980, 'Friedrich Ratzel and the Origins of Lebensraum', *German Studies Review*, vol. 3, no. 1, February, pp. 51–68.

Stanhope, D 2010, 'Doug Stanhope: Voice of America – Abortion is Green', Youtube, 9 February.

Stanton, E 2007, *The Human Development Index: A History*, Working paper series no. 127, Political Economy Research Institute, University of Massachusetts, Amherst, US.

Stanton, J 2015, 'Italy's Birth Rate Drops to Its Lowest Level in 150 Years as Economy Tanks', *Daily Mail Australia*, 14 February.

Steffen, W, Crutzen, P and McNeill, J 2007, 'The Anthropocene: Are Humans Now Overwhelming the Great Forces of Nature?' *Ambio*, vol. 36, no. 8, December, pp. 614–621.

Streatfield, P and Kamal, N 2013, 'Population and Family Planning in Bangladesh', *Journal of the Pakistan Medical Association*, vol. 63, no. 4 (Suppl. 3), April, pp. S73–S81.

Tertullian, 1951 [c.200 AD], 'Treatises on Marriage and Remarriage: To His Wife: An Exhortation to Chastity: Monogamy', in *Ancient Christian Writers*, Quaten, J and Plumpe, J (eds), no. 13, Catholic University of America, Paulist Press, New York, US.

UN: United Nations 2014a, *The Millennium Development Goals Report 2014*, United Nations, New York, US.

UNDESA: United Nations Department of Economic and Social Affairs, Population Division 2014, World Urbanisation Prospects: The 2014 Revision, highlights.

UNDESA: United Nations Department of Economic and Social Affairs, Population Division 2015, World Population Prospects 2015.

UNU-IHDP and UNEP 2012, *Inclusive Wealth Report 2012: Measuring Progress toward Sustainability*, Cambridge University Press, Cambridge, UK.

Wang, X and Lau, S 2002, *Pursuing New Urban Living Environment in the New Millennium: Projecting the Future of High-Rise and High Density Living in Hong Kong.*

White, L 1967, 'The Historical Roots of Our Ecological Crisis', *Science*, vol. 155, no. 3767, pp. 1203–1207.

Williams, E 1944, *Capitalism and Slavery*, University of North Carolina Press, Chapel Hill, US.

Worldometers 2014. Viewed 28 June. www.worldometers.info/world-population/#growthrate.

Yun, J 2012, 'The Next Black Swan: Global Depopulation', *Forbes,* 7 December.

5 Affluence

It is commonly asserted that the average contemporary Westerner enjoys greater affluence than the kings and emperors of the pre-industrial era. Certainly an investigation of the palace of Suleiman the Magnificent in Istanbul is instructive: while between 1520 and 1566 he led an empire controlling much of South Eastern Europe, Western Asia, the Caucasus and Northern Africa, Suleiman's personal lavatory, although lined in marble may as well have been of concrete. It essentially consisted of a hole in the floor. He possessed no transport apart from that which meandered at the pace of a horse or with the lassitude of sail. He knew nothing of flight, except that of birds, bats and insects. He communicated at a distance only by messenger. His entertainment was necessarily live and his literature limited to inscriptions made by hand.

While it may be true of the average Westerner, it is probably not true of the average Southerner that she is better off than an ancient emperor. Her lavatory will not be lined in marble. It will at best be a hole on the floor. There may well be no floor. It will probably not be a flushing system connected to a sewer. It may be at some distance from her dwelling, or not a facility at all. It may be dangerous for her to even venture there for fear of violence (McCarthy 2014). She may know of crowded trucks and buses (especially in Africa) and trains (especially in India). Aircraft trails may be seen far above in the sky. She may possess a mobile phone and watch films using a communal diesel generator to power a DVD player. Neither, however, will protect her from the risk of death in childbirth or from infection by malaria, tuberculosis or HIV. She may well be one of billions who are vulnerable to multi-dimensional poverty (UNHDR 2014, p. 19).

The affluent Westerner and the impoverished Southerner are contemporary extremes in geography as well as recent history. Arguably they depend on each other: As Marshall McLuhan famously wrote in 1964, "affluence creates poverty" – the poor are poor because the rich are rich; wealth depends on the exploitation of others. However, the observation that the sun causes poverty does not hold throughout history. Ancient Egypt, the Khmer empire, and the Mayans were all civilisations of immense wealth, yet arose in tropical settings.

Arguably all three failed to adapt to changing circumstance; the Egyptians failed to innovate,[1] the Khmer suffered degradation of their irrigation system and Mayan agriculture did not cope with drought (Diamond 2003). Today, the tropical city-states of Singapore and Hong Kong are icons of wealth. If it was not the sun that caused the poverty of the South, then what is it that causes the wealth of the West? Perhaps it is that the 'North wind made the Vikings', or that overexploitation of natural resources – the denuded forest – led to the use of coal, in which power is concentrated (Nef 1977, pp. 140–141).

In stark contrast with people of the South, it was only during the twentieth century that it became possible for the majority of ordinary people in the West to enjoy unprecedented affluence – on a scale unimaginable before the modern era. This was an affluence free from past drudgery and the daily struggle for survival. It was an affluence that included the universal availability of flushing lavatories. For the first time in human history vast numbers of people could live fully rather than subsist, to have leisure, for art and for knowledge beyond any immediate need for food, shelter and companionship. Although interrupted by war and economic depression, this new widespread affluence was a product of three key factors – a new economic system, the exploitation of energy reserves and mass production.

The new economic system was based on "the lure of gain" and effected through the market system, "the most important revolution, from the point of view of shaping modern society, that ever took place" (Heilbroner 1983, p. 17). This system – capitalism – superseded traditional (and largely agricultural) societies, in which tasks were allocated according to custom, where son followed father, daughter followed mother, where occupation was often attached to caste. It proved superior to systems of command that involve the whip of authoritarian rule, whereby the pyramids were built and more recently, Soviet farms collectivised. Neither of these other economic systems necessarily involved the transfer of money throughout society. But it was this new system that introduced the novel idea of individual gain that, with widespread use of money both as wages and as wealth, could build upon earlier gains and involve whole societies in the advance of affluence. The reinvention of money during the Middle Ages was critical in this new system because, instead of barter, money required acceptance of the concept of value. Once established, this abstract concept led to its expression in a market that could be leveraged and exploited (Sattin 2014).

The exploitation of energy reserves remains central to this new economic system and to the affluence it has created.[2] During the nineteenth century, the primacy of coal was established in fuelling new secondary industries, which could be located where labour was cheap, rather than only where waterwheels could be powered (Klein 2014, p. 172). The conversion of coal to heat allowed steel production, mass transport (railways), mass electrification and new factories producing affordable consumer goods. During the twentieth century, oil and gas made individual autonomous transport (cars) and aircraft feasible.

Fossil-based energy is highly concentrated, considerably more so than primitive windmill, waterwheel or animal power. Accordingly, without the extensive use of concentrated energy, modern affluence would not be possible. Metals, electricity and consumer goods cannot be provided on a mass scale by a horse plodding around the axis of some crude turbine. Rather, billions of tonnes of coal and billions of barrels of oil – or their equivalent – are needed each year to maintain the elements of mass affluence.

Mass production techniques involved new factories producing affordable consumer goods. When applied to complex manufactures and the assembly line, this technique became known as 'Fordism', after Henry Ford the car manufacturer,[3] a term coined by Antonio Gramsci in his 1934 essay *Americanism and Fordism*. But the most important aspect of Fordism was its virtuous re-cycling of profits through increased workers' wages, that then enabled workers to buy the cars they made, that enabled higher production, lower unit costs through economies of scale and so on. Gramsci was sceptical about this virtuous cycle and questioned "the so-called 'high wages' paid", rather suggesting that Fordism was "the ultimate stage in the process of progressive attempts by industry to overcome the law of the tendency of the rate of profit to fall" (Gramsci 1999 [1934], pp. 562–563). Growing affluence based on reduced costs of complex manufactures remains with us today. However, much of the lower costs of, for example, cars, apparel and homewares in the West is now due to exploiting the lower wage structures of the East, foreshadowed by Gramsci as part of an 'international division of labour' (ibid., p. 607).

There are parallels between the economies of the early twentieth and early twenty-first centuries. In both eras the prices of iconic items continually fell while quality improved, due to scale:

> Economies of scale allowed the robber barons to keep reducing prices and improving quality. Henry Ford cut the price of his Model T from $850 in its first year of production to $360 in 1916. In 1924 you could buy a much better car for just $290. The silicon sultans performed exactly the same trick. The price of computer equipment, adjusted for quality and inflation, has declined by 16 percent a year over the five decades from 1959 to 2009.
> (*The Economist* 2015a)

Definition

The word 'affluence' appeared in English around the mid-fourteenth century, according to the Oxford English Dictionary, and derives from the Old French *affluence* and Latin *affluentia*, "a plentiful flowing, an abundance, rich, copious", including a sense of 'wealth' from about 1600, from the notion of "a plentiful flow" of the gifts of fortune. The word 'wealth' is from the thirteenth century

Middle English 'wele' or 'well-being' and is based on an analogy of 'health'. In economics, 'wealth' means 'the sum of all goods and services that have an exchange value' but it originally meant 'happiness', as well as its contemporary meaning of 'prosperity in abundance of possessions or riches'. The association of affluence with wealth and thus with happiness may have been lost because objectively the connection is not entirely clear. Rather, it is a paradox:

> the happiness – income paradox is this: at a point in time happiness varies directly with income, but over time happiness does not increase when a country's income increases.
>
> (Easterlin and Angelescu 2009, p. 2)

This finding is based on an analysis of responses to questionnaires about life satisfaction and income in 37 countries at different stages of development.

The same study shows that people are more dismayed about loss of income than are elated by an equivalent gain (ibid., p. 12). Similarly, at the intersection of economics and psychology, there is the profound observation that affluence more relative than absolute. It is distressing if we are less affluent than our friends and neighbours and comforting if we are wealthier, irrespective of absolute levels of affluence (Veblen 1899, ch. 4; Solnick and Hemenway 1998, pp. 380–381). Thorsten Veblen, who coined the term 'conspicuous consumption', detailed the display behaviours of the wealthy that both symbolised and underscored their relative material superiority during the Golden Age of the late nineteenth century.[4] However, the abundance now generally enjoyed in the West is conspicuous relative to the scarcity of the South. And it appears to be built on fragile foundations.

At the individual psychological level, western affluence has produced the dysfunctional behaviour of 'hoarding'. Cherrier and Ponnor (2006, p. 26) distinguish between three groups: While 'collectors' enjoy societal approval for their organised collections based on aesthetic or historical value, 'functional hoarders' risk disapproval because they find it difficult to part with many useless items for sentimental reasons or perceptions of wastefulness. 'Non-functional hoarders', however, risk both societal intervention and their own well-being due to compulsive accumulations of valueless items. Some of this hoarding behaviour may be due to perceptions of shortage originating during the Depression or the oil crises of the mid-1970s. Nevertheless, it is facilitated by affluence. Basic material goods in recent times are cheap and diverse. In many cases they are made to be disposable, but instead, they are hoarded for psychological security. Sometimes its manifestation is extreme:

> The Collyer brothers, for example, died in the United States in 1947 due to over-cluttered space. It is reported that their house contained 136 tons

> of refuse. One brother was buried alive when piles of rubbish collapsed on him, leaving his blind brother starving to death.
>
> (Ibid., p. 6)

The world's first shopping mall, 'Southdale' in Minnesota, started in 1954 in a building complex purpose-designed by Victor Gruen and "spread like an epidemic across the USA and the rest of the world". Such malls embody mass consumerism, and despite the current prevalence of online shopping continue to exert a powerful attraction throughout the world. The term 'Gruen transfer' refers to the disorientation many people feel when they first enter these spaces as they forget their original intentions and buy more than initially planned (Sloterdijk 2005, p. 274).

There is stress associated with material choice that consumerism encourages:

> Once upon a time in Springfield, the Simpson family visited a new supermarket. *Monstromart's* slogan was "where shopping is a baffling ordeal". Product choice was unlimited, shelving reached the ceiling, nutmeg came in 12lb boxes and the express checkout had a sign reading, "1,000 items or less". In the end the Simpsons returned to Apu's Kwik-E-Mart.
>
> (Jeffries 2015, para. 1)

Aristotle viewed wealth as either 'unnatural' if it were attained through trade, or 'natural' if attained through "skilful management of house and land". Trade is unnatural because things are used for other than their natural purpose, and because it involves money, which has no use other than exchange. Thus the most unnatural form of wealth is that produced from money itself – usury – the lending of money at interest (Aristotle 1952 [c. 330 BC], 1258b).

Aristotle may have only been reflecting his class interests, however. Throughout history debtors have disapproved of interest while creditors have the opposite attitude. Because Greek philosophers were aligned with the landowning class who were often in debt, they did not like usury (Russell 1980 [1946], p. 198). Nevertheless, Aristotle presented a more profound insight: wealth of the household is limited, but monetary wealth, the spurious kind, has no limit and nor has the desire for it. As "money supplants other values and becomes their only measure", no matter how affluent a person becomes, there is always a desire for more (Harvey 2014, p. 277).

Measurement

While the invention and widespread use of money enabled the easy storage, measurement and accumulation of wealth, only since the 1930s were detailed national accounts constructed for government, especially in the United Kingdom and the United States, due to pressures of the Depression and in preparation for World War II (Van Dieren 1995, p. 39; Coyle 2014, p. 12). Up until that point, 'the economy' was essentially the private sector only. Government was excluded because it had minimal input into production. With the 'New Deal' and the supporting *General Theory* of J.M. Keynes (1936) government became significant to economic growth, especially in the West. In the Communist states of course, government was already pre-eminent.

The universal use of GDP to measure national wealth and economic growth is even more recent. While the measure was formally established in 1953 through the UN statistical system, its universal use did not occur until towards the end of last century. Up until then GDP was one of several similar but competing concepts, in particular Gross National Product (GNP), which measures the net wealth generated by national citizens and corporations, irrespective of their location. GDP, however, is the market value of everything produced within a country's borders. Gross National Income (GNI) is the same concept as GNP but uses an income rather than a production measure. The persistence of GNP and GNI to measure the US economy may be because, unlike most countries, US income tax is payable irrespective of in which country its citizens reside. At the global level of course, all three measures are theoretically equal since total exports[5] should equal total imports and production is accounted for but once in each system.

Interestingly, Simon Kuznets, who was awarded a Nobel Prize for developing the comprehensive system of national accounts during the Roosevelt years, wanted them to measure human *welfare*, not just total output or consumption. Accordingly, he argued – unsuccessfully – that the new national economic measure should subtract all expenditure on armaments, advertising, "speculative activities" and "necessary evils", like subways. In this, however, he was at odds with the spirit of the time, and particularly with Roosevelt who wanted government spending to be included as a positive economic contributor and thus demonstrate that the US economy was growing again post-Depression (Coyle 2014, pp. 13–15). Today, the resultant GDP measures without moral perspective. A billion dollars' worth of nuclear bombs is counted as exactly equal to a billion dollars' worth of baby food.

But irrespective of Kuznets' or FDR's views on what should be included, any such quantitative indicator tends to lead to an over-emphasis on the notion of 'growth' at the expense of 'development'. Former World Bank ecological economist Herman Daly (1990, p. 1) argues that the distinction between 'growth' and 'development' is that the former is about an increase in size, whereas the latter is about qualitative improvement. Thus ultimately the human global economy cannot grow to be more than the finite global ecosystem of which it is part. Yet while economic growth cannot be sustainable in the longer term, it is possible for economic development to be sustainable, because the global ecosystem develops without growing. In measuring wealth as a quantitative GDP, ironically we tend to value that which leads to unsustainability and its own negation.

Robert Kennedy would have agreed. In a speech at the University of Kansas in 1968 he observed that:

> The gross national product does not allow for the health of our children, the quality of their education or the joy of their play. It does not include the beauty of our poetry or the strength of our marriages, the intelligence of our public debate or the integrity of our public officials. It measures neither our wit nor our courage, neither our wisdom nor our learning, neither our compassion nor our devotion to our country, it measures everything in short, except that which makes life worthwhile.

Before the point of negation is reached, however, measuring relative affluence between countries can be complicated The United States has the world's largest economy and China the second largest in nominal terms. Yet in purchasing power parity (PPP) terms, where the value of goods and services consumed is weighted according to the cost of living in each country, the result is different. If ten dollars (or its exchange currency) buys one Happy Meal in the United States, but two Happy Meals in China, then each dollar equivalent of GDP in China is worth twice each dollar of GDP in the United States, at least as far as Happy Meals are concerned. The difficulty in PPP measurement is ensuring that the basket of goods used to indicate purchasing power is representative of consumption in both countries. In nominal terms, US GDP was about USD16 trillion compared with China's GDP of about USD9 trillion in 2013. But in PPP terms China's GDP was nearly the same as that of the United States, according to the World Bank (2014b). However, in terms of personal affluence, both are distorted comparisons. China has more than four times the US population; therefore, the average American is more than four times as affluent as their Chinese counterpart in PPP terms. It is this measure – GDP per person (PPP) – that is more relevant when discussing relative national affluence as well as its impact on the entire planet.

It is also important to be aware that GDP is a measure of production, consumption or income. It is a gauge of flow. It does not measure the stock of accumulated wealth. Thomas Piketty (2014, p. 463) points out that more and more countries are becoming owned by their own billionaires, including especially the established billionaires of Europe, rather than being taken over by foreign wealth. Thus, consistent with the fears of Aristotle, wealth is accumulating and concentrating within national borders as much as it is concentrated in the West compared with the South.

Hazel Henderson (2014) believes that the measurement of GDP contributes to a debt imbalance and therefore artificially weakens national economies.

Unlike business accounts with flows of income and expenditure and stocks of assets and liabilities, national accounts lack assets:

> So, the Grossly Distorted Picture in current GDP only records levels of public debt for vital infrastructure and public services (police, fire protection, teachers, etc.). Omitted is an asset side to account for valuable taxpayer investments in public infrastructure: transport, ports, railways, schools, etc. ... Imagine trying to run a company this way!

Governments focus on GDP, however, because not only is it a simple measure that correlates with other indicators of well-being (such as employment and education levels), but also it can be influenced by government policy in the medium term. Other indicators can be influenced by government policy, but few combine all these apparent virtues as well as does GDP.

Wealth creation

The concept of national (and now global) wealth is central to political economy and economics generally, and is generally attributed to Adam Smith's *Wealth of Nations* (1776). In fact, the notion of national wealth as a capital stock appeared in texts that predate Smith's works. In *A Discourse of Money* for example, John Briscoe (1696, p. 198) writes of "the capital stock of national treasure". Andrew Hooke, in his *Essay on the National Debt and National Capital* (1750), treats "national capital" as consisting of (1) "cash, stock, or coin," (2) "personal stock" or "wrought plate & bullion, jewels, rings, furniture, apparel, shipping, stock-in-trade, stock for consumption, and live-stock of capital," and (3) "land stock" – the value of all lands in the kingdom.

During the eighteenth century the French physiocrats[6] developed an economic theory whereby national wealth is composed of the value of agricultural land and its produce, due to productive work. According to its proponents such as François Quesnay (1694–1774) and Anne-Robert-Jacques Turgot (1727–1781), capital (derived from surplus saved) is necessary for economic growth and economies function best where each participant is free to pursue their own wants (*'laissez-faire'*).

When Adam Smith took the term 'capital' in hand, he began by distinguishing capital from interest. In his *Lectures on Jurisprudence* (Smith 1766, p. 248), it first appears as a sum of money lent, as opposed to the interest paid on the loan.

Much of a century after Smith, Engels was more concerned with how the concept of national wealth masked poverty:

> The term national wealth has only arisen as a result of the liberal economists' passion for generalisation. As long as private property exists, this term has

> no meaning. The "national wealth" of the English is very great and yet they are the poorest people under the sun.
>
> (Engels 1844, para. 13)

More generally, wealth is created by human labour, according to Adam Smith, and this was the original basis of exchange value. The exchange value of a good, Smith said, is determined by the amount of labour it contains in its production. In a primitive society it is the only way of determining how goods can be exchanged. If it takes twice the labour to kill a beaver as it does a deer, then one beaver is worth two deer (Smith 1952 [1776], p. 20). But in advanced societies, the price of a good includes three factors – labour, the owner's profit, plus rent. However, the real value of all three components, "is measured by the quantity of labour which they can, each of them, purchase or command" (ibid., p. 21). Consistent with Smith's insight, machinery and technologies that facilitate the production of goods in modern economies are simply stored labour, since it is human labour that enabled their conception and creation. Skill is the embodiment of earlier labour. Services, as distinct from goods, are still more directly composed of labour.

Others, notably Karl Marx, based much analysis on this concept, in particular that the surplus value inherent in a good derives from the value of the labour it embodies beyond that needed for the subsistence of the labourer (Wolff 2011, s.3). Marx's model of perfect contemporary capitalism demonstrated that the only source of profit was this surplus value, because the labourer must sell his labour at a subsistence rate, considerably less than the value of labour he makes available to the capitalist. The difference is the profit that fuels capitalism (Heilbroner 1983, pp. 120–121).

While the labour theory of value has been described as "an appalling jumble of ideas" (Whitaker (2001 [1904], p. 6), it underlined the central importance of labour to political economics and the creation of wealth. The creation of wealth, according to Smith in his early *Lectures on Jurisprudence*, progressed in four societal stages:

> ...hunting, pasturage, farming, and commerce. If a number of persons were shipwrecked on a desert island their first sustenance would be from the fruits which the soil naturally produced, and the wild beasts which they could kill. As these could not at all times be sufficient, they come at last to tame some of the wild–beasts that they might always have them at hand. In process of time even these would not be sufficient, and as they saw the Earth naturally produce considerable quantities of vegetables of its own accord they would think of cultivating it so that it might produce more of them...The age of commerce naturally succeeds that of agriculture. As

> men could now confine themselves to one species of labour, they would naturally exchange the surplus of their own commodity for that of another of which they stood in need.
>
> (Smith 2005 [1766], pp. 522–523)

Smith was writing at the brink of the industrial era, which was unknown to him. However, affluence, in the sense of wealth accumulation, increasingly depended on possession of the means of production. In line with his societal stages, the means of production was for hunters the spear, for herders the stock, for farmers the land and for entrepreneurs the firm. Further, the associated development from communal to private ownership enabled wealth to be concentrated in fewer hands. In the industrial era, factory ownership was key to affluence. In the post-industrial era it is more the ownership of a technology or 'intellectual property' that is the source of affluence, as no doubt Bill Gates and the geeks of Silicon Valley would concur. Especially in its patent-protected recent forms, technology tends to concentrate affluence.

Thus in both classical and Marxist political economy, nature might appear peripheral. Human intervention, invention and organization is presumed more significant. Smith's desert island illustration, however, did recognise the importance of the natural world. It is the source of sustenance from gathering and cultivation, from hunting and husbandry. Human labour does not produce wealth from nothing. It produces wealth from the natural world. Marx too, was aware of nature and its relationship to human wealth (Bellamy Foster 2000). Engels also knew that wealth ultimately derives from nature: "labour is the source of all wealth … next to nature, which supplies it with the material that it converts into wealth" (Engels 1925 [1883], p. 452).

That ultimate foundation of wealth, however, "based on ever-more alarming environmental reports"[7] is now dramatically "sinking under the weight of demands to supply more resources and absorb more wastes" (Gale 2014, p. 1). Indeed, capitalism is built on unending economic growth:

> A purely private enterprise system can only function if companies can obtain sufficient profits which in turn requires that the selling price of goods exceeds the costs of production. This means that the selling price must exceed the spending power that has been distributed through payments to factor inputs. Hence, to ensure sufficient "aggregate demand" to clear the market, additional spending power is required from some other source…Investment therefore serves the dual role of increasing productive capacity and creating additional demand to clear the market of whatever has already been produced.
>
> (Sorrell 2010, p. 1797)

As far as consumption is concerned, the economist Kenneth Boulding was one of the first of that discipline, to point out that "the closed Earth of the future requires economic principles which are somewhat different from those of the open Earth of the past" (Boulding 1966, p. 10), a challenge to economists to re-think the centrality of growth, which has yet to be fully accepted. This may be because he was uncomplimentary to others of his own profession: "Anyone who believes exponential growth can go on forever is either a madman or an economist". Redclift (1993, p. 19) accuses "modern economics" of causing "unsustainable development" due to the pursuit of growth at the expense of ecological consequence. This "has its roots in the classical paradigm which informed both market economies and state socialist ones".

And economic growth does not even translate to more happiness. Easterlin and Angelescu's (2009, p. 2) study of 37 countries showed, as with income relativities, there is no relationship between the rate of growth of GDP per person and increase in happiness. Whilst there may be short-term associations between the growth of income and happiness due to fluctuations in macroeconomic conditions, the long-term relationship "is nil" (ibid.). That our happiness has little to do with economic growth is an encouragement for environmentalists who advocate restraint in production and consumption (Trainer 1985; Alexander 2015), and an obstacle for development economists who advocate continuous growth.

But wealth is not necessarily material. Karl Marx's notes in the *Grundisse* approvingly quotes a view of Charles Dilke – that true national wealth is not a matter of money or other capital, but rather consists of more free time for everyone:

> Truly wealthy a nation, when the working day is 6 rather than 12 hours. Wealth is not command over surplus labour time' (real wealth), 'but rather, disposable time outside that needed in direct production, for every individual and the whole'.
>
> (Dilke 1821, p. 6, quoted in Marx 1973 [1861], p. 706)

Dilke's view of wealth as free time is in turn probably based on that of the utopian socialist, William Godwin[8] (1793): "Is there not a state of society beyond that needed for production … in which leisure shall be made the inheritance of every one of its members?". Indeed, there have been recent studies of the phenomenon of "downshifting", whereby significant numbers of people in affluent societies choose simpler lives that consume less while providing leisure time beyond 'the rat race'. Involving around 20 percent of adults in the United States and Australia, these people forgo what they regard as unneeded income and material goods, and are mostly happy with their decisions. The most prominent reason for downshifting is that trope of retiring politicians – 'to spend more time with their family' (Hamilton and Mail 2003, p. 20), but probably meant more sincerely.[9]

This profound view of wealth as free time is arguably now within the grasp of contemporary advanced economies as automation replaces unskilled labour. To a significant extent, even the skilled labour of journalists, accountants and marketing executives is demonstrably replaceable. In ancient times it was the few who enjoyed the wealth of leisure, the freedom to create rather than the drudge of repetitive unskilled labour. Now that machines are capable of that production as well as routine labour of the mind, the political issue again is distribution.

Wealth distribution

In Plato's Republic (1952 [c. 380 BC], s.744) its distribution would result in an inequality of wealth no more than four to one, otherwise it threatened social order. That ratio has been since far exceeded in modern republics, but there remains a certain wariness about its implications for unrest. The 1789 French and the 1917 Russian revolutions are obvious examples of its menace.

While there were minimum income payments by the state in eighteenth century England via the 1795 Speenhamland Law, abandoned because it inhibited the creation of an industrial labour market (Polanyi 2001 [1944], p. 81), industrialisation at first brought with it greater inequality, justified as virtue. During the Victorian era various political and economic 'iron laws' were asserted to imply their universality, as if they were the enduring *principia* of some economic Isaac Newton. They include the 'iron law of wages' attributed to Malthus and Ricardo, which states that wages of labourers will tend to fall to the minimum for human subsistence, such is the unfortunate tendency for the poor to reproduce themselves creating an oversupply of their kind. This is associated with an 'iron law of population', which states that rapidly increasing population inevitably leads to poverty.

Hence it became virtuous for harsh measures to be taken against the poor – as if they were an undesirable separate species. If the poor were provided with more than the barest of necessities via wages or social welfare measures, then they would simply increase their numbers through ill-discipline and fornication. Wages would thus fall still further, below subsistence levels, to their greater detriment. How kind it was therefore to be cruel; how much better for the poor to be contained in a misery that made possible their subsistence, but no more. There was no affluence for the masses to be found in Victorian economic theory cast as enduring truth.

However, the beginnings of the modern era began to bring leverage to the cousins of these early wage slaves in the former colonies of Australasia and North America. Universal suffrage, trade unions and the creation of political movements entwined to result in wages that provided "reasonable comfort" in Justice Higgins' decision of 1907 in Australia (Robbins et al. 2005, p. 488) and the beginnings of welfare measures in the United States during the Depression. Winston Churchill, who observed in 1945 that capitalism is good at accumulating wealth, but poor at distributing it, nonetheless opposed welfare measures in the post-war United Kingdom and was dismissed by electors in the same year.[10]

As welfare measures took hold in the West, a degree of universal affluence was identified mid-century in the work of Simon Kuznets. After his involvement in quantifying national income, Kuznets found evidence for an inverted U-shaped curve that described the relationship between inequality and economic development. As economic development proceeds and more people take advantage of greater opportunities, average incomes rise. At first there is increasing inequality as only a few benefit, which then peaks when the entire population is involved in the more developed economy. Inequality then begins to fall and return to earlier levels as progressive taxation and social welfare programs are implemented.

Margaret Thatcher and Ronald Reagan, who both championed the resurrection of economic liberalism during the 1980s, oversaw rising inequality in the West as wealth generation was regarded as more important than distribution, their national economies having suffered a decade of stagnation.

The predominance of neoliberalism has since been undermined by the global financial crisis of 2008 and the subsequent long global recession. Recently, French economist Thomas Piketty's *Capital in the Twenty-First Century* (2014) created a scholarly storm with its thesis that in the longer-term since the Industrial Revolution, inequality will continue to increase, as the return on capital normally exceeds income from labour. Inequality grows especially in times of political stability, but can be reversed in times of crisis such as the Depression and the two world wars. This contradicts earlier assumptions and narrow evidence that increasing economic growth produces greater equality (Kuznets 1955).

Piketty's research greatly widens the scope of Kuznets' earlier inquiry into inequality and encompasses it in a manner that recalls Einstein's subsumption of Newton. Kuznets had found that inequality tended to decline over time in developed economies. Piketty, by analysing a much greater range of data covering most major economies over the past 300 years, shows that Kuznets was accurate, but that the period he chose (1913–1948) was an aberration in the long-term trend of growing inequality. What appeared to Kuznets as a compression of wealth disparities was due to the disruption of the two world wars and the effects of reconstruction. When less chaotic times returned, wealth – including income — again resumed its long march to widening disparity. Piketty's findings have since been dramatically extended to encompass even ancient times. The only factors that reduce otherwise inevitable inequality are 'the four horsemen of levelling' – mass warfare, revolution, state failure, and pandemics (Scheidel 2017).

Importantly, Picketty surmises that economists became uninterested in inequality and its risks after Kuznets mid-century. Although it was a central concern of nineteenth century political economists, such as Marx, Engels and Ricardo, inequality became no longer an issue because of his reassurance. But Picketty goes further in showing how Kuznets brought his immense prestige as a Nobel laureate and president of the American Economics Association (AEA) to bear on the issue. For the first time the dire predictions of Marx – growing inequality, overproduction, collapse of demand, impoverishment, revolution – were apparently disproven, based on Kuznet's detailed research of then newly available data. In future, according to Kuznets' enthusiastic 1954 speech to the Association, providing that countries followed the US model, economic growth would not only diminish poverty, but society would become more equal as development matured. In the middle of the Cold War, for America and the West this was good news indeed.

Kuznets was careful to decorate his crucial news with qualification, but any proviso was swept away behind the word 'however':

> No adequate empirical evidence is available for checking this conjecture of a long secular[11] swing in income inequality; nor can the phases be dated precisely. However, to make it more specific, I would place the early phase in which income inequality might have been widening, from about 1780 to 1850 in England; from about 1840 to 1890, and particularly from 1870 on in the United States; and, from the 1840's to the 1890's in Germany.
>
> (Kuznets 1955, p. 19)

These three countries, England,[12] the United States and Germany, are the only three that Kuznets was able to study in detail. Picketty's sources are not only much more extensive in time, but also many more countries are covered.

It is hard to overstate Kuznets' prestige before Picketty arrived. Not only was he a Nobel laureate and president of the AEA, he was also central to the development of national accounts and the very concept of GDP. He had been trusted by Franklin Roosevelt to provide the data constructs for the recovery of the United States and the world from the Great Depression (Coyle 2014, p. 12). He was, at least until 2014 and Picketty, regarded as one of the most important authorities on the effects of economic development and poverty reduction. The 'Kuznets curve' that describes economic development and inequality, the 'Kuznets ratio' that relates income of the highest-earning households against the lowest, and the 'environmental Kuznets curve' that relates environmental degradation against economic development are all indications of his regard. It is a tragic irony that one of the conclusions of his watershed 1954 speech concerns Marx and an "overgeneralization" of tendencies not properly understood:

> much of Marxian economics may be an overgeneralization of imperfectly understood trends in England during the first half of the nineteenth century when income inequality may have widened; and that extrapolations of these trends (e.g. increasing misery of the working classes, polarization of society, etc.) proved wrong because due regard was not given to the possible effects upon the economic and social structure of technological changes, extension of the economic system to much of the then unoccupied world, and the very structure of human wants.
>
> (Kuznets 1955, p. 27)

Inequality tends to fray society, reduce opportunity and innovation and produce social unrest. Picketty suggests measures including wealth and property taxes to redress the situation. World Bank president Jim Yong Kim mentioned Picketty's findings soon after the book was published, suggesting that growing inequality affected capitalism's legitimacy:

> As an economic system, global market capitalism has produced affluence and innovation. These are very good things. However, an economic system's legitimacy is also tied to its ability to make two things accessible to all: the riches it generates and the social benefits that arise from that wealth. Unfortunately, national income gains from growth tend not to be shared among a population in anything close to equal measure. In his 2014 best seller *Capital in the Twenty-First Century*, French economist Thomas Piketty showed that, in developed economies, these gains generally flow at substantially higher rates to owners than to workers. Ultimately, we want to ensure the global economic system's gains are distributed in a fashion that creates opportunity and respects human dignity.
>
> (Kim 2014)

This sort of observation may be valid as far as it goes, but it restricts its focus to inequalities within nation-states. Yet as mentioned at the beginning of this chapter, the differences in affluence between the global West and the global South are immense. In human terms the contrast is probably far greater as almost everyone in the West enjoys standards of infrastructure, health and nutrition that are out of reach for the majority in the South. Further, contemporary global capitalism relies on the poverty of the South to drive low cost resource extraction and the manufacture of consumer goods.

But what is the relationship between increasing inequality and humanity's impact on the environment? The relationship is positive between the two variables – and is therefore negative for the environment. Apart from the plausibility that the rich might favour nature, clean air and clean water in their own interests, concentration of wealth implies a concentration of political power. Political rules and legislation therefore tend to favour the interests of the wealthy, who are less concerned with the environmental impact of their own consumption because they are insulated from its effects. There is extensive research evidence that supports such findings, including that of the UN's Department of Economic and Social Affairs (Islam 2015). The wealthy can choose where they live, which will not be the deltas of Bangladesh or the atolls of Kiribati that will be inundated as sea levels inexorably rise. The wealthy can pay more for increasingly scarce resources to fuel their limousines and their business jets. The wealthy can heat their swimming pools and surround themselves with security walls as others might become increasingly desperate for food, shelter and fuel, as in the gated communities of South Africa, the United States and Brazil. They can afford organic food that is free from contaminants as do the elite of a polluted China. The wealthy and the powerful will continue to meet together at Davos[13] to ski, to dine finely together and to reinforce their own perceptions, not to prioritise the environment.

Unless of course those with children and grandchildren may worry about the world their descendants will inherit. The media baron, Rupert Murdoch, for instance, says he worries about increasing inequality because it leads to social polarisation (Blutstein 2014), although he does not appear to have environmental concerns. Yet inequality is based on unequal benefit from the exploitation of natural resources. In Australia, for example, much wealth is concentrated amongst the owners of iron ore and coal mines and it is not that elite that is concerned about the effects of the dredging and infrastructure needed to enable its transport through the Great Barrier Reef. But to extend Murdoch's point: inequality tends to reduce societal cohesion. It is harder to debate political issues when society is polarised, and certainly more difficult to reach a consensus on what needs to be done, especially on environmental matters. The wealthy, who benefit from less regulation, tend to resist it. The impoverished, who often rely on environmental regulation to maintain their access to natural resources, may advocate more. As the World Bank notes:

> environmental income shares are higher for low-income households than the rest. This is because the poor are more reliant on subsistence activities and products harvested from natural areas such as forests and lakes.
>
> (World Bank 2015a, p. 173)

Other interests, such as blue collar labour unions, may give priority to short-term jobs over long-term environmental protection, especially where jobs are in short supply. Growing inequality thus implies dwindling sustainability.

Better measures

Apart from Dilke's concept of wealth as free time, there may be better measures of affluence than GDP and its clones – at least some that involve environmental impact. However, this is tricky because "the ecological and social sciences have developed independently and do not combine easily" (Ostrom 2009). As yet, none has been able to establish the value of an irreplaceable songbird (Funtowicz and Ravetz 1994, p. 197).[14]

Most measures that relate to sustainable affluence are either monetary or physical. Most monetary indicators are expressed as flows (such as GDP), and most physical indicators describe stocks.

There is Gross National Happiness (GNH) for example, which results from household surveys in the Kingdom of Bhutan to measure a "multidimensional" happiness index that includes "subjective well-being". However, it is much wider than this. GNH uses 124 variables ultimately grouped into nine domains. People are considered happy when they have "sufficiency" in two thirds of the variables. The nine equally-weighted domains are: psychological well-being, health, education, culture, time use, good governance, community vitality, ecological diversity and resilience, and living standards (Ura et al. 2012, p. 1). 'Ecological diversity and resilience' are based on Article 5 of the Constitution of Bhutan, whereby citizens shall:

> contribute to the protection of the natural environment, conservation of the rich biodiversity of Bhutan and prevention of all forms of ecological degradation including noise, visual and physical pollution.
>
> (Ibid., p. 30)

It attempts to measure such variables as traffic congestion as well as "wildlife damage to crops", which may appear inconsistent with environmental protection, but is important to an agricultural economy.

Under the 'living standards' domain, Bhutan's GNH comes closest to 'material well-being' and Western notions of GDP. This includes such variables as consumption, income and expenditure, including household per capita income, assets and 'housing conditions' (ibid., p. 33). Some of the results are fascinating. For example, there appears to be no relationship between per capita income and the GNH index for each district. The district with the highest per capita income is the capital (Thimphu), but the district with the highest GNH is Paro, which has considerably less per capita income. Also there is much more equality in GNH across the twenty districts than there is in per capita income.

The United Nations Development Program's (UNDP's) 'Human Development Index' (HDI) is a related concept that, like GDP, also enables cross-country comparisons to be made. The Index is based on three factors – health, education and wealth, specifically, life expectancy at birth, mean and expected years of schooling and gross national income (GNI) per capita ($PPP). According to the UNDP, it was created because "people and their capabilities should be the ultimate criteria for assessing the development of a country, not economic growth alone" (UNDP 2014b). But while life expectancy is indirectly related to the environment, there is no attempt to measure environmental quality as there is with GNH.

The *Sarkozy Report* (Stiglitz, Sen and Fitoussi 2009) arose from the *Commission on the Measurement of Economic Performance and Social Progress* (CMEPSP). It was instigated by the former French President to assess the feasibility of alternative measurement tools to GDP, which he regarded as a distorted measure of economic and social progress. Richard Easterlin (2010, p. 1) points out that the Report's recommendations concerning the inclusion of subjective (as well as objective) measurements of human well-being are economically "revolutionary", almost "heresy", especially considering that almost all its 25 members were trained in the era of behavioural economics. With regard to the environment, however, the Report finds that:

> Choices between promoting GDP and protecting the environment may be false choices, once environmental degradation is appropriately included in our measurement of economic performance.
>
> (Stiglitz, Sen and Fitoussi, p. 7)

Further, the *Sarkozy Report* assesses the measurement of sustainability, which it regards as aspects of economic, environmental, and social dimensions of well-being over time. This implies the measurement of wealth, or stocks of physical, natural, human and social capital that carry over into the future. It considers two approaches to measuring sustainability. One kind estimates changes in each stock separately with a view to keeping each one above its critical threshold. The other converts all assets into a monetary equivalent, which implies that different types of capital can be substituted for each other. However, the lack of market values for some items and the validity of the assumption are problematic. As well as the songbird conundrum above, for example, Donald Worster (1993, p. 133). takes exception to it in his collection of essays *The Wealth of Nature*. Rather than the market, "ecological harmony is a nonmarket value that takes a collective will to achieve".

According to the *Sarkozy Report*, this suggests a more limited method, involving monetary measurement of stocks that already have valuation techniques available. In so doing, it should be possible to assess the economic component of sustainability, that is, whether or not countries are over-consuming their economic wealth (Stiglitz, Sen and Fitoussi 2009, p. 17).

Inclusive wealth

The value of Rapa Nui's last palm tree was never quantified, which may be connected to its demise. Achim Steiner, writing in the UN *Inclusive Wealth Report 2012,* points out that the very lack of measurement of nature contributes to environmental damage:

> conventional indicators such as gross domestic product (GDP) or the Human Development Index (HDI) are failing to capture the full wealth of a country. These limitations may be in part fuelling environmental decline and degradation because changes in natural or "nature-based" assets are not factored into national accounts, rendering those accounts less useful as an indicator of changes in human well-being.
>
> (UNU-IHDP and UNEP 2012, p. xi)

The Report begins to rectify this lack of measurement and resultant focus by constructing an 'inclusive wealth index' (IWI) that shows how the full range of productive wealth changes over time. 'Inclusive wealth', according to the Report, includes three types of productive assets – capital, human and natural. So not only does it broaden the accountant's view of wealth to include both human and natural capital (the environment), but it also focuses on the stock (wealth) rather than the flow (income or production) of GDP-style accounting, as recommended by the *Sarkozy Report*. This is similar in concept to the adjusted net savings (ANS) approach (World Bank 2013).

Therefore, importantly, positive growth rates in 'inclusive wealth' represent sustainability, because the total productive base is not eroding. As a result, the asset base can continue to produce similar or better levels of output for future generations (UNU-IHDP and UNEP 2012, p. 11). By contrast, positive growth rates in GDP accounting often correlate with loss of sustainability because they are based on the exploitation of limited resources. The HDI, because it includes GNI per person as one component, is similarly limited, although to a lesser degree.

The *Inclusive Wealth Report* shows how natural resources, along with human and productive capital, can be monitored through calculation of the Index over time. However, the representation of the three sub-categories as well as the total wealth index in the report facilitates monitoring of each separately and shows how they are inter-related. For example, there are five countries that had reductions in their IWI over the period 1990–2008, (Colombia, Nigeria, Russia, Saudi Arabia, and Venezuela) but showed an increase in GDP per capita. Most of these countries were drawing down their oil reserves without sufficient compensatory increases to their produced and human capital bases. Thus the negative IWI growth rates "suggest an unsustainable track and most of the

GDP growth has come at the expense of the natural capital base" (UNU-IHDP and UNEP 2012, p. 43).

The contrast between IWI and GDP becomes clearer in Chapter 4 of the Report, where each of the contiguous states of the US are compared for both measures. In summary, states with high GDP growth tend to have much lower rates of inclusive wealth growth, again suggesting that GDP is generated by running down capital stocks (ibid., p. xxiv).

There are three main problems with GDP. First, well-being is associated with income alone, which may be a necessary condition for well-being, but not a sufficient one. 'Peace of mind', belonging to a community, safety and good health also support well-being. Second, GDP ignores the environmental externalities that result from the production process or the scarcity of dwindling natural resources, which are often public goods with no market prices. Third, GDP represents flows over a short time period, but does not provide information on the state of the capital stocks necessary to generate the income measured, or if they are sufficient for future generations. It masks production based on an unsustainable exploitation of natural resources.

The key variables used to measure inclusive wealth are shown in Table 5.1. Of the 54 categories; more than half relate to natural capital *(A–E)* and how it is adjusted.

As shown in Table 5.1, natural capital assets are grouped as fossil fuels, minerals, forest resources, agricultural land, and fisheries. The IWI attempts to measure the value of these natural assets through the technique of 'shadow pricing', whereby a monetary value is assigned to the asset and compared with the stock of the asset at a later time. For non-renewable resources such as oil, the stock diminishes according to how much has been extracted during the period. The shadow price of each unit is the market price minus the marginal cost of its extraction[15] (ibid., p. 53).

There are also 'externalities' involved. Because using the oil also damages the environment through oil spills and carbon dioxide accumulation for example, the total value of the asset must also be reduced to account for the cost of rectifying the damage. These 'externalities' are listed in the section '*Adjustments in IWI' (A–C)* in the table. Some are positive adjustments, such as productivity increases arising from new technology and the discovery of new oil reserves, while some, such as *carbon damages* of fossil fuels are negative. While parts of the index could be calculated differently – oil and mineral reserves could be simply diminished by extraction and increased by discovery of new deposits for example – the index is reasonably comprehensive without being overly detailed.

The IWR points out that its *'inclusive wealth'* is different from the *'comprehensive wealth'* concept that the World Bank developed in the late 1990s. Because *inclusive wealth* is the social worth of an economy's capital asset base, crucially

Table 5.1 Variables used to measure inclusive wealth

Natural capital	*Human capital*
A. Fossil fuels	Population by age and gender
Reserves	Mortality probability by age and gender
Production	Discount rate
Prices	Employment
Rental rate	Educational attainment
	Employment compensation
B. Minerals	Labour force by age and gender
Reserves	
Production	**Produced capital**
Prices	Investment
Rental rate	Depreciation rate
	Assets lifetime
C. Forest resources	Output growth
Forest stocks	Population
Forest stock commercially available	Productivity
Wood production	
Value of wood production	**Health capital**
Rental rate	Population by age
Forest area	Probability of dying by age
Value of non-timber forest benefits (NTFB)	Value of statistical life
Percentage of forest area used to extract NTFB	Discount rate
Discount rate	
	Adjustments in IWI
D. Agricultural land	*A. Total factor productivity*
Quantity of crops produced	Technological change
Rental rate	
Price of crops produced	*B. Carbon damages*
Harvested area in crops	Carbon emission
Permanent cropland area	Carbon price
Permanent pastureland area	Climate change impacts
	GDP
E. Fisheries	Discount rate
Fishery stocks	
Value of capture fishery	*C. Oil capital gains*
Quantity of capture fishery	Reserves
Rental rate	Oil production
	Oil consumption
	Prices
	Rental rate

Source: UNU-IHDP and UNEP 2012, *Inclusive Wealth Report*, p. 31.

its accounts do not assume sustainability of consumption. Changes in wealth are measured directly from the changes in the asset base. *Comprehensive wealth*, however, "inadvertently assumes that consumption is always on a sustainable path" (ibid., p. 24). On the other hand, according to the World Bank itself, its *comprehensive wealth* allows changes in wealth "to measure the sustainability of development" (World Bank 2011, p. 4). The IWR claim about the Bank may be debatable, but the single index it produces is an outstanding and valuable feature.

Importantly for this book, however, the IWR also says that:

> population has to be acknowledged as a critical factor in sustainability. Although the comprehensive wealth accounts do provide per capita figures, the underlying assumption is that population is kept constant. In the case of the inclusive wealth estimates, population growth is intrinsically captured in the framework and the growth rate has been factored in the analysis. Not surprisingly, results show significant differences between estimates with and without population growth.
>
> (UNU-IHDP and UNEP 2012, p. 23)

But while the *Inclusive Wealth Index* has advantages as a measure of well-being and sustainability, it also has other weaknesses. Most importantly, its breadth of data is limited. While they are reasonably representative, it covers only twenty countries,[16] and only four of them in relation to fisheries due to lack of comparable data for the rest. It also covers only ten types of minerals,[17] albeit generally the more valuable (ibid., p. 18).

However, the strength of the IWI is the singular clarity it brings to sustainability. None of its weaknesses are insurmountable, but if the IWI is to gain the wider acceptance of measures such as GDP, its incorporation into broader programs and the securing of institutional political clout are indicated.

As with the *Sarkozy Report*, a key issue with assessing inclusive wealth is how to measure the value of environmental assets. Many natural assets are not traded in a market and therefore have no price. Also, if natural assets did have a price, it would tend to rise with scarcity, so that environmental destruction may be rewarded in the index as a result of higher values for parts of the environment that remain. For example, if the value of natural forest (or 'non-timber forest benefits' (NTFB)) is, say, a thousand dollars per hectare, and there are a million hectares of it, then the total value of natural forest is a billion dollars. But if half of that forest area were destroyed through logging, mining and fire for example, then each remaining hectare would be worth much more, conceivably double its previous value. The total value of forest assets thus (absurdly) remains the same, even though half has been destroyed.

Microcosms

In this light, one wonders what price could have been put on the last palm tree on Rapa Nui. While that Pacific outpost often serves as a microcosm of the whole Earth, there is conjecture about the alleged collapse of its population and its link to over-exploitation of its resources, especially tree cover. The most famous of these conjectures is in Jared Diamond's *Collapse*, which does draw such a conclusion (Diamond 2005, p. 118). Others, however, point out that it would have been possible to know that the last tree was being destroyed – if it happened – as the whole island is visible from its centre. An alternative and plausible reason for deforestation is that rats arrived with the Polynesians around 1000 AD and ate the seeds of the palm trees, and that during the nineteenth century its population was devastated by Western diseases and the practice of blackbirding to service the mines of Chile (Hunt and Lipo 2009; Lynas 2011).

Perhaps Nauru is a more accurate microcosm of a future degraded planet, as Naomi Klein suggests in *This Changes Everything* (2014, p. 161). A lone Pacific island that was once per person the richest country in the world is now reduced to the status of a gaol for people seeking refuge in Australia. Its modern history began with its annexation by Germany in 1880 with an eye to its one main natural resource. Allocated to Australia by the 1919 Paris Peace Conference (Bashford 2012, p. 120) and made rich due to the extraction and export of its rich phosphate deposits, those deposits are now long exhausted and much of the island is a pitted moonscape of a dug-out resource. The capital from phosphate mining was badly invested. Its few thousand people now appear to maintain a diet that maximises weight and diabetes. Before the arrival of refugees in detention, its only village area – a place I myself have visited several times – appeared desolate and windswept like some ghost town; its only sources of income the sale of territorial fishing rights and fees for 'processing' unnecessary transit visas.

Banaba, formerly *Ocean Island* and now politically part of the Republic of Kiribati in the central Pacific, is a variant of the Nauru story. Like Nauru it was extensively mined for phosphate, but was depopulated during World War II due to massacre by the Japanese and forcible relocation by the British (Hindmarsh 2002, pp. 9–10). Ultimately a trust fund of $10 million was established sourced from its profits, prior to Independence in the 1970s. That fund has been hardly touched since its establishment and now totals about $650 million, a vast sum representing several thousand dollars for each inhabitant of the republic, let alone the island, making the republic debt free. However, the ownership of the fund is now a matter of dispute between the Banaban diaspora, who live mainly in Fiji where they were relocated by the British, and the government of Kiribati, that incorporates the island as part of its territory. The population of the island itself is now only about 300 and it is almost completely isolated as there is no regular transport link between the island and anywhere else, either by sea or air. Thus even though the (natural) capital of the island was at least partly converted to the (financial) capital of the trust fund, neither the island itself, its inhabitants nor their descendants have benefited to any degree. The island remains derelict, its inhabitants are reduced to subsistence and the Fijian diaspora face an uncertain future.

Makatea in French Polynesia was the third Pacific phosphate island, mined between 1917 and 1964. Like its more westerly neighbours it too now lies ravaged and desolate.

In the Indian Ocean, Christmas Island's phosphate deposits were exhausted during the late twentieth century and grandiose proposals for an international casino and a spaceport have yet to eventuate. Instead, like Nauru, its main industry is now as detention centre for asylum seekers to Australia. Its namesake, Kiritimati[18] Island, the world's largest coral atoll, located in the Central Pacific, escaped phosphate mining only to become a site for nuclear testing by the United States and the United Kingdom during the 1950s and 1960s.

Whether or not these six examples foretell a global calamity, they at least demonstrate the importance of maintaining, or at least offsetting the loss of natural capital. The possession of a considerable natural resource can be a curse as much as a blessing.

Footprint

If the IWI has yet failed to gain much political traction, the now-familiar concept of the *ecological footprint* (EF), is an alternative that has several advantages. The EF is a relatively simple aggregate indicator, developed by Wackernagel and Rees in 1996. It calculates the productive land area needed to sustain the consumption and assimilate the waste generated by populations. It is expressed as the land area required to meet consumption levels and can be compared to a country or region's natural carrying capacity. If the EF exceeds the natural carrying capacity, then the population is living beyond the carrying capacity of the land they occupy. It must be either depleting its own natural resources unsustainably, or living off the natural capital of other nations or regions. Of course globally, apart from fanciful ideas of asteroid mining, the latter option is not available. If the global EF is greater than the carrying capacity of the Earth, then it is unsustainable. Wackernagel has calculated estimates of the EF for 52 countries or 80 percent of the world's population (Anielski 2001, p. 7).

Other indicators

Anielski mentions other attempts to measure sustainable well-being in line with Kuznet's earlier dashed hopes in constructing the national accounts for GDP. These include the *Index of Sustainable Economic Welfare* (ISEW) developed by

Herman Daly, John Cobb Jr and Clifford Cobb in the late 1980s. It is a flow measure, based on consumption expenditure as for GDP, but it adds benefits such as unpaid work and deducts social costs such as crime and environmental costs such as pollution and depletion of non-renewable energy. The *Genuine Progress Indicator* (GPI) is a modification of the ISEW concept that originated by Cobb in the United States and developed by Clive Hamilton in Australia (Hamilton and Denniss 2005). There is also the '*dashboard of sustainability*', developed by the International Institute for Sustainable Development (IISD), which aggregates the three clusters – economic, environmental and social – as dials on a car dashboard (Anielski 2001, p. 6). In this respect, it has been observed that "the idiot lights on our planet's dashboard are flashing" yet we continue with business as usual (Sutton and Costanza 2015).

Prompted by the activity surrounding the SDGs, the fifth OECD World Forum held in Mexico in 2015, centered on the limitations of using GDP as a measure of progress and discussing alternatives. Within its 34 developed-country members there is increasing acknowledgment that well-being does not automatically flow from economic growth and that the natural environment must be preserved (Shaw 2015).

Environmental Kuznets curve

The environmental Kuznets curve (EKC) is a construct named for him, rather than ascribed to him, that emerged during the 1990s (Stern 2004, p. 1420) and which crucially addresses the relationship between affluence and environmental damage. The EKC concept says that, like Kuznet's inequality curve, environmental damage will at first rise with economic development, then peak at a certain level of income, and then decline again as the economy matures into a 'developed' state. As Grossman and Krueger (1995 p. 353) put it in their influential study *Economic Growth and the Environment*:

> Will continued economic growth bring ever-greater harm to the Earth's environment? Or do increases in income and wealth sow the seeds for the amelioration of ecological problems? The answers to these questions are critical for the design of appropriate development strategies for lesser-developed countries. Exhaustible and renewable natural resources serve as inputs into the production of many goods and services. If the composition of output and the methods of production were immutable, then damage to the environment would be inextricably linked to the scale of global economic activity.

The study involved matching national-level GDP with local air and water quality data (urban air pollution, oxygen levels, fecal contamination and heavy metal contamination of river basins) in urban areas of 42 different countries between 1977 and 1988. Grossman and Krueger confirm and extend earlier findings, especially including those of the World Bank (1992, p. 11). They say that:

> while increases in GDP may be associated with worsening environmental conditions in very poor countries, air and water quality appear to benefit from economic growth once some critical level of income has been reached.
>
> (Grossman and Krueger 1995, p. 370)

That critical level of income varies again by indicator or pollutant but averages about USD8000 per person per year in 1985 dollars, or around USD14,000 in 2015 terms – coincidentally about the mean global GDP per person. Thus the underlying concept of the EKC has significant support.

However not everyone favours growth in whatever form. For example, the former Professor of Economics at the London School of Economics, EJ Mishan, lamented that:

> The 'Age of Abundance', it transpires, is abundant with pre-packaged and chemically processed foodstuffs, with plastic knick-knacks, with plug-in electric gadgets and stereo equipment. And a part of the price that people in the West pay for this unending procession of shiny assembly-line products is the concomitant loss of those now rarer things that once imparted zest and gratification.
>
> (Mishan 1992 [1967], pp. 125–126)

Similar attitudes are evident in Kenneth Galbraith (1958), *The Affluent Society* and Hamilton and Denniss (2005), *Affluenza*.

More specifically, Grossman and Krueger (1995) do not address two critical findings of the 1992 World Bank report on economic development and the environment. While the Bank found that increasing levels of income were associated with better access to clean water and sanitation, and ultimately to cleaner air, it also found that *municipal waste* rose dramatically with income per person as did *carbon dioxide* emissions per person. Municipal waste is of course a major issue. However it is not insurmountable: there is plenty of scope for profit in its minimisation, as the fortunes of Veolia, the French multinational, demonstrate.[19] Further, the impact of carbon dioxide emissions on climate change is now probably the most important global issue. In 1992 the Bank suggested several policy measures aimed at improving the situation, especially reduction of fuel and energy subsidies (World Bank 1992, p. 12). Today it is clear that much more is required (IPCC 2014a).

Consumption and capitalism

The Nobel laureate astrophysicist, Brian Schmidt has summed up the situation of affluence that confronts us. Affluence depends on the availability of energy, but using readily available energy has negative consequences:

> the great world challenge is figuring out how to transition to an Earth of ten billion people who all want what we have, which turns out to be seven times what the average person has right now and 80 times more than the median person has. My guess is that we rich are not keen to drop our standard of living, so we need to raise the standards of everyone else...prosperity correlates better worldwide with the amount of energy consumed... we already have a ready supply – coal and gas – but the problem is that it leads to climate change...the greatest scientific challenge for the world is figuring out how to use technology to develop the energy necessary to sustain in relative harmony the needs of ten billion people.
>
> (Schmidt 2014)

There are other perspectives. For example, apart from Fowler and Hobbs' (2003, p. 2579) finding that human consumption differs from other species by orders of magnitude, Vitousek et al. (1986, p. 368) had earlier assessed the total use of the world's food supply by humans, then numbering about five billion. They found that humans were appropriating 40 percent of that available through photosynthesis on land, crowding out that available to the other "5–30 million animal species on Earth". A more recent study found that humans consume 25 percent of the production of all land plants, twice the rate of replacement (Charlton 2011, p. 6).

The contemporary Marxist perspective is that there is an ecological contradiction at the heart of capitalism (Foster 2002). A particularly jaundiced view of this is from the US political scientist Michael Parenti in *Against Empire* (1995):

> The essence of capitalism is to turn nature into commodities and commodities into capital. The live green Earth is transformed into dead gold bricks, with luxury items for the few and toxic slag heaps for the many. The glittering mansion overlooks a vast sprawl of shanty towns, wherein a desperate, demoralised humanity is kept in line with drugs, television, and armed force.

The psychologist, Oliver James (2008), tends to support such a view. Writing on the brink of the Great Recession, he finds that misery and distress especially in the English-speaking West has become widespread since the 1970s as inequality has increased and the wealth of a tiny minority has ballooned. He ascribes this situation to the extreme neoliberal capitalism that afflicts the Anglosphere and the inequalities it has produced. Devin Nordberg (2002, p. 15) writes that the issue is based on "serious, structural injustices on a global scale". Sustainable futures should not be built on "technical imperatives", but rather "on political values", he says. While hungry people will always support growth in the hope "it will relieve their misery", "as long as production occurs for profit rather than for human needs, growth will continue", irrespective of any human values of sufficiency.

The philosopher Clive Hamilton believes that there are forces that tell us it is our fault and that it is up to each of us to save the environment through our own personal habits of consumption. But the forces that produce unsustainability need to be approached socially and politically: "In the end we cannot consume our way to sustainability" (Hamilton 2005).

Thus affluence – ultimately material wealth rather than happiness – is measured by both human production and human consumption. It is created originally from nature using human labour and technologies that define both an age and our relationship with the environment. Affluence is accumulated and distributed increasingly unevenly both within and between countries and, because it tends to deplete stored sources of energy, it will have to become less dependent on fossil sources if its distribution is to be made more equal. Its effects on species and their habitats have been highly destructive, especially in recent decades as consumption is amplified through pervasive technologies and population increase.

To the extent that economics' worship at the singular altar of growth is a barrier to sustainability in the developed world, it is trumped by the basic need for survival. A better economics would emphasise sustainability. 'Growth' would emphasise deep quality, not brute quantity. More could be better off if durability, energy conservation and zero waste were revered more than obsolescence, extravagance and effluent.

Lip service

The World Bank's 1992 World Development Report Development and the Environment, showed promise in its awareness and policy prescriptions for sustainability:

> The main message of the Report is the need to integrate environmental considerations into development policymaking. The value of the environment has been underestimated for too long, resulting in damage to human health, reduced productivity, and the undermining of future development prospects.
>
> (World Bank 1992, p. iii)

But despite the positive developments of the 1990s, the measurement options and analyses produced since, and the climate imperatives of the early twenty-first century, recent World Bank reports indicate that sustainability has yet to be mainstreamed within the Bank's collective consciousness. The issue of environmental impact is almost entirely ignored in its 2015 report, *A Measured Approach to Ending Poverty and Boosting Shared Prosperity*, for example.

The opening statement of the report mentions that prosperity must "fully account for environmental degradation and natural resource depletion" (World Bank 2015a, p. 1) and later in a faint echo of Brundtland, stresses that "the path toward [growth and prosperity] must be environmentally, socially, and economically sustainable over time" (ibid., p. 14). However, the document is otherwise all about growth. Apart from the risk of climate change and access to natural resources treated cursorily in Chapter 4 – and the back cover declaration that it is printed on environmentally-friendly recycled paper – these examples are the only times the environment is mentioned in the entire 280-page report. Thus for this major Bank report, the issue is still how to get economic growth to reduce poverty, while it pays lip service to the environment in the face of overwhelming concern about how that growth is achieved. The Bank's 2017 *Atlas of Sustainable Development Goals* does mention decoupling environmental degradation from economic growth, which is one of the 169 SDG targets.

The IMF also appears to lack focus on environmental sustainability in its endorsement of growth. While in a particular statement about the SDGs, there is a mention that the Fund is "deepening policy advice on aspects of inclusion and environmental sustainability and bringing this advice to its operational work" (IMF 2017a), this is the last and least specific of several initiatives outlined. Its World Economic Outlook (IMF 2017b) also mentions the term 'sustainable' – but only in the context of debt repayment.

However, where economic growth has been spectacularly successful in raising affluence, yet disastrous in environmental terms, there is renewed emphasis on environmentally sustainable economic growth (Liu 2010; Zhang 2012; Cohen 2016). But this is because pollution in China is so pervasive that it is impossible to ignore, including the political consequences of failing to address it. Elsewhere, the environmental price of growth may be more conveniently ignored because it is out of sight. Few governments – and major international institutions like the Bank and the Fund – risk dilution of the 'growth and jobs' paradigm because political success demands it. Governing elites tend to avoid complicating the message. Economic growth delivers prosperity. Any environmental nuance that takes nature into account threatens political defeat.

Species perspective

As José Mujica, then President of Uruguay, said to the UN General Assembly on 28 September 2013, we tend to think and reason as individuals or as countries, but "poverty could be eliminated from the planet if we could begin to reason as a species. The current form of civilisation cannot be maintained. We must understand that we are a species and we must govern ourselves as a species" (Mujica 2013). Pope Francis echoes this in his encyclical *Laudato si'*:

> We need to strengthen the conviction that we are one single human family. There are no frontiers or barriers, political or social, behind which we can hide, still less is there room for the globalisation of indifference.
>
> (Pope Francis 2015, s.52)

Mujica occupied no palace as president of his country. At his insistence, his dwelling was little more than a shack. It did, however, have a Victorian water closet; at least in this small way he was more affluent than Suleiman the Magnificent.

Notes

1 Staying with archaic bronze weapons when their enemies, the Hittites, used iron, for example.
2 Indeed, the maintenance of societal wealth is directly proportionate to the generation of energy (Garrett 2014).
3 The Ford Motor Company was one of the first multi-national corporations (MNCs). It expanded production to Canada in 1904, to Europe in 1917,and to Australia in 1925.
4 Although he was anticipated by Adam Smith (2005 [1776], p. 202): "With the greater part of rich people, the chief enjoyment of riches consists in the parade of riches".
5 In practice, total global exports are higher than imports, probably due to political pressures to produce favorable statistics.
6 Physiocracy: from the Greek 'government of nature'.
7 Reports about climate change (IPCC 2013), forestry and fisheries depletion (FAO 2010; FAO 2014), ocean acidification (UNEP 2010), and species extinction (MEA 2005).
8 Husband of Mary Wollstonecraft and father of Mary Shelley.
9 Although a recent (albeit limited) study did not show that 'post-materialist values' influence consumption in Germany, Canada and Sweden (Eklund 2012, p. 27).
10 Churchil's pointed corollary was *"The inherent virtue of socialism is the equal sharing of miseries"*, House of Commons, 22 October 1945.
11 The word 'secular' refers to 'long-term' rather than cyclical trends.
12 Presumably Kuznets is referring to the United Kingdom here.
13 Swiss village site of the annual World Economic Forum.
14 They suggest that ecological economics is a 'post-normal science' in which the songbird has a qualitative, beyond ordinary, value.
15 Also known as the 'rental value'.
16 The twenty countries are: Australia, Brazil, Canada, Chile, China, Colombia, Ecuador, France, Germany, India, Japan, Kenya, Nigeria, Norway, the Russian Federation, Saudi Arabia, South Africa, United Kingdom, the United States and Venezuela.

17 Bauxite, copper, gold, iron, lead, nickel, phosphate, silver, tin, and zinc.
18 Pronounced similarly – the 'ti' is pronounced as an 's' in Gilbertese, which uses only 13 letters.
19 According to its website, *Veolia Environment* in 2012 it had more than 300,000 employees in 48 countries and revenue of nearly 30 billion euros.

References

Alexander, S 2015, *Prosperous Descent: Crisis as Opportunity in an Age of Limits*, Simplicity Institute, Melbourne, Australia.

Anielski, M 2001, *Measuring the Sustainability of Nations: The Genuine Progress Indicator System of Sustainable Well-Being Accounts*, Paper for the Fourth Biennial Conference of the Canadian Society for Ecological Economics: Ecological Sustainability of the Global Market Place, August, Montreal, Canada.

Aristotle 1952 [c. 330 BC], 'Politics', in *The Works of Aristotle*, Vol. 2, University of Chicago/Encyclopedia Britannica, Chicago, US.

Bashford, A 2012, 'Karl Haushofer's Geopolitics of the Pacific Ocean', in Fullagar, K (ed.), *The Atlantic World in the Antipodes: Effects and Transformations since the Eighteenth Century*, Cambridge Scholars, Newcastle Upon Tyne, UK.

Blutstein, H 2014, 'Murdoch Discovers Inequality, But He's Not on 'Team Australia'', *The Conversation*, 9 October.

Boulding, K 1966, 'The Economics of the Coming Spaceship Earth' in Jarrett, H (ed.), *Environmental Quality in a Growing Economy*, Johns Hopkins University Press, Baltimore, US, pp. 3–14.

Briscoe, J 1696, *A Discourse of Money: Being an Essay on That Subject, Historically and Politically Handled, With Reflections on the Present Evil State of the Coin of This Kingdom, and Proposals of a Method for the Remedy, in a Letter to a Nobleman, &c.*, Sam. Briscoe, London, UK.

Charlton, A 2011, 'Man-Made World: Choosing between Progress and Planet', *Quarterly Essay*, vol. 44, pp. 1–72.

Cherrier, H and Ponnor, T 2006, *Hoarding Behavior and Attachment to Material Possessions*, Luis Guido Carli University, Rome, Italy.

Cohen, S 2016, 'Sustainability Policy Is Taking Hold in China', *Huffington Post*, blog, 7 June 2015, updated 6 July 2016.

Coyle, D 2007, *The Soulful Science*, Princeton University Press, Princeton, US.

Coyle, D 2014, *GDP: A Brief but Affectionate History*, Princeton University Press, Princeton, US.

Daly, H 1990, 'Toward Some Operational Principles of Sustainable Development', *Ecological Economics*, vol. 2, pp. 1–6.

Diamond, J 2003, 'The Last Americans: Environmental Collapse and the End of Civilisation', *Harper's Magazine*, June, pp. 43–51.

Diamond, J 2005, *Collapse: How Societies Choose to Fail or Succeed*, Penguin, London, UK.

Dilke, C 1821, *The Source and Remedy of the National Difficulties, Deduced from Principles of Political Economy*, in a Letter to Lord John Russell, published as a pamphlet, London, UK.

Easterlin, R 2010, *Policy Implications of the Sarkozy Report*, Conference Paper, University of Southern California, Los Angeles, US. www.aeaweb.org/aea/2011conference/program/retrieve.php?.

Engels, F 1996 [1844], 'Outlines of a Critique of Political Economy', *Deutsch-Französische Jahrbücher*, Marxists.org.

Engels, F 1925 [1883] 'The Dialectics of Nature', in *Karl Marx Frederick Engels Collected Works*, Vol. 25, Word Press, pp. 313–588.

Easterlin, R and Angelescu, L 2009, *Happiness and Growth the World Over: Time Series Evidence on the Happiness-Income Paradox*, Discussion Paper No. 4060 IZA, Bonn, Germany.

The Economist 2015a, 'Self-Made Wealth in America: Robber Barons and Silicon Sultans', 3 January.

Foster, J Bellamy 2000, *Marx's Ecology: Materialism and Nature*, Monthly Review Press, New York, US.

Foster, J Bellamy 2002, 'Capitalism and Ecology: The Nature of the Contradiction', *Monthly Review*, vol. 54, no. 4, 1 September.

Fowler, C and Hobbs, R 2003, 'Is Humanity Sustainable?', *Proceedings of the Royal Society*, vol. 270, pp. 2579–2583.

Funtowicz, S and Ravetz, J 1994, 'The Worth of a Songbird: Ecological Economics as a Post-Normal Science', *Ecological Economics*, vol. 10, pp. 197–207.

Gale, F 2014, *On the Deep Unsustainability of Actually Existing Liberal Democracy*, Conference Paper, APSA, Sydney, Australia.

Godwin, W 1793, *Enquiry Concerning Political Justice and Its Influence on Morals and Happiness*, Robinson, London, UK.

Gramsci, A 1999 [1934], 'Americanism and Fordism', in Hoare, Q and Nowell Smith, G (eds and trans), *Selections from the Prison Notebooks of Antonio Gramsci*, Vol. 1, Electric Book, London, UK, pp. 568–622.

Grossman, G and Krueger, A 1995, 'Economic Growth and the Environment', *The Quarterly Journal of Economics*, vol. 110, no. 2, May, pp. 353–377.

Hamilton, C 2005, *Speech at the Launch of 'In Search of Sustainability'*, 18 January, Sydney, Australia.

Hamilton, C and Denniss, R 2005, *Affluenza: When Too Much Is Never Enough*, Allen and Unwin, Crows Nest, Australia.

Hamilton, C and Mail, E 2003, *Downshifting in Australia: A Sea-Change in the Pursuit of Happiness*, Discussion Paper No. 50, The Australia Institute, January. ISSN 1322–5421

Harvey, D 2014, *Seventeen Contradictions and the End of Capitalism*, Oxford University Press, New York, US.

Heilbroner, R 1983, *The Worldly Philosophers*, Pelican, Suffolk, UK.

Henderson, H 2014, 'Grossly Distorted Picture: GDP Still Misleading Governments, Banks and Investors by Omitting Asset Accounts', *Ethical Markets*, 7 July.

Hindmarsh, G 2002, *One Minority People: A Report on the Banabans, Formerly of Banaba (Ocean Island) Who Were Relocated to Rabi Island in Fiji*, Report commissioned by UNESCO (Apia).

Hooke, A 1750, *Essay on the National Debt and National Capital*, W Owen, London, UK.

Hunt, T and Lipo, C 2009, 'Revisiting Rapa Nui (Easter Island) "Ecocide"', *Pacific Science*, vol. 63, no. 4, pp. 601–616.

IMF: International Monetary Fund 2017a, *The IMF and the Sustainable Development Goals*, Factsheet, 14 April.

IMF: International Monetary Fund 2017b, *World Economic Outlook, April 2017: Gaining Momentum?*

IPCC: Intergovernmental Panel on Climate Change 2014a, *Climate Change 2014, Mitigation of Climate Change*, WMO-UNEP, Geneva, Switzerland.

Islam, S 2015, *Inequality and Environmental Sustainability*, Working Paper No. 145, August, ST/ESA/2015/DWP/145, United Nations Department of Economic and Social Affairs (DESA).

James, O 2008, *The Selfish Capitalist: Origins of Affluenza*, Vermilion, London, UK.

Jeffries, S 2015, 'Why Too Much Choice Is Stressing Us Out,' *The Guardian*, Manchester, UK, 22 October.

Keynes, J 1936, *The General Theory of Employment, Interest, and Money*, University of Adelaide Library Electronic Texts Collection, Adelaide, Australia.

Kim, J 2014, *Boosting Shared Prosperity*, Speech by World Bank Group President at Howard University, World Bank Group, Washington, DC, US, 1 October.

Klein, N 2014, *This Changes Everything: Capitalism vs. the Climate*, Penguin Group, London, UK.

Kuznets, S 1955, 'Economic Growth and Income Inequality', Presidential address delivered at the sixty-seventh Annual Meeting of the American Economic Association, Detroit, Michigan, December 29, 1954, *The American Economic Review*, vol. 45, no. 1, March, pp. 1–28.

Kuznets, S 1971, 'Modern Economic Growth: Findings and Reflections', Nobel prize lecture, in Lindbeck, A (ed.), *Nobel Lectures, Economics 1969–1980*, World Scientific Publishing, Singapore.

Liu, J 2010, 'China's Road to Sustainability', *Science*, Galley policy forum, vol. 328, no. 5974, 2 April, p. 50.

Lynas, M 2011, *The Myth of Easter Island's Ecocide*, website, 19 September. Viewed 6 November 2014. www.marklynas.org/2011/09/the-myth-of-easter-islands-ecocide.

Marx, K 1973 [1861], *Grundrisse der Kritik der Politischen Ökonomie (Outlines of the Critique of Political Economy)*, Economic Works of Karl Marx 1857–1861, Nicolaus, M (trans.), Penguin, London, UK.

McCarthy, J 2014, *How a Lack of Toilets Puts India's Women at Risk of Assault*, NPR: National Public Radio, Washington, DC, US, 9 June. www.npr.org/blogs/parallels/2014/06/09/319529037/indias-rape-uproar-ignites-demand-to-end-open-defecation.

Mishan, E J 1992 [1967], *The Costs of Economic Growth*, Oxford University Press, Oxford, UK.

Mujica, J 2013, *Address to the 68th General Assembly of the United Nations*, 13 September. http://gadebate.un.org/68/uruguay#sthash.Bv285VHR.dpuf.

Nef, J 1977, 'An Early Energy Crisis and Its Consequences', *Scientific American*, vol. 237, no. 5, November, pp. 140–151.

Nordberg, D 2002, *The IPAT Equation, Global Modelling and the Search for Sustainable Development*, University of Hawaii, Honolulu, US.

Ostrom, E 2009, 'A General Framework for Analyzing Sustainability of Social-Ecological Systems', *Science*, vol. 325, no. 5939, 24 July, pp. 419–422.

Parenti, M 1995, *Against Empire*, City Lights Books, San Francisco, US.

Piketty, T 2014, *Capital in the Twenty-First Century (Le capital au XXI siècle)*, Goldhammer, A (trans.), Harvard University Press, Cambridge, US.

Plato 1952 [c.380 BC], *Laws*, Book V, University of Chicago/Encyclopedia Britannica, Chicago, US.

Polanyi, K 2001 [1944], *The Great Transformation*, Beacon Press, Boston, US.

Pope Francis 2015, *Laudato si'*, Libreria Editrice Vaticana, 24 May.

Redclift, M 1993, 'Sustainable Development: Needs, Values, Rights', *Environmental Values*, vol. 2, no. 1, pp. 3–20.

Robbins, W, Harriss, I and Macklin, R 2005, *Fact and Myth: Reflections on Why Higgins Made the Harvester Decision*, Association of Industrial Relations Academics of Australia and New Zealand Conference paper, Charles Sturt University, Perth, Australia, pp. 487–494.

Sattin, A 2014, 'The Edge of the World Review – A Radical Perspective on the Modern World: Michael Pye's Argument That the North Sea, Not the Roman Empire, Has Been At The Heart Of Europe's Greatest Recent Advances Is Persuasive and Eloquent', *The Observer*, London, 9 November.

Scheidel, W 2017, *The Great Leveler: Violence and the History of Inequality from the Stone Age to the Twenty-First Century*, Princeton University Press, Princeton, US.

Schmidt, B 2014, *Science and Society: Highlights from the 2014 Kenneth Myer Address*, Radio National Big Ideas, Australian Broadcasting Corporation, 28 August.

Shaw, T 2015, 'Beyond GDP – How Australia Could Help Redefine Well-Being', *The Conversation*, 13 October.

Sloterdijk, P 2005, 'The Crystal Palace', *Im Weltinnenraum des Kapitals: Fur eine philosophische Theorie der Globalisierung* [In the Global Inner Space of Capital: For a Philosophical Theory of Globalisation], Suhrkamp, Frankfurt am Main, Germany, chapter 33, pp. 265–276.

Smith, A 1952 [1776], *An Inquiry into the Nature and Causes of the Wealth of Nations*, University of Chicago/Encyclopedia Britannica, Chicago, US.

Smith, A 2005 [1766], *Lectures on Jurisprudence*, Glasgow Edition of Works, Vol. 5 (1762–1766), Online Library of Liberty, Liberty Fund.

Solnick, S and Hemenway, D 1998, 'Is More Always Better? A Survey on Positional Concerns', *Journal of Economic Behavior and Organization*, vol. 37, pp. 373–383.

Sorrell, S 2010, 'Energy, Economic Growth and Environmental Sustainability: Five Propositions', *Sustainability*, vol. 2, pp. 1784–1809.

Stern, D 2004, 'The Rise and Fall of the Environmental Kuznets Curve', *World Development*, vol. 32, no. 8, pp. 1419–1439.

Stiglitz, J, Sen, A and Fitoussi, J-P 2009, *Report by the Commission on the Measurement of Economic Performance and Social Progress*, Eurostat.

Sutton, P and Costanza, R 2015, 'Beavers Are Worth $1b a Year, Yet Still Our Economy Grossly Undervalues Nature', Fleming, R 1942, *The Oceans, Their Physics, Chemistry, and General Biology*, Prentice-Hall, New York, US.

Trainer, T 1985, *Abandon Affluence*, Zed Press, London, UK.

UNDP: United Nations Development Program 2014b, *Human Development Index*. Viewed 22 October 2014.

UNHDR: United Nations Human Development Report 2014, *Sustaining Human Progress: Reducing Vulnerabilities and Building Resilience*, United Nations Development Programme, New York, US.

UNU-IHDP and UNEP 2012, *Inclusive Wealth Report 2012. Measuring Progress toward Sustainability*, Cambridge University Press, Cambridge, UK.

Ura, K, Alkire, S, Zangmo, T and Wangdi, K 2012, *A Short Guide to Gross National Happiness Index*, Centre for Bhutan Studies.

Van Dieren, W (ed.) 1995, *Taking Nature into Account: A Report to the Club of Rome*, Copernicus Books, Springer-Verlag, New York, US.

Veblen, T 1899, *The Theory of the Leisure Class: An Economic Study of Institutions*, Project Gutenberg, eBook #833.

Vitousek, P, Ehrlich, P, Ehrlich, A and Matson, P 1986, 'Human Appropriation of the Products of Photosynthesis', *BioScience*, vol. 36, no. 6, pp. 368–373.

Wackernagel, M and Rees, W 1996, *Our Ecological Footprint: Reducing Human Impact on the Earth*, New Society Publishers, Gabriola Island, BC, Canada.

Whitaker, A 2001 [1904], *History and Criticism of the Labor Theory of Value in English Political Economy*, Columbia University, Batoche Books, Kitchener, ON, Canada.

Wolff, R 2011, 'The Thought of Karl Marx Part Seventeen', *The Philosopher's Stone*, blog, 9 February. http://robertpaulwolff.blogspot.com/2011/02/thought-of-karl-marx-part-seventeen.html.
World Bank 1992, *World Development Report: Development and the Environment*, Oxford University Press, New York, US.
World Bank 2011, *The Changing Wealth of Nations: Measuring Sustainable Development in the New Millennium*, The World Bank, Washington, DC, US.
World Bank 2014b, GDP, PPP (Current International $) Data. http://data.worldbank.org/indicator/NY.GDP.MKTP.PP.CD.
World Bank 2015a, *A Measured Approach to Ending Poverty and Boosting Shared Prosperity: Concepts, Data, and the Twin Goals*, Policy Research Report, Washington, DC, US.
World Bank 2017a, *World Development Report 2017: Governance and the Law*, overview, Washington, DC, US. doi: 10.1596/978-1-4648-0950-7.
Worster, D 1993, *The Wealth of Nature*, Oxford University Press, New York, US.
Zhang, J 2012, *Delivering Environmentally Sustainable Economic Growth: The Case of China*, Asia Society.

6 Technology

Humanity has an ancient and intense relationship with technology. Such is our mutual embrace that it seems unique to our species and is definitive of us. But in its relationship to sustainability, contradictions are apparent. It is technology that has led to resource depletion and degradation of the environment, whereas it is in technology that there lies hope for a more sustainable future. Technology has enabled the creation of modernity that briefly shields our species from its long-term consequences. It has both allowed us to become more crowded and consume ever more, yet it may also enable us to become happily fewer and for each of us to tread more lightly on the planet. Technology is associated with industrial pollution, but at the same time it can help reduce pollution. Its impact differs according to the nature of technology and how it is deployed. As the technology historian Melvin Kranzberg (1986, p. 545) said in his first law, "technology is neither good nor bad; nor is it neutral". His fifth law underlines its importance to humanity: "all history is relevant, but the history of technology is the most relevant" (ibid., p. 550).

There is a yawning chasm between extreme views of the relationship concerning technology and humanity, both positive and negative polar opposites. How that void is bridged may well determine our continued existence as a species, as well as the fate of the other life forms that cohabit the Earth with us.

Warnings

Warnings about technology include the wings of Icarus that enabled him to fly to the heavens but brought about his death. Significantly the wings were built for him by his father Daedalus, the artisan who had also constructed the labyrinth that imprisoned the Minotaur. Daedalus' creation of a means to achieve the god-like power of flight has destructive consequences, and the gift of this technology results in a double disaster: the death of a son and a father's grief. Similarly, the titan Prometheus stole fire to benefit the humanity he had created, and was condemned to eternal torture for his efforts by Zeus, because it was a challenge to the power of the gods. One version of the myth includes the vesting of the troubles of an archetypical Pandora onto humanity by an angered Zeus: "From her the tribe of women comes – for men a grievous bane" (Hesiod 2010 [c. 800 BC], p. 590). Hephaestus was the god associated with the technologies of the forge, but he was deformed due to poisoning from the arsenic used in bronze metallurgies in ancient times.

During the Middle Ages mechanical technologies involving clockwork and other means spread through Europe from the East. They were regarded as magical, much as nature was regarded as a powerful entity to be feared rather than understood. According to the historian E. R. Truitt (2015), eventually around the seventeenth century these mechanical wonders were generally regarded as operating according to natural laws, instead of by magic. At about the same time, nature was further diminished by Newton's *Principia* (1952 [1686]), a work itself made possible by new optical technologies. Rather than a powerful entity, 'nature' became abstract, predictable and subject to law much like a mechanical clock.

During the same era the morality tale of Dr Faustus, who sold his soul to the Devil for the chance to to command nature, was widespread throughout northern Europe. This fable was disseminated by the new technology of print and became available to the English playwright, Christopher Marlowe who wrote it as a play in 1602. Faustus' authority over nature was:

> To do whatever Faustus shall command,
> Be it to make the moon drop from her sphere,
> Or the ocean to overwhelm the world.
>
> (Marlowe 2009 [1604], p. 23)

At the beginning of the nineteenth century, Johan von Goethe's *Faust* (1870 [1828]) was published on a similar theme.

Still during the early Industrial Revolution, Mary Shelley's fear of technology was evident in her immensely popular novel *Frankenstein: The Modern Prometheus* (1818), which involved the creation of life through technology, as the titan had created humanity. In the introduction to the 1831 edition, Shelley mentions Erasmus Darwin, Charles' grandfather, and Luigi Galvani as sources of inspiration. Darwin had written of experiments concerning life developing from apparently lifeless forms and Galvani's theories of animal electricity had already famously been applied to the corpse of a murderer (Simili 2014). Dr Frankenstein's god-like creation of life was a warning about new technology, but Shelley linked technology and humanity to total catastrophe in her later book, *The Last Man* (1826). In the book, at the end of the twenty-first century, masculine belief in technological mastery over nature is proved false when rapid climate change, famine and finally plague exterminate humanity on Earth (Shelley 1997 [1826]).

Apocalyptic visions of the future based on technological catastrophe are now commonplace, such as Margaret Atwood's 2003, *Oryx and Crake* in which a genetically modified virus wipes out the entire population. Earlier works include René Barjavel's 1943 novel *Ravage*, in which a future France is devastated by the sudden failure of electricity as well as E. M. Forster's 1909 novella *The Machine Stops*, in which humanity is entirely dependent on a god-like machine that deteriorates and eventually stops, ending the lives of everyone.

Less cataclysmic but more poignant is Hans Christian Andersen's 1843 tale *The Nightingale* in which the emperor of China banishes his living nightingale in preference to a bejewelled mechanical bird, which "could sing like the real one, and could move its tail up and down, which sparkled with silver and gold", sent to him by the emperor of Japan. But the emperor is overwhelmed with despair when the mechanical nightingale can no longer sing due to a broken spring. Nearing death from his misery, finally the emperor hears the song of the living nightingale and is revived (cited in Johnson 1990, p. 71). Mechanical technologies can be alluring but are ultimately only imperfect imitations of nature.

From a subtler perspective, German existentialist philosopher Martin Heidegger was a major twentieth century critic of technology. In *The Question Concerning Technology* (1953), Heidegger said that the modern technological "mode of being" saw the natural world only as a resource to be exploited, as a means to an end. He illustrated this with what might be regarded as a rather benign form of technology – a hydroelectric plant on the Rhine – but which changed the river from an unspoiled natural wonder to a mere power supply (Heidegger 1953, p. 321). In Heidegger's view technology is not just a collection of tools, but rather a way of understanding the world, which is both "instrumental and grotesque" (Wheeler 2013). Whether Heidegger would have preferred his electricity to be supplied from a coal-fired power station is not made clear in the essay. Yet his insight remains: existing technologies, especially those that exploit natural resources, reduce nature to that singular measure. In this sense, humanity too is diminished by our technologies.

Technology equates to pollution in some views. As well as the poisoning of Hephaestian bronze workers and lead pollution from mining in ancient times (Hong et al. 1994), atmospheric pollution from fires and furnaces has been a problem since at least the Middle Ages. Before World War II, "smoke, sewage, and soot were the main environmental concerns" (Heaton, Repetto and Sobin 1991, p. 5).

However, something changed after 1945 and it concerned new technologies. There was the new threat of radiation and radioactive waste from nuclear technology, as well as concern about the new synthetic chemical technologies highlighted by Carson's *Silent Spring* (1962). Barry Commoner, "the Paul Revere of ecology" in *The Closing Circle* (1971, p. 146) identified the "technological suicide" caused by the scale and nature of pollution that had engulfed the United States, also since the 1940s. Commoner showed how these unregulated new technologies (especially synthetic organic chemicals) caused great harm to ecosystems, polluting at a rate ten times that of economic growth. Commoner engaged in intense public debate with population alarmists such as Paul Ehrlich, arguing that humanity's negative impact on the planet was largely due to these uncontrolled new technologies rather than due to too many people who consumed too much. In 1974 Ariyoshi Sawako,[1] in the novel *Fukugo osen* (*Compound Pollution*), tapped into deep concerns about the effects of organic chemicals in Japan. The destruction of the Earth's protective ozone layer by chlorofluorocarbons from refrigerants and propellants was recognized in the 1970s, leading to identification of the 'ozone hole' over the Antarctic.

Positives

Carson's book led to the banning of DDT[2] and to far greater scrutiny of the effect of pesticides. Commoner contributed to an awareness of ecosystems, while Ariyoshi's work led to consumer demand for organic products in Japan (Moen 1997). Protecting the ozone layer was effected through the world's first universally signed treaties – the *Vienna Convention for the Protection of the Ozone Layer* (1985) and its *Montreal protocol* (1997) – both endorsed by 197 states including the European Union.

However, recent science has shown that technology was decisive in humanity's ultimate survival. In *When the Sea Saved Humanity*, anthropologist Curtis Marean (2010) explains how the development of advanced composite stone tools, including "sophisticated implements such as microblades" involving heat-treated quartz to make it easier to shape and attached to wooden shafts to form spears, was a factor that enabled the last remnants of humanity to stay alive during an Ice Age around 150,000 years ago in southern Africa. Humanity, then reduced to a total of only a few hundred breeding individuals literally clinging to the edge of the continent, would likely have become extinct without these technologies that enabled successful fishing when there was no game and little vegetation. If so, we owe our very existence to that relationship – all of us are descended from those lucky few.

Gore (2013, p. 357) asserts that this critical time for humanity before migration from Africa occurred was one of three occasions when humanity did not appear sustainable. The other two were in the 1960s when nuclear war between the United States and the USSR appeared imminent, and now when we are changing the biosphere irreversibly to the detriment of ourselves as well as other species. In the first case technology is positive. But the latter two cases show that complex technologies can have profoundly negative effects.

Nevertheless, technology as a positive force is defined by the future. Without technological change the future would consist of successive generational replacement and little else. In prehistoric times there might be tomorrow or even next winter, but there was no future. A better future only became possible with the development of complex technologies. How those technologies might be applied became the stuff of science fiction. For example, Jules Verne anticipated submarines in *Twenty Thousand Leagues Under the Sea* and spaceflight in *From the Earth to the Moon* during the nineteenth century. The Internet was invented in fiction long before it existed in reality. H. G. Wells publicised the idea of a 'world brain' or permanent encyclopedia using microfilm in 1936, so that anyone, anywhere could examine any book or document that had been assembled within it (Wells 1938). Arthur Clarke presciently suggested in 1962 that this encyclopedia or library could be accessed by personal computers by the year 2000, but that its extension to involve artificial intelligence might take another century.

Isaac Asimov in his *Foundation* series, beginning in the 1940s, conceived of a foundation located at the end of the galaxy. This foundation enabled human knowledge to be preserved and advanced during a period of galactic chaos, which was predicted by the invention of a mathematical sociology, *psychohistory*. The foundation's technology is regarded by its neighbouring primitive planets as a form of religious magic, which anticipated Arthur Clarke's famed third law: "any sufficiently advanced technology is indistinguishable from magic" (Clarke 1962, p. 14). Clarke's own later work, *Imperial Earth* predicts desktop "communications consoles" with screens and keyboards as well as "minisec" mobile computing devices that can communicate with the consoles, albeit only within visual proximity (Clarke 1975, pp. 126–128). And the once ubiquitous Samsung flip-phone of the early twenty-first century was reputedly modelled on the 'Beam me up Scotty' communicator device of the 1960s *Star Trek* television series, set in the twenty-third century.

Meanings

The word 'technology' was first recorded in English in the early 1600s, meaning "a discourse or treatise on an art or the arts," from the Greek 'tekhnologia' – the "systematic treatment of an art, craft, or technique". The meaning "study of mechanical and industrial arts" such as "spinning, metal-working, or brewing" was first recorded in the mid 1800s. Its former component, 'techno' is from the Greek 'tekhne' "art, skill, craft in work; system or method of making or doing," originally related to weaving. Its latter component '-logy' means "a speaking, discourse, treatise, doctrine, theory, science," from the Greek '-logia' from root of legein "to speak" (Online Etymology Dictionary 2014).

However, the contemporary meaning of the word 'technology' is much broader. As a result of the so-called 'second Industrial Revolution' during the latter nineteenth and early twentieth centuries, when steel, electrification, the telegraph and the production line defined the era, the term came to mean both the study of the useful arts and crafts, as well as the arts and crafts themselves. The phrase 'high technology' is from 1960s while its short form 'high-tech' (later 'hi-tech') is from the 1970s. *New* technology arises from the application of innovative systems, whether for the same or different ends.

The word 'technology' also incorporates the applied sciences in its meaning, although not all technologies are derived from science. Most generally, it is the "practical application of knowledge" (Merriam-Webster 2014). In fact, many significant technological developments have been contributed by non-scientists – people with practical skills such as James Watt, an instrument maker; Henry Ford, a metal machinist; Samuel Morse, a professional artist; Thomas Edison, a telegrapher; Guglielmo Marconi, an aristocrat inventor; and Steve Jobs, a geek and college drop-out. Technological innovations depend more on the skills of technicians than on scientific theories (Coan 2012), and, rather than science driving new technology it is more likely that new technologies are "the force that move science forward" (Wise 1985, p. 229), as no doubt Galileo, Pasteur and the Wright brothers would agree. Much of Galileo's science depended on his use of the telescope, Pasteur's on the microscope and the Wright brothers' aeronautical miracle of powered flight depended on the technology of internal combustion and the practical observation of kites.

Technologies have been with us since pre-historic times, whereas science is a newcomer. Technology may incorporate aspects of art and aesthetics (as the functional beauty of an Apple computer), but neither is intrinsic to technology. Importantly, what is intrinsic to technology is the systematic use of particular methods.

It is tempting also to associate technology with mass production, as it has continued to be so since the Industrial Revolution and even earlier with the use of pottery moulds in ancient times. This is because the use of systematic method often leads to reproducibility and hence to mass production – that multiplier of investment capital. Certainly the rise of modern technology is connected with capitalism and the Industrial Revolution, especially through mass production. Early capitalism encouraged new technologies by simply rewarding them. Arkwright's spinning jenny was probably the first use of technology that resulted in mass production (of cloth) and its associated concentration of a workforce alongside factories. However, these early factories at first used water from fast flowing streams to power the looms. Newcomen's invention of the steam engine to pump water from coal mines enabled coal production to feed new furnaces and factories, while later developments (Watt, Trevithick, Stevenson) enabled the railways that began to connect cities and ports, and powered ships that linked ports around the globe. Edison's breakthrough technologies such as grid electricity, the electric light and the phonograph resulted from an intensive and systematic approach to innovation itself, propelled by the motive for profit.

From a philosophical point of view, Clive Hamilton (2013, p. 30) points to a definition of technology as 'a phenomenon captured and put to use':

> ...where phenomena are mostly physical effects such as the release of energy when carbon-based molecules are oxidised and heated, the way light is refracted through a lens, the way wind energy turns a propeller that can drive a turbine, and so on. These myriad phenomena—mechanical, electrical, photonic, biological, nuclear, etc. — are waiting to be discovered by humans and then orchestrated to our benefit.

Engels described man as the 'tool-making animal'. We have not evolved the specialised organs Darwin refers to and instead we develop "specially prepared instruments". Technology is the tools we develop that shape our relationship with nature – and that in turn shape us in both hand and mind (Engels 1940 [1876], p. 281). Similarly, Karl Marx in *Capital,* wrote that:

> Technology reveals the active relation of man to nature, the direct process of the production of his life, and thereby it also lays bare the process of the production of the social relations of his life, and the mental conceptions that flow from those relations.
>
> (Marx 1952 [1883], p. 181)

This are important observations. The creature at the computer screen thinks and acts differently from a tiller of the soil, and so whole societies are shaped.

However, in other definitions humans are not alone in technological behaviour. Other creatures construct technological artefacts such as nests, webs, traps, complex termite mounds and beaver dams. The artefacts that characterise humanity are rather machines (Polanyi 1968, p. 1308), complex manufactures with more than one part (such as the bow and arrow)[3] that give us added power over nature. Machine-making was both enabled by and enabled specialisation, exchange value and the creation of ever more complex technologies. It has meant that now in the contemporary West, we are entirely surrounded by machines, networks and specialised technologies that few understand enough to recreate, but which are in universal use (Aunger 2010). Further, as major technologies are rolled out, there is a tendency to overestimate their effects in the short run, but more importantly, to underestimate their effects in the long run.[4]

Slavery and capitalism

Slavery, capitalism and technology are interrelated. The economic historian, C. Knick Harley, says it is commonly asserted that the West Indies slave trade financed the Industrial Revolution in England (Harley 2013, p. 5), based on

Eric Williams' seminal work *Capitalism and Slavery* (1944). Certainly the timing is consistent: slavery and the slave trade was at its profitable height immediately before the Industrial Revolution, yet the slave trade was abolished in 1807 and slavery itself eventually abolished in 1833 within the British Empire (Harley 2013, p. 8). But while the transatlantic slave trade and its profits were based on sugar during the eighteenth century, it was the technological and resultant organisational changes in the cotton industry "that were central to the emergence of a modern economy based on mechanised factory production" (Harley 2010, p. 2). Further,

> there is general agreement that technological change lies behind historical economic growth and that the creation of knowledge and technology must be seen as a part of the economy, i.e. endogenous.
>
> (Harley 2013, p. 21)

While the British cotton industry depended on slavery for its raw material at the beginning of the Industrial Revolution, by the mid-nineteenth century it was less dependent on slavery due to sourcing from India and the Levant, and the industry itself was much less significant. At that time, the export of manufactured goods to the Empire and North America was the source of new profits. The Industrial Revolution in Britain arose from industrial research and development resulting from the relatively high wages and cheap energy at the time. This provided incentives to find manufacturing techniques that substituted fuel and capital for labour, and this in turn created new knowledge that further enhanced the process of technological change (ibid., pp. 21–23). Thus, plausibly, slavery (in the sugar industry) was one source of capital for the Industrial Revolution. Slavery also was associated with the production of an important raw material (cotton) that was exploited by new technology.

The divergent costs of labour and energy pressed the technological innovations that defined the era. That there was no Industrial Revolution in ancient times indicates that slavery is a negative factor in technological advance. The ancients devised many technological innovations, including steam engines, pumps and watermills, a railway across the Corinthian neck, as well as canals in Egypt, Greece and in China. However, none of these innovations spread far. One reason was the presence of slaves in all ancient societies, which obviated the need for labour-saving devices (Cowell 1964, p. 368). Thus, conceivably, the slavery that helped produce the technological advances of the Industrial Revolution did so because it was at a distance from it. The slavery that prevented an ancient Industrial Revolution did so because it was part of the societies that would have been affected. In the ancient world, there was already an army of intelligent robots and thus no need for the technologies of the Victorian era.

However, capitalism is not essential to technological development. During much of the last century, for example, the command economy of the Soviet Union at least mimicked much of the technological development of the West and in significant ways exceeded it. The indisputably revolutionary *Sputnik* of 1957 is the most obvious example of this. Others include the *Voskhod* space capsule that could land on the hard earth, rather than at sea as the US capsules had to; the world's most popular assault rifle, the Kalashnikov AK-47; superior aircraft such as the MiG 15 fighter of the 1950s, the MiG 25 Foxbat of the 1970s that could fly at Mach 3, and the giant Antonev transport plane that is still in use in many parts of the world. Perhaps, as C. P. Snow asserts, it was partly because the Soviet education system managed to blend the arts and the sciences:

> An engineer in a Soviet novel is as acceptable, so it seems, as a psychiatrist in an American one. They are as ready to cope in art with the processes of production as Balzac was with the processes of craft manufacture.
>
> (Snow 1959, p. 19)

And whilst technological innovation was certainly not frequent before the Industrial Revolution and capitalism, several pre-modern economic epochs are discernable from around 7000 B.C. that are based on it. The innovations that drove economic (and population) growth include the invention of agriculture itself (c.7000–4000 B.C.), bronze metallurgy (c. 4000–1900 B.C.), the iron plough (c. A.D. 300–930) and the spectacle lens (c. A.D. 1340–1600) (Šmihula 2011, p. 66).

In fact, modern economic growth may be more dependent on technological innovation than it is on capitalism. Simon Kuznets' Nobel speech on modern economic growth illustrates this in his discussion of economic epochs:

> we may proceed on the working assumption that modern economic growth represents such a distinct epoch – growth dating back to the late eighteenth century and limited (except in significant *partial* effects) to economically developed countries. These countries, so classified because they have managed to take adequate advantage of the potential of modern technology, include most of Europe, the overseas offshoots of Western Europe, and Japan.
>
> (Kuznets 1971)

Interestingly, the rest of this lecture – presumably an example of the best thinking of the time – has but one dismissive reference to the sustainability of economic growth: "even if we disregard the threatening exhaustion of natural resources, a problem that so concerned Classical (and implicitly even Marxian) economics…" (ibid.), but goes no further on the subject, suggesting that the question of resource depletion was then thought no longer an issue.

Kenneth Boulding, too, was in agreement with the link between economic growth and technology, insofar as technology is applied knowledge. His 1966 paper *The Coming Spaceship Earth* is explicitly about limits. But as far as economic development is concerned:

> The cumulation of knowledge, that is, the excess of its production over its consumption, is the key to human development of all kinds, especially to economic development.
>
> (Boulding 1966, p. 5)

Kondratieff waves

Long-term 50- to 60-year economic cycles, known as Kondratieff[5] waves, have been associated with technological innovation since the early twentieth century. The Soviet economist Nikolai Kondratieff identified these phenomena in his book *The Major Economic Cycles* (1925), although some of his observations had been anticipated by Dutch economists, Jacob van Gelderen and Samuel de Wolff, in 1913 (Narkus 2012, p. 12). Kondratieff believed that the crises that punctuate periods between cycles in capitalism were indicative of a clearing away of debris to allow for new technological and economic growth, rather than leading to the inherent doom predicted by Marx. These views were in obvious conflict with Stalinist orthodoxy, and after working in agricultural reform for a time, he was sent to a Gulag labour camp and subsequently executed in 1938. In 1939, the economist Joseph Schumpeter, whose views on "the perennial gale of creative destruction" in capitalism (Schumpeter 2011 [1942], p. 85) and new growth based on "the swarming of technological innovations" (Phillimore 2001, p. 28) benefited from this work, suggested calling the cycles 'Kondratieff waves' in his honour.

Several other possible factors that contribute to long economic waves have also been identified, including demographic "baby booms" and debt deflation as part of the credit cycle. However, according to the technological innovation theory championed by Schumpeter, these waves arise from the clustering of innovations that launch technological revolutions, which in turn create leading industrial or commercial sectors.

A modification of Kondratieff's theory was recently developed by the Slovak, Daniel Šmihula, who identified six long-waves within modern capitalist economies. Each of these was initiated by technological revolution (Šmihula 2011, p. 51). The financial-agricultural wave occurred between 1600 and 1780. The Industrial Revolution wave occurred between 1780 and 1880, especially involving textiles, iron, coal, railways and canals. The technical wave occurred between 1880 and 1940, especially involving the chemical, electro-technical and machinery industries. The scientific-technical wave between 1940 and 1985 involved aviation, nuclear technologies, space, synthetic materials, oil and cybernetics. The information and telecommunications revolution, from 1985 until 2015, involved telecommunications, cybernetics, "informatics" and the Internet. The sixth wave to 2035 is a 'post-informational technological revolution', including biomedicine, nanotechnology and alternative fuel systems such as hydrogen.

Unlike Kondratieff and Schumpeter, Šmihula believes that each new cycle is shorter than its predecessor, but although it is his central conjecture, he does not advance any particular reason as to why this may be so. However, it seems evident that the increasing mass of all preceding innovations would result in compression of the cycles that spring from later innovations. Earlier technologies facilitate newer technologies, as the printing press spread ideas on heat and steam engines, as smelting produced steel, as railways linked industries that enabled complex manufactures. The accumulation of technical knowledge is geometric and thus technological progress accelerates. Richardson (2013, p. 161) tends to support this view in his paper on holarchies and technological evolution, in which he says that the idea of a "continual acceleration of change" is credible. The concept of 'holarchies', or nested layers of systems, he says, is common in many disciplines, including physics[6] and biology,[7] but also applies especially to technology. Imagine what could happen if the technological systems of our world were "integrated into an organic whole" supporting both "humanity and nature in an elegant and sustainable way", he writes (ibid., p. 167).

Šmihula (2011, p. 53) stresses technological progress as a decisive factor in any long-time economic development. Each of these waves has an innovation phase, which is described as a technological revolution, and an application phase in which the number of new innovations falls and the focus is on exploiting existing innovations. Each wave of technological innovations is characterised by the sector in which the most revolutionary changes took place. Every wave of innovations lasts until the profits from the new innovation or sector fall to the level of older, more traditional sectors. Šmihula (p. 61) goes on to identify the important difference between invention and innovation:

> The invention is a making-up of something new which did not exist before. But as such, it may finish at the bottom of a drawer. Only an invention which is applied in practical life can be a real innovation. The source of such innovations may be not only inventions but also imitations taken, for example, from some other society [such as Arabia and China to Europe before the seventeenth century].

Carbon

While economic development may occur in technological waves, as many have observed (Commoner, Boulding, Gilding), the technological icons of modernity are all based on the combustion of carbon – coal, oil and gas – that is in effect the stored sunlight from plant life laid down during the Carboniferous period of the late Paleozoic era. Those icons – grid electricity, cars and aeroplanes, as well as more lately the computer – enable us to be shielded from the environment and isolated from the harm that has been mounting since the Industrial Revolution. The everyday carbon technology of the West provides more comfort than concern in our daily lives, yet this luxury is fragile because it is based on a limited resource that in its use is ultimately detrimental. In fact, "all of our current environmental problems are unanticipated harmful consequences of our existing technology" (Diamond 2003, p. 44).

The significance of a link between system, technological innovation and economic growth may be overdrawn. Some, such as George Monbiot (2014a), deny the force of ideology as compared with the real 'meta-trend':

> it was neither capitalism nor communism that made possible the progress and the pathologies (total war, the unprecedented concentration of global wealth, planetary destruction) of the modern age. It was coal, followed by oil and gas. The meta-trend, the mother narrative, is carbon-fuelled expansion. Our ideologies are mere subplots. Now, as the most accessible reserves have been exhausted, we must ransack the hidden corners of the planet to sustain our impossible proposition.
>
> (Ibid.)

Monbiot also says that the failure of our society is inescapable because it is built on compound growth and the destruction of the Earth's living systems. He is not alone in this assessment (Meadows et al. 1972, Huesemann and Huesemann 2011, Higgs 2015, for example).

One key aspect of the carbon meta-trend is its relationship to that gift of Prometheus: fire. The idea of the Anthropocene prompts us to see humanity as a geological agent, involving the use of pyrotechnology to transform materials and thus "shape our social and physical worlds" (Clark 2014). Pre-humans, the hominids of East Africa, may have mastered the use of fire for cooking and for simple tool-making more than 500,000 years ago. But it was only around 10,000 years ago that the technology of the containment of fire led to the transformation of materials such as metals from ore, bricks and pottery from clay and glass and ceramics from silica, as well as the baking of bread from grain. This containment – ovens, hearths, furnaces, boilers – made possible much higher temperatures and led to the very fabric of ancient civilisations (stone cutting tools, metal spears and armour, bricks and concrete) and ultimately thence to the Industrial Revolution.

This is supported by the insight that two of the three key factors in the domination of the West identified by Jared Diamond (1999) are made possible by the furnace – guns and steel. Germs already existed and were transported to the new worlds largely by accident. In this sense, the European settlement of Australia can be viewed as a clash between the indigenous land-altering technology of open fire and the European metals technology based on contained fire (Clark 2014).

But as Stephen Pyne (2015) observes, industrial combustion is changing the nature of diversity:

> The new energy is rewiring the ecological circuitry of the Earth. It has scrambled ecosystems and is replacing biodiversity with a pyrodiversity – a bestiary of machines run directly or indirectly from industrial combustion.

Alternatives

A feminist counter-view may be to put the technology of weaving before fire in terms of importance, which is certainly arguable as it is supported by the etymology of the word itself.[8] Such a view is also consistent with the fact that little of this ancient technology remains – fabric is soft and decays, metal is hard and durable. Nevertheless, Penelope's weavings and re-weavings in Homer's *Odyssey* establish it in ancient times as do sophisticated Egyptian relics from much earlier periods. Various forms are observable in many indigenous societies today, depending on the plant fibre available. Weaving enables not only the production of clothing from plant or animal fibre, but also enables lightweight containers (baskets) for the transport and security of food, the making of shelters and the production of portable fencing for herding as was practised in Neolithic times. It is a technology that is not fuelled by carbon and, ironically, runs counter to the 'mother narrative'.

A broader alternative to the mother-narrative imagines a modern world where fossil fuels do not exist (Le Page 2014, pp. 34–39). He suggests that such a world would be dependent on hydroelectricity, with that technology possibly originating in Norway and Switzerland rather than in England. Such a world may have more highly developed electrical motors and batteries, cleaner air, and more electric powered public transport. Long-distance travel would be by sailing ship, electric railway and hydrogen-filled airship. However, he points to an apparent consensus that at best, there would be much slower growth and such a civilisation could not support a similar population to the present, even if industrialisation did proceed in this manner. This would be due to limiting factors such as an impossible number of dams required, as well as lack of materials required for construction. The intensity of fossil energy sources is very hard to replace (ibid., p. 39).

Silicon

Kenneth Boulding (1966) would have agreed about the carbon meta-trend. However, he was optimistic about the possibility of technology to provide answers to the exhaustion of carbon reserves:

> Failing this [development of atomic fusion], however, the time is not very far distant, historically speaking, when man will once more have to retreat to his current energy input from the sun, even though this could be used much more effectively than in the past with increased knowledge. Up to now, certainly, we have not got very far with the technology of using current solar energy, but the possibility of substantial improvements in the future is certainly high.

In line with Boulding's prophetic observation, the most recent 'meta-trend' or 'mother narrative' arises less from the exploitation of carbon and more from its fraternal element, silicon, to produce the knowledge explosion resulting from interlinked computers.

The history of computer development really starts in Bletchley Park in England during the World War II, when Tommy Flowers and Alan Turing led teams that were trying to break German military codes and applied computing theory to the problem. In so doing they constructed ten electronic programmable computers, each appropriately called 'Colossus'. Turing himself pointed out two key advantages of electronic computer technology when he delivered a lecture to the London Mathematical Society in 1947 on the automatic computing engine (ACE). Digital computers were unrestricted in both their accuracy and their applications:

> the property of being digital should be of greater interest than that of being electronic. … That the machine is digital however has more subtle significance. It means firstly that numbers are represented by sequences of digits which can be as long as one wishes. One can therefore work to any desired degree of accuracy. … A second advantage of digital computing machines is that they are not restricted in their applications to any particular type of problem. … A good working rule is that the ACE can be made to do any job that could be done by a human computer, and will do it in one ten-thousandth of the time.
>
> (Turing 1947, p. 509)

Just as the exploitation of carbon had an enormous impact across a vast range of industries, so has the exploitation of silicon soon after the Bletchley Park era. This was when vacuum-tube valves were replaced with transistors[9] and then microchips in much smaller computers through exploiting the semiconducting properties of silicon. It was also when photovoltaic cells began to be developed that exploited the capacity of silicon to translate photonic energy into electronic energy.[10] But while carbon is exploited directly in vast quantities for both energy and materials, silicon is exploited indirectly and in much lesser quantities for the manipulation of vast amounts of data, as well as for harnessing energy from the sun. Compared with carbon, our relationship to silicon is not, and need not be, voracious.

The silicon-based development of linked computers has since enabled data gathering and manipulation across all fields and industries as well as creating new endeavours such as those based on chaos theory and the information sciences. The origin of the DIKW (data-information-knowledge-wisdom) hierarchy[11] concept is uncertain (Boulding 1955?), but linked computing has enabled a new universe of possibilities within the first three categories. The fourth category, wisdom, may be enhanced by the availability of better information and knowledge on which to base good judgement.

Again, exponential growth is apparent, but unlike carbon technologies, silicon-based information technology uses a virtually limitless resource – data – that is created and processed at increasing rates. Moore's Law, that computing power doubles approximately every 18 months, is looking a little ragged now that processing chips are at near sub-atomic levels. Yet the law has held for the past several decades. And in any case, the growth of the amount of information available is independent of growth in computing power; it relates more closely to storage and retrieval capacity. This most recent technological trend forms a kind of symmetry with the major technological revolution of the mid-millennium: printing, which produced a radical increase in the sharing of knowledge and ideas, and which therefore made the Industrial Revolution possible.

Other technologies

Brundtland (1987) first mentions (new) technology at page 13 as a "mainspring of economic growth" which has the potential to reduce "the dangerously rapid consumption of finite resources", but which risks new forms of pollution and new life forms that could change the direction of evolution. Later the Brundtland report also asserts that technology will continue "to change the social, cultural, and economic fabric of nations and the world community" but that new technology encompasses both promises and risks. The promises outlined in the report include information and communication technologies that can benefit productivity as well as energy and resource efficiency.

The Brundtland report also highlights the advantages of new materials including composites that require less energy and less matter to manufacture. She mentions how biotechnology could dramatically improve human and animal health and how renewable plant-sourced energy may substitute for non-renewable fossil fuels, as well as how engineered crop varieties "could revolutionise agriculture". Biotechnology could be an answer to polluting products as well as to the issue of hazardous waste disposal. New satellite remote sensing technology could ensure better short-term agricultural forecasting and longer-term "optimal use of the Earth's resources" by monitoring climate change, marine pollution, soil erosion and plant cover (Brundtland 1987, pp. 150–151).

Some of these examples did not quite work out as forecast, however. Brundtland also mentions "new plant strains that can fix atmospheric nitrogen" and so reduce pollution from chemical fertilisers. The use of nitrogen-based fertiliser started to peak in 1988, around the same time as the Brundtland report was issued, after its invention early in the twentieth century had "incited fears of a runaway technology causing a rain of N" (Frink, Waggoner and Ausubel 1999, p. 1179). While the Food and Agriculture Organisation (FAO) cautions that the method is not effective in all plants, a 1995 technical report on the concept of biological nitrogen fixation using bacteria that naturally occur in the roots of some plants, summarises such a development as suitable for farming of tropical legumes:

> Since nitrogen is commonly the most limiting plant nutrient in arable farming in the tropics and also the most expensive element as a mineral fertilizer, biological nitrogen fixation (BNF) holds great promise for smallholder farmers in sub-Saharan Africa. Alley farming systems which use leguminous woody species in the hedgerows can reduce or eliminate farmers' needs for commercial N fertilizer.
>
> (Mulongoy 1995)

However, in 2013, 26 years after the forecast in the Brundtland report, a different method for fixing atmospheric nitrogen was announced by researchers at the University of Nottingham after a decade of research. Rather than using new plant strains, the method uses bacteria that naturally occur in sugar cane that are harvested to provide a sustainable biological coating for any plant to enable it to fix nitrogen (Kurzweil 2013).

While this alternative nitrogen-fixing technology appears to be beneficial, the Brundtland report warns that other technologies may have negative consequences. The report says developments in seed technology could lead to greater dependence on fewer crop varieties owned by even fewer transnational companies. Also, new materials may create previously unknown hazards to health. Further, the report anticipates the genetically modified (GM) food debate by pointing out that genetically engineered organisms may have unforeseen effects on the environment.

Pessimism

One recent text stands out as a denial of the possibility that technology can be an answer to unsustainability. With *Technofix*, Canadians Huesemann and Huesemann (2011), who aptly live on a small island in British Colombia, have drawn praise from some ecologists for their "mythbusting"[12] explanation of "why technology won't save us or the planet". After showing that all new technologies cause significant unanticipated problems, the Huesemanns suggest that there are basically three types of technological remedy: (1) counter-technologies that neutralise the negative effects of other technologies, such as environmental remediation (2) social fixes, which use technology to solve social, economic, political, or cultural problems, such as industrialised agriculture to solve world hunger and (3) efficiency improvements to technological processes, such as better fuel efficiency of cars (Huesemann and Huesemann 2011, p. 73).

Although lacking clarity about which category of remedy it falls, the Huesemanns are pessimistic that even renewable energy technologies contribute to sustainability, contrary to the opinions of many ecologists, as "all renewable energy technologies are expected to have significant environmental impacts" when deployed on a sufficiently large scale. They base this view on "land limitations, severe environmental impacts and public opposition" as well as the greater expense of particularly solar technology compared with coal (ibid., p. 132). This view is questionable. Certainly there has been public opposition to renewable wind farming on both land and sea, where there has been visual, auditory and birdlife impacts. But this has been observably less than the sort of opposition engendered by coal and nuclear power plants. As far as solar energy is concerned, there has been opposition to large-scale solar farms in the United Kingdom, for example, where they may have been sited "insensitively" (Gosden 2014), but this could not be said to be a wave of discontent that has prevented such harvesting. Moreover, the future of solar energy is as much about decentralised domestic and local production as it is about large scale grid-based energy delivery, and the cost of renewables at the point of delivery is already lower than coal (Parkinson and Gilding 2014), only three years after the Huesemanns' assertion to the contrary.

Nonetheless, more ominously Huesemann and Huesemann outline three general preconditions for environmental and societal collapse "(i) rapid growth in resource use and pollution, (ii) limited resource availability and waste absorption capacity, and (iii) delayed responses by decision-makers when limits have already been exceeded or soon will be" (Huesemann and Huesemann 2011,

p. 139). While this text tends to accept that unanticipated problems are inevitably of a great magnitude compared with the benefits of a technology, these three preconditions are cause for concern. The first two are largely met – partly due to the vastly increased growth and consumption in both China and India in the past few decades, combined with the foreseeable exhaustion of global oil supplies and the obvious incapacity of the atmosphere to absorb more carbon dioxide without dramatically affecting climate. With respect to the third precondition (delayed response by decision-makers), the planetary scale of the problem suggests that any delay may render any action too late.

To illustrate this issue of scale, it has been observed that in the time leading up to Australia's 1964 'Voyager' naval disaster, which involved the collision between a destroyer and an aircraft carrier, there was an interval of 45 seconds during which no avoidance manoeuvre whatever could have averted the tragedy because the speed and course of the ships could not be varied enough in that time (Hickling 1965, p. 141). The momentum of ships is large. Yet it is infinitesimal compared with the momentum of the planet's biosphere and the destructive course it is following. Over how many times longer than 45 seconds will it matter not at all whatever corrective action is taken? Unless of course, that point in time has already passed.

Optimism

Whether or not it is indeed too late, technology is involved in yet another paradox, but with potentially positive consequences. The Harvard biologist Edward Wilson (2013) draws attention to the "inconvenient truth" that no ecosystem under human pressure can be made sustainable without humanity knowing all the species that compose it. The paradox is that the more species that we extinguish, the more we discover. We currently identify less than two million, whereas there may be at least five million or even possibly 100 million species on Earth. At microbial level, 99 per cent of such organisms remain unknown to us: the soil remains a "*terra incognito*" to us (van der Hout 2014, p. 435). Nevertheless, the technology now helping to identify and preserve species, which involves a kind of barcoding of DNA sequences, is some distance from Heidegger's lament.

Technology optimists include the US industrial ecologist Jesse Ausubel, whose article *Can Technology Spare the Earth?* (1996) outlines how improving efficiencies in resource use through especially silicon-based technology can restore the environment even as the population grows:

> Families named Smith, Cooper, and Miller people our nation because until not long ago most of us beat metal, bent casks, and ground grain. … So far, except in video, we are not named Programmer, Sub-Micron, and

> Genesplicer. … We forget the power of compounding our technical progress, even at one or two percent per year. Knowledge can grow faster than population and provide abundant green goods and services. The message from history is that technology, wisely used, can spare the Earth.

This message echoes that of Marx and Engels more than a century and a half ago. While Marx attacked Malthus' views of the limits to growth and the inevitable impoverishment of the working class (cited in Dresner 2008, pp. 13–14), Engels asserted that any geometric increase in population would be matched by a geometric increase science and its application:

> science increases at least as much as population. The latter increases in proportion to the size of the previous generation, science advances in proportion to the knowledge bequeathed to it by the previous generation, and thus under the most ordinary conditions also in a geometrical progression. And what is impossible to science? (Engels 1996 [1844])

Incongruously, it is a similar argument that now attaches to its opposite ideology of neoliberal economics (there will be no resource crisis because applied science and technology will be "incentivised" to find and develop substitutes as resources become harder to extract and their price rises). However, the argument is also consistent with the key advantage of a silicon-based technology – that it uses vastly less material resource and more an inexhaustible virtual resource: it grows geometrically, building upon the 'knowledge bequeathed to it by the previous generation' as Engels suggested so long ago.

Jevons paradox

Much technological development has been driven by the quest for efficiency, highlighted by Watt's improvements to Newcomen's steam engine and the continuing pursuit of more efficient internal combustion. The IPCC endorses efficiencies through improved technologies as a way of mitigating climate change and promoting sustainable development (IPCC 2014a, s.4.3.6). But all is not as it seems; efficiency can be a mirage that recedes as it nears. The Jevons paradox (or 'rebound effect') is an observation made about coal consumption in the nineteenth century and has much wider application for all forms of technological efficiency. Jevons (1865) projected that the consumption of coal in Britain would follow an exponential curve so that it would be almost infinite by 1950 if its then rate of increase were maintained. At that time many in the United Kingdom worried that coal reserves were rapidly dwindling, but some believed that improving technology would reduce coal consumption. Jevons argued that this view was incorrect, as further increases in efficiency would tend to increase the use of coal, as Watt's improved steam engine had generated more demand for the resource. More generally, the proposition is that as technology progresses, efficiency increases with which a resource is used tends to increase the rate of consumption of that resource.[13]

This tendency is apparent because greater efficiency makes the resource effectively cheaper. However, its degree does depend on the elasticity of demand for the resource, so there is not a rigid link between efficiency and resource use. Demand elasticity is partly a function of the availability of substitutes, and this also must be taken into account. According to energy economist Harry Saunders (1992, p. 131), there is a further factor linking efficiency with increased energy consumption: at the macroeconomic level, more efficient technology also leads to faster economic growth, which in turn increases energy use throughout the economy. Thus technological progress that improves energy efficiency will tend to increase overall energy use.

Nevertheless, the Jevons paradox assumes an unregulated market, whereas in reality it is possible to tax undesirable consumption. As Joe Stiglitz (2014b) has said in respect of the carbon tax, "it's absolutely clear that it's better to tax bad things than good things; that way you get a double dividend". Finally, on this efficiency issue, the Jevons paradox obviously does not result in the same concern when the energy source is renewable: as photovoltaics, for example, are becoming more efficient, they are able to generate more electricity per unit. But there is no concern about depletion of the energy resource. That concern only applies where the energy source is not renewable, where the fuel is not free. Technology and sustainability is less about fuel efficiency and more about the nature of the technology and the form of energy it uses.

Automation, artificial intelligence and development

For both the developed and developing worlds, an important technological issue is associated with automation and artificial intelligence (AI). Al Gore, for example, believes that:

> The technological extension of the ability to think is therefore different in a fundamental way from any other technological extension of human capacity … [thus the impact of AI] will be far greater than that of any other technological revolution.
>
> (Gore 2013, p. 40)

David Roden (2015, pp. 16–17), in discussing the concept of 'posthumanism', speculates on the implications of a robot or computer system with better than human reasoning ability. Such a possible super-intelligence may be ambivalently good or evil. He cites the cryptographer Irving Good as saying that since the design of machines is an intellectual activity, then an intelligent machine could design even better machines, resulting in an "intelligence explosion", leaving human intelligence far behind. "Thus the first ultra-intelligent machine is the last invention that man need ever make" – as long as it can be kept under control (Good 1966, pp. 33–4). The cosmologist Stephen Hawking agrees. In a BBC interview of 2 December 2014 discussing new anticipatory software that enables him to communicate more easily, he said he was worried about the idea of creating machines that think because an artificial intelligence would be able to independently re-design itself at an ever-increasing rate, so that humans would be "superseded". While Isaac Asimov anticipated and attempted resolution of such an issue in the 1940s with his three laws of robotics,[14] it is conceivable that a sophisticated AI could re-program itself or its progeny beyond such constraints, or become dangerously dysfunctional as did HAL in Clarke's *2001: A Space Odyssey*.

Asimov's laws are an example of highly distilled AI regulation. According to *The Economist* (9 May 2015), coping with AI will be relatively easy because we are used to controlling autonomous bureaucracies, markets and armies, all of which can do great harm without appropriate regulation, oversight and means of accountability. It points out, though, that the most important safeguard for controlling a future AI will be an *off* switch.

The current approach to AI though is to ignore the goal of human-like intelligence or understanding and rather teach machines through vast numbers of example, rather than by adopting certain principles. Known as 'machine learning', this is how the first successful machine translation system, *Candide*, worked.[15] *Google Translate*, with its vast database uses the same approach. According to one of its software engineers:

> you can take one of those simple machine-learning algorithms that you learned about in the first few weeks of an AI class, an algorithm that academia has given up on, that's not seen as useful—but when you go from 10,000 training examples to 10 billion training examples, it all starts to work. Data trumps everything.
>
> (Estelle quoted in Somers 2013, para. 60)

The technique is so effective that Google Translate people do not speak most of the languages that the application translates. Engineers are much more important to the system than linguists. Translation through machine learning is "data-mining at a massive scale" (ibid., para. 61). It is not understanding, nor is it anything like self-awareness.

But while AI is yet to achieve "singularity" – the point at which machine (general) intelligence exceeds that of humans – nevertheless automation, or the replacement of human labour by embodied capital, is progressing at pace, exemplified by the explosion of computerisation and robot factories of recent decades. It is not just manufacturing employment that is at risk in both rich and poor countries; more skilled jobs are susceptible to replacement by machine intelligence, including routine legal work. Already journalism is often the product of software rather than human effort, as are translations due to *Google Translate*.

Some are skeptical of the degree of automation possible due to Polanyi's paradox[16] and the recurring phenomenon of "automation anxiety" since the time of the Luddites (Autor 2014, pp. 2–3). Nevertheless, the futurist Martin Ford's (2009 p. 237) view is that, as technology accelerates, machine automation may create a 'downward economic spiral' as predicted by Marx, due to over-production coupled with unemployment and hence lack of consumer demand. Automation is already established even in developing economies and there is no possibility of arresting the trend: "the world economy is a closed system", he says (ibid., p. 97). However, Ford does not accept Marxist solutions, instead valuing the consumer more than the worker-producer. He proposes that as automation accelerates, people in developing countries should be paid incentives for conserving resources and protecting the environment, thus maintaining consumer demand while at the same time "decoupling economic prosperity from negative ecological impact" (ibid., p. 198).

More recently, but for different reasons, the Intergovernmental Panel on Climate Change (IPCC) is in line with such thinking. The Panel finds that the basic clean energy and nutritional needs of billions of people remain unmet. Therefore, there is a need to adopt non-conventional incentives "for technology innovation and diffusion processes that respond to social and environmental goals" (IPCC 2014a, ch. 4, p. 25). The World Bank too, believes that innovation and technology can lead to better development models. Its former Chief Innovation Officer, Chris Vein, says that:

> technology evolves and the possibilities of movements like the 'Internet of Everything' come in to focus, new and innovative approaches become the antidote to development in the usual manner.
>
> (Vein 2013)

This is noticeably different from the sort of approach suggested by earlier sustainable development advocates. Schumacher (1989 [1973], p. 61) of 'small is beautiful' fame, advocated "maximum well-being for minimum consumption" and small-scale, local "appropriate" technologies for the developing world. As John Phillimore (2001, pp. 24–25) points out, Schumacher's ideas, once influential, fell from favour as Schumpeterian big-scale technological innovation ideas showed success in the economic development of East Asia. Although the two economists are usually seen as opposites, they actually have much in common, not least that Schumacher was once Schumpeter's pupil — in Bonn in 1929. Both were especially interested in technology and its relationship to economic growth. Both favoured entrepreneurial innovation, although politically the two were rather further apart (Narkus 2012).

Transitions

A century or so ago, a very different – albeit distant – technological future from that based on coal was envisaged by the scientist JBS Haldane (1923):

> ...four hundred years hence the power question in England may be solved somewhat as follows: The country will be covered with rows of metallic windmills working electric motors which in their turn supply current at a very high voltage to great electric mains. At suitable distances, there will be great power stations where during windy weather the surplus power will be used for the electrolytic decomposition of water into oxygen and hydrogen.

Such a transition appears to be happening faster than Haldane predicted. As at 2017, twice as much electricity in the United Kingdom was generated by wind power than by coal (Wilson and Staffell 2018).

A related transition question is how to encourage more sustainable 'green' (Schumacherian) technological development on a Schumpeterian scale. This essentially depends on energy source (Phillimore 2001, p. 27). Bill McKibben (2008, p. 206), while favouring small-scale, local and appropriate technologies in the Schumacherian tradition, also allows that centralised 'clean-coal' and even nuclear power generation should not be ruled out in the "desperate" fight to slow carbon emissions (ibid., p. 144). At the same time, Phillimore points out that information and communication technology (ICT) is not inherently green because it has aspects that – shades of Jevons – encourage more energy use, such as increased travel, materials and energy consumption based on the production shift to Asia. Ultimately he laments that the ICT paradigm depends on fossil fuels as its main source of energy (ibid.). As Haldane forecast, this is still true, but increasingly less so, as wind and silicon replace carbon in the generation of electricity in both grid and off-grid applications, leading to a transformational 'tipping point' (Parkinson and Gilding 2014; CEA 2016; Forsythe 2017).

There is a widespread view that coal, oil and gas will be with us for many more decades, despite concerns about climate change. This is based on a world view of an energy system defined by large, long-life assets like power stations and coal mines, whereas there is an emerging distributed system that more closely resembles the growth of mobile phones – a consumer-driven, rapidly shifting market with diverse players. It has been argued that, driven by dramatic price falls in solar, the energy system may be on the verge of a revolutionary transformation. This would involve a shift from fossil fuels to renewables; a transition from centralised utilities to distributed house scale generation; a car industry that might enable storage of renewable energy; and perhaps global energy price deflation in stark contrast to forecasts of peak oil and spiralling prices (Parkinson and Gilding 2014). The challenge to regulatory frameworks and institutional arrangements is already felt. In many countries policy makers are caught between the demands of consumers testing the current boundaries of institutional structures, and utilities that want to defend their standard business models.

In Africa, renewable energy has already made large inroads at the non-grid household level. Kerosene lamps have been described as the "vampires" of developing communities on the continent because although used in almost every home their fuel is expensive. These lamps also provide poor light, emit noxious fumes and cause house fires and child poisonings. As of 2014, it was estimated that 110 million households in Africa did not have electricity. But recently, new solar powered lights manufactured in China have begun to replace kerosene lighting. These lights cost eight dollars each, cost nothing to run and have none of the other disadvantages of kerosene lamps. *SunnyMoney*, a supplier of the lights, which is now the biggest solar lighting company in Africa, has sales of more than one million units per year (*Science Show* 2014). Because kerosene uses still more fuel in its transport, this solar lamp project has the potential to cut global oil consumption by three per cent (Miller et al. 2013).

Already, within key economic sectors energy demand is not what might be expected. According to *The Economist* (2014), agriculture is now far more energy-intensive than manufacturing. While this appears counter-intuitive, considerable energy is needed to produce fertilisers, and in many countries much electricity is needed to pump water from deep aquifers and distant rivers, for example. Each dollar of agricultural output needs four or five times as much energy to produce as the same value of manufactured goods. The transition to a future sustainable agriculture to feed ten billion people will demand a lot of energy, and how it is sourced and distributed are already important issues.

The cloud

The 'Internet of Everything' (IoE), or 'Internet of Things' (IoT) concept mentioned above is based on cloud data storage and the wireless interconnection of myriad computing devices through the Internet. Increasingly it is powered by sunlight directly, as illustrated by Apple's decision to convert all its data centres as well as some of its manufacturing to solar power (Goldenberg 2014b). Of the IoT's many potential applications, two in particular stand out for sustainability: its potential for environmental monitoring and remediation, and its potential for energy management. Monitoring of the environment is made vastly more sophisticated through applications of the IoT. There are potentially limitless data available from environmental sensors[17] already, and when wirelessly connected through the cloud, both real-time and historical information is made available with which to link with social and economic data, and therefore on which to target remediation and mitigation decisions (Sense-T 2014).

With respect to energy management, cloud connectivity makes it possible to remotely manage energy-producing and energy-using devices, thus minimising energy production through matching it more exactly to real-time demand. While this does not yet drive a stake through Jevons' resilient heart, this oblique approach must help reduce energy needs.

Of course the inventor and futurist, Buckminster Fuller (1969, p. 44), believed this half a century ago:

> Man has ever-increasing confidence in the computer; witness his unconcerned landings as air transport passengers coming in for a landing in the combined invisibility of fog and night. While no politician or political system can ever afford to yield understandably and enthusiastically to their adversaries and opposers, all politicians can and will yield enthusiastically to the computers safe flight-controlling capabilities in bringing all of humanity in for a happy landing.

Thermodynamics

The relevance of thermodynamics to the issue of the sustainability of the anthropocene Earth is that the planet is, in effect, a 'closed' thermodynamic system. However, in the language of applied physics, this is different from a completely *isolated* system: a closed system does exchange energy (in the form of sunlight and other radiation) with its surroundings while an isolated system exchanges neither energy nor mass.

The laws of thermodynamics have been described by the Colombia theoretical physicist, David Albert, as "one of the monumental achievements of the physics of the nineteenth century", a "beautiful and breathtakingly concise summary of the behaviours of macroscopic material systems" (Albert 2015, p. 3).

The Earth contains the means for the creation of energy, such as the fossil deposits of ancient sunlight. However, those reserves are limited to perhaps a century or two. Other apparently internal energy sources such as hydroelectricity, wave and wind power are also the indirect result of sunlight, albeit with a virtually inexhaustible supply. And while there are two genuinely internal energy sources – geothermal and radioactivity (from uranium and potentially thorium deposits) – geothermal supplies are probably available for the very long-term, supplies of uranium are estimated to last perhaps only 200 years even 'at current rates of consumption' (Fetter 2009). Thus energy sources are either external or internal to the Earth's thermodynamic system. Most energy sources are based on the input of sunlight, whether over the shorter or longer term. The key point is that sustainable energy must balance the flow from external sources with its use over time – that is, power. Because internal sources are a limited stock, we must rather focus on matching the external flow to become sustainable.

That leaves other energy sources that are as yet little more than theory. In particular, this means nuclear fusion, rather than the controlled fission reaction now used in nuclear power stations. Fusion research, however, although currently funded by an international government consortium centered on France is typically "ten years" from any practical development and is frequently announced as such, as recently by the US military manufacturer, Lockheed Martin (Tollefson 2014).

Still, the US science writer Dawn Stover disputes that so-called renewable forms of energy are as renewable as they are thought to be. She says that 'renewables' need non-renewable resources to be effective, including the stuff of photovoltaic panels, precious groundwater for solar and geo-thermal, steel and rare Earths[18] for wind turbines as well as vast amounts of concrete for hydroelectric dams. In many countries, biomass for heating and cooking results in "severe deforestation and air pollution". Further, meeting the world's energy demands from such sources in the foreseeable future is completely impracticable as it would require the construction of an impossible number of solar farms, wind turbines and the like (Stover 2011, p. 1). Her critics, however, point to the recyclability of photovoltaics and steel, as well as the fact that the "rare Earths" of generator magnets are not actually so rare, as they are distributed in the Earth's crust at about the same volume as copper.

And also a comparison between renewable and finite energy reserves shows that of the finite (non-renewable) reserves, coal has estimated total reserves of 900 TW-yrs,[19] petroleum 240, natural gas 215 and uranium 90–130. The total world energy use is about 16 TW-yrs each year. Thus if coal were the *only* source of energy it would last for only a further 56 years. Likewise, petroleum would last only 15 years, natural gas 13 years and uranium around ten at current usage rates. By contrast, potential hydro and biomass energy could each account for about a quarter of energy needs indefinitely, wind for two or three times current consumption indefinitely, ocean thermal about half, and geothermal about one-tenth. However direct solar potentially could provide more than *1400 times* our current world energy consumption – indefinitely (Perez and Perez 2009).

The US ecological economist Eric Zencey's (2013, pp. 73–74) view is that the economic system is subsumed by energy deficits that will result in its demise:

> Seen through the thermodynamic lens, what has been called the Industrial Revolution is, more properly, the Hydrocarbon Revolution, a once-in-planetary-history drawdown of stored sunlight to do work and make wealth in the present. The petroleum era will most likely depart as suddenly as it came; in the grand sweep of geologic time, our use of petroleum is just an instant, a brief burst of frantic activity that has produced exponential growth in wealth and human population—and in humanity's impact on planetary ecosystems.

He goes on to sketch the history of ecological economics through the mathematician-economist Georgescu-Roegen and his pupil Herman Daly, who developed a thermodynamic critique of economics and the non-market "ecosystem services" it takes for granted (ibid., p. 77).

Entropy

The concept of entropy relates to matter, biology and knowledge, as well as to energy. An isolated system is more restrictive than a closed system as it does not interact with its surroundings in any way. Mass and energy remain constant within the system, and no energy or mass transfer takes place across the boundary. As time passes in an isolated system, internal differences tend to even out and pressures and temperatures tend to equalise, as does density. A system in which all equalising processes have gone practically to completion is in a state of thermodynamic equilibrium – one of maximum entropy.

According to Georgescu-Roegen, understanding what entropy is relates to the distinction between *available* and *unavailable* energy. "This distinction is anthropomorphic" because energy is available or unavailable according to whether it is suitable for human use. Within "an isolated system, the amount of energy remains constant (the first law of thermodynamics), while the available energy continuously and irrevocably degrades into unavailable states (the second law). Therefore, irrespective of technical detail, "entropy is an index of the amount of available energy relative to the absolute temperature of the corresponding isolated system" (Georgescu-Roegen 1986, pp. 3–4).

Georgescu-Roegen also outlines how matter is similarly affected by entropy and in so doing denies that that continuous recycling, or re-creation of ordered states of matter is feasible:

> It is an elementary fact…that matter also exists in two states, available and unavailable…just like energy, it degrades continuously and irrevocably from the former to the latter state. Matter, just like energy, dissipates into dust, as is best illustrated by rust, by wear and tear of motors [or] of automobile tyres.
>
> (Ibid., p. 7)

More specifically this will lead to scarcity of materials for, as he puts it, "materials vital for the current hot technology will sooner or later become extremely scarce…even scarcer than the available energy from fossil fuels", which then exposes the "logical weaknesses" of Herman Daly's (1973) promise of salvation through steady-state economics (ibid., p. 8).

As to biological entropy, the Nobel laureate, Erwin Schrödinger, who was invited to live in Ireland by Eamon de Valera after the Anschluß, lectured in Dublin in 1943 about life from a physicist's perspective. The lectures centred on the notion of entropy and the laws of thermodynamics. Usefully for this book, Schrödinger defines 'life' as a kind of 'entropy border'. Inside the boundary of a living organism is order, or 'negative entropy'. This order within is achieved at the expense of increasing entropy without, that is in its environment:

> entropy, taken with the negative sign, is itself a measure of order. Thus, the device by which an organism maintains itself stationary at a fairly high level of orderliness (= fairly low level of entropy) really consists continually sucking orderliness from its environment.
>
> (Schrödinger 1944, p. 26)

James Lovelock picked up this theme when working at the Jet Propulsion Laboratory, pondering how to identify life for the first Mars lander during the 1960s. His first step was to look for "entropy reduction" as it must be a characteristic of every life form. He describes the key but elusive nature of entropy in the following passage:

> …few physical concepts can have caused as much confusion and misunderstanding as that of entropy. It is almost a synonym for disorder and yet, as a measure of the rate of dissipation of a system's thermal energy, it can be precisely expressed in mathematical terms. It has been the bane of generations of students and is direfully associated in many minds with decline and decay, since its expression in the second law of thermodynamics (indicating that all energy will eventually dissipate into heat universally distributed and will no longer be available for the performance of useful work) implies the predestined and inevitable run-down and death of the Universe.
>
> (Lovelock 2000 [1979], p. 2)

Zencey (2013, p. 73) makes the notion of entropy still clearer in his article *Energy as Master Resource*. The first law of thermodynamics says that both matter and energy cannot be created or destroyed, only transformed. The second law says that when energy is transformed, its capacity to do useful work is diminished. While the total amount of energy remains the same (the first law), some of it is no longer useful. This degraded energy is known as entropy. Further, Albert (2015, p. 3) describes the second law – the famous one – as meaning that "entropy can never decrease as time goes forward". Smoke spontaneously spreads from but does not collect in a cigarette. Soup cools not heats in a cool room. Chairs slow, not accelerate, when sliding on a floor. Eggs hit a rock and break, not jump from a rock and reassemble. All tends to less order, energy ultimately disperses as heat, temperatures equalise.

The US historian Henry Adams (1910) used entropy as a metaphor to interpret history at around the end of the nineteenth century, thus bringing the concept to the humanities. Kenneth Boulding writing in the 1960s, thought to apply the concept to culture and knowledge:

> The question of whether there is anything corresponding to entropy in the information system is a puzzling one, though of great interest. There are certainly many examples of social systems and cultures which have lost knowledge, especially in transition from one generation to the next, and in which the culture has therefore degenerated [such as in Appalachia and with American Indian culture]. On the other hand, over a great part of human history, the growth of knowledge in the Earth as a whole seems to have been almost continuous.
>
> (Boulding 1966, pp. 6–7)

More recently, it has been noticed that the stock of useful knowledge tends to decline due to technological change. Knowledge of how to handle a square rigged sailing ship is no longer needed. Nor is how to produce antibiotics rendered ineffective by overuse. At the same time, useful knowledge is replenished and advanced only due to research and invention (Matson, Clark and Andersson 2016, p. 49).

From yet another angle, the astronomer John Barrow (1999, p. 153) cites Clarke's cautionary fable of technology and entropy *Superiority* (1951), in which an advanced civilisation is defeated in a space war by its technologically inferior opponents. The advanced group keeps developing ever more sophisticated high-tech weapons that require downtime to upgrade delivery systems. In their final iteration the new systems bamboozle each other because each deployment warps space-time and gradually increases entropy so that electronics and communications are mismatched. "The more sophisticated and powerful a technology becomes, the more susceptible it is … to breakdowns and subtle malfunctions".

Ultimately, the second law of thermodynamics has been described as "the supreme" law of nature, whereby if any new theory is found to contradict it, then the new theory must collapse (Eddington 1928, pp. 73–75). It is only in the field of economics that its force is ignored (Zencey 2013, p. 83). Therefore, it is only when economics takes into account present 'externalities' such as energy return on energy invested (EROEI) can human civilisation hope to be sustainable. A "bare minimum" of 3:1 EROEI would be required, while at least 5:1 EROEI would be needed to maintain "anything like what we call civilisation". But while renewables are well above this figure, they still require a net energy input to build, as well as for their infrastructure. Promised energy does not build the windmill (ibid., p. 80; Sgouridis, Bardi and Csala 2015, p. 2).

Exponential growth

The opposite face of entropy is exponential growth, a growth built on technology, especially since the World War II. The curves have the same shape, but are inverted. In fact, it is the converse interactions of these theoretical twins that describes our possible futures. Schmidt and Cohen (2014, p. 253), citing Kurzweil (1999), assert that "technology is a continuation of evolution by other means" and is itself evolutionary, "building on its own increasing order, leading to exponential growth and accelerated returns over time".

If it is increasing entropy that ultimately limits growth (Ekins 1993, p. 272), then there are only two ways of raising that limit. One way is by exploiting the continuous flow of incoming energy available from the sun. While economic activity increases entropy by depleting resources and creating wastes, sunlight can have the opposite effect as it is both a direct energy source and the fuel of nearly all life on Earth.[20] Physical production might then be increased, but would still be limited by the quantity of that energy available – at a rate of about 1.37 kilowatts per square metre, known as the 'solar constant'. Economic growth may still continue beyond these physical limits to the extent that it is non-physical – which it increasingly is. This form of 'decoupling' and depends on adding value while using fewer physical resources, such as Internet applications, recycling and provision of services rather than physical goods.

Systems theory

Systems theory was developed from different perspectives by such thinkers as Ludwig von Bertalanffy, Anatol Rapoport, Talcott Parsons and Kenneth Boulding during the mid-twentieth century. It involves both the physical and the social sciences, and concentrates on "the dynamics that define the characteristic functions, properties, and relationships that are internal or external to the system", rather than so much on its components (Laslo and Krippner 1998, p. 49).

Systems theory is useful because disciplinary worldviews are increasingly fragmented. While the natural sciences have tended to synthesis through unified theories, in contrast the social sciences have tended to relativism and to deny normative behavioural views (ibid., p. 50), as well as being mutually incomprehensible (Boulding 1956; Snow 1959).

This concept is relevant to technology and sustainability in several ways. The technology to model Earth systems such as climate is now well-developed and involves vast numbers of sensors throughout the biosphere combined with mathematics, binary algorithms and the power of cloud computing. This approach is also applicable to the development of possible solutions to human impact. It enables evolutionary systems designers to align their simulations with the dynamics of civilisation change and with patterns of environmental sustainable development.

To an evolutionary systems designer, a 'situation' is a system of interconnected problems. One such situation is the evolution of economics. Another is the development of energy-intensive technologies. And a third is the destruction of the Earth's ecosystems. Humanity is in a conundrum of interconnected problems that relate to sustainability in which technology is interwoven. Their dynamics may be more important than their static components. Nevertheless, as Donella Meadows cautions, it is unwise to assume that a dynamic systems approach brings with it the prospect of all-conquering control. Rather than control, or even predict, a systems approach can help envision and bring into being, as systems can be designed and redesigned: "we can't control systems or figure them out. But we can dance with them!" (Meadows 2008, p. 170). Systems thinking implies humility.

Ethics

The Dutch eco-philosopher, Sanne van der Hout (2014, pp. 424–429), approaches the question of sustainable technologies from an ethical framework. She describes new technologies that increasingly mimic natural systems from the molecular level and larger — '*biomimetic*' or '*homeo*' technologies — which have been associated with a more sustainable co-existence between humans and nature. These technologies contrast with the *classical* technologies, which are fundamentally different from natural organisms, such as the wheel and combustion engines. Classical technologies tend to "counteract or disturb the dynamics of nature"; they are "radical simplifications" that enable some degree of human control over our environment.

But van der Hout disputes that biomimetic technologies are necessarily ecologically benign. In their very mimicry they may enable subtler and more pervasive ecological interference than classic brutalism. To fulfil the potential of biomimetic technologies to enhance the carrying capacity of the Earth, humanity's predilection for luxury and consumption, for dominance over and separation from nature, must be contained by an ethic of peaceful coexistence and of humility towards our environment. In this hope she cites Norman

Cousins (1979, p. 31) who put the beginnings of a changing world view most eloquently: "What was most significant about the first lunar voyage was not that men set foot on the moon, but that they set eye on Earth".

In this discussion, the different reach of our current technologies, whether classical or biomimetic, compared with earlier technologies is also apparent. In the ancient past, human ethics were based on the immediate – the family, the neighbour, the locality — because it was only the immediate that could be affected by our technologies. However now that technology affects the entire planetary biosphere, ethics must encompass this dimension because we are now responsible for it, both in time and space (Jonas 1984; Plumwood 1993, 2002; Sloterdijk 2009). Thus the scale of human technological reach brings with it an ethical imperative of technological accommodation with the natural world.

Relationships

This then is where humanity is poised in its ancient and intensifying relationship with technology. How that relationship is wrought in the near future will have a decisive impact on sustainability. But there remains a gulf between attitudes to technology and sustainability. Those who regard technology as dangerous, potentially disastrous and to be kept under firm control (Goethe, Shelley, Heidegger, Atwood, Carson, Commoner, the Huesemanns, and van der Hout) are distant from the technological transformationalists (Marx, Engels, Verne, Schumpeter, Turing, Clarke, Fuller, Ford, Richardson, Miller, Gilding, Gore and Vein), for whom nothing is impossible. Others see technology as useful in particular contexts, but that economic relationships have to change to achieve sustainability (Jevons, Schumacher, Brundtland, McKibben, Dresner, Stiglitz, Wilson, Dietz and O'Neill, the IPCC). It is this inter-relationship with economics that is unresolved. Technology may be used as a shield from the reality of increasing environmental degradation for the few, or it may be used to improve the environment for the many.

Technology's relationship to population is already clear. During the Industrial Revolution, technology implied more people due to higher birthrates and lower death rates. It now implies the opposite: modern technology means fewer people due to declining birthrates because fewer people are needed. Technology's relationship to affluence has always been one of association. With a world connected through its latest expression it becomes possible that such affluence can be more evenly distributed as well as less destructive for the planet.

Notes

1 Ariyoshi won the Nobel Prize for literature in 1968.
2 India, however, continued its use as a malaria control agent (Kranzberg 1986, p. 546).

3 Which in turn are composed of shaped wooden lengths, stone, bone or metal tips, gut or hemp stringing and fibrous binding.
4 This is known as Amara's Law, after Roy Amara c. 1985.
5 Also written as 'Kondratiev'.
6 Atoms > molecules > cells > organelles > organs > individual > species > planet > solar system > galaxy > universe.
7 Such as Fibonacci sequences in branching trees, the fruitlets of a pineapple, the flowering of artichoke, an uncurling fern, and the scales of a pine cone.
8 The Greek word root of 'technology is from 'teks' – 'to weave' or 'fabricate'.
9 Initially germanium rather than silicon was used for transistors.
10 Although Alexandre Becquerel noticed the conversion of light into electricity in 1839, practical photo-voltaics were first developed at the Bell laboratories in 1954.
11 Often expressed as 'data is not information, information is not knowledge, knowledge is not wisdom'.
12 Rees, W, University of British Colombia, quoted in the book frontispiece.
13 The Jevons paradox is in principle similar to the Downs–Thomson paradox, that increasing road capacity can make traffic congestion worse.
14 The first law is that a robot must protect its human master from harm. The second law is that a robot must obey the commands of its human master, unless it conflicts with the first law. The third law is that it must protect itself unless it conflicts with the first or second laws. In later stories there is also a 'zeroth law', that a robot must primarily protect humanity.
15 *Candide* used 2.2 million pairs of sentences, mostly from the bilingual proceedings of Canadian parliamentary debates for its English-French translations (Somers 2013, para. 56).
16 Polanyi's paradox: our tacit knowledge of how the world works often exceeds our explicit understanding. We understand how to use our body or drive a car, but can do so without any knowledge of physiology or the dynamics of transport objects. It is therefore very difficult to create a robot to emulate human capability (Michael Polanyi 1968).
17 According to the World Economic Forum there will be one trillion sensors connected to the Internet by 2022 (WEF 2015, p. 16).
18 Rare earths such as neodymium are used in wind turbine generator magnets.
19 One terawatt equals one trillion (or 10^{12}) watts.
20 Except for some life forms in deep sea volcanic vents.

References

Adams, H 1910, *A Letter to American Teachers of History*, Washington, DC, US.

Albert, D 2015, *After Physics*, Harvard University Press, Cambridge, US.

Andersen, H C 2015 [1843], *The Nightingale (Nattergalen)*, Jean Hersholt (trans.), SDU, Hans Christian Andersen Centre, Odense, Denmark.

Asimov, I, 1951+, The Foundation series: *Prelude to Foundation* (1988), *Forward the Foundation* (1993), *Foundation* (1951), *Foundation and Empire* (1952), *Second Foundation* (1953), *Foundation's Edge*, 1982), *Foundation and Earth* (1986), Ballantyne, New York, US.

Atwood, M 2003, *Oryx and Crake*, Bloomsbury, London, UK.

Aunger, R 2010, 'Types of Technology', *Technological Forecasting and Social Change*, August.

Ausubel, J 1996, 'Can Technology Spare the Earth?', *American Scientist Magazine*, vol. 84, no. 2, March-April, pp. 166–178.
Autor, D 2014, *Polanyi's Paradox and the Shape of Employment Growth*, Working Paper No. 20485, National Bureau of Economic Research, Cambridge, US.
Barjavel, R 1972 [1943], *Ravage*, Editions Flammarion, Paris, France.
Barrow, J 1999, *Impossibility: The Limits of Science and the Science of Limits*, Vintage, Random House, London, UK.
Boulding, K 1955, 'Notes on the Information Concept', *Exploration*, vol. 6, pp. 103–112.
Boulding, K 1956, 'General Systems Theory – The Skeleton Of Science', *Management Science*, vol. 2, pp. 197–208.
Boulding, K 1966, 'The Economics of the Coming Spaceship Earth', in Jarrett, H (ed.), *Environmental Quality in a Growing Economy*, Johns Hopkins University Press, Baltimore, US, pp. 3–14.
Brundtland, G 1987, *Report of the World Commission on Environment and Development: Our Common Future*, United Nations, Geneva, Switzerland.
Buckminster Fuller, R 1969, *Operating Manual for Spaceship Earth*, Design Science Lab, Buckminster Fuller Institute, San Francisco, US.
Carson, R 2002 [1962], *Silent Spring*, Houghton Mifflin Harcourt, New York, US.
CEA: Central Electricity Authority 2016, *Draft National Electricity Plan*, Volume 1 Generation, December, Government of India, Ministry of Power.
Clark, N 2014, *Nigel Clark*, Lancaster Environment Centre, Lancaster University, UK.
Clark, N 2018, 'Pyropolitics for a Planet of Fire', in Peters, K, Steinberg, P and Stratford, E (eds), *Territory Beyond Terra: Geopolitical Bodies, Material Worlds*, Rowman and Littlefield, London, UK.
Clarke, A 1962 [rev. 1973], 'Hazards of Prophecy: The Failure of Imagination', in *Profiles of the Future: An Enquiry into the Limits of the Possible*, Macmillan, London, UK.
Clarke, A 1968, *2001: A Space Odyssey*, Hutchinson, London, UK.
Clarke, A 1975, *Imperial Earth*, Pan, London, UK.
Coan, R 2012, 'What's the Theory Behind Innovation and the Knowledge-Based Economy', *Journal of Applied Research in Economic Development*, vol. 9, August.
Commoner, B 1972, *The Closing Circle*, Knopf, New York, US.
Cousins, N 1979, 'Rendezvous with Infinity', *Cosmic Search*, vol. 1, no. 1, January, pp. 30–34.
Cowell, F 1964, *Cicero and the Roman Republic*, third edition, Pelican, London, UK.
Diamond, J 1999, *Guns, Germs and Steel*, Norton, New York.
Diamond, J 2003, 'The Last Americans: Environmental Collapse and the End of Civilisation', *Harper's Magazine*, June, Harpers, New York, US, pp. 43–51.
Dresner, S 2008, *The Principles of Sustainability*, second edition, Earthscan, London, UK.
The Economist 2014, 'Cheaper Oil: Winners and Losers, America and Its Friends Benefit from Falling Oil Prices; Its Most Strident Critics Don't', 25 October.
Eddington, A 1928, *The Nature of the Physical World*, Macmillan, New York, US.
Ekins, P 1993, ''Limits to Growth' and 'Sustainable Development': Grappling with Ecological Realities', *Ecological Economics*, vol. 8, pp. 269–288.
Engels, F 1940 [1876], *The Dialectics of Nature*, International Publishers, New York, US.
Engels, F 1996 [1844], 'Outlines of a Critique of Political Economy', *Deutsch-Französische Jahrbücher*, Marxists.org.
Fetter, S 2009, 'How Long Will the World's Uranium Supplies Last?' *Scientific American*, 26 January.

Ford, M 2009, *Automation, Accelerating Technology and the Economy of the Future,* e-book, Acculant Publishing, New York, US.
Forster, E M 2008 [1909], *The Machine Stops*, Dodo Press, Gloucester, UK.
Forsythe, M 2017, 'China Cancels 103 Coal Plants, Mindful of Smog and Wasted Capacity', *New York Times*, 18 January.
Frink, C, Waggoner, P, and Ausubel, J 1999, 'Nitrogen Fertilizer: Retrospect and Prospect', *Proceedings of the National Academy of Sciences of the United States of America*, vol. 96, no. 4, 16 February, pp. 1175–1180.
Georgescu-Roegen, N 1986, 'The Entropy Law and the Economic Process in Retrospect', *Eastern Economic Journal*, vol. X11, no. 1,.
Gilding, P 2011, *The Great Disruption*, Bloomsbury, London, UK.
Goldenberg, S 2014b, 'Apple Eyes Solar to Power the Cloud and Iphone 6 Sapphire Manufacturing' *The Guardian,* 26 July.
Good, I 1966, 'Speculations Concerning the First Ultraintelligent Machine', *Advances in Computers*, vol. 6, January, pp. 31–88.
Gore, A 2013, *The Future*, Random House, London, UK.
Gosden, E 2014, 'Energy Minister Vows to Curb the Spread of Solar Farms', *The Telegraph*, London, 4 April.
Haldane, JBS 1923, *Daedalus, or, Science and the Future*, Paper read to the Heretics, Cambridge, UK, 4 February.
Hamilton, C 2013, *What Would Heidegger Say about Geo-Engineering?*, 11 September.
Harley, C 2010, *Prices and Profits in Cotton Textiles during the Industrial Revolution*, Discussion paper in Economic and Social History, no. 81, Department of Economics and St. Antony's College, University of Oxford, Oxford, UK.
Harley, C 2013, *Slavery, the British Atlantic Economy and the Industrial Revolution*, Discussion paper in Economic and Social History, no. 113, Department of Economics and St. Antony's College, University of Oxford, Oxford, UK.
Heaton, G, Repetto, R and Sobin, R 1991, *Transforming Technology: An Agenda for Environmentally Sustainable Growth in the 21st Century*, World Resources Institute, Washington, DC, US.
Hesiod, 2010 [c. 800 BC], *Works and Days*, Evelyn-White H G (trans.), e-text, The Theoi Project, Auckland, NZ.
Heidegger, M 1953, 'The Question Concerning Technology', in Krell, D (ed.), *Martin Heidegger: Basic Writings*, Harper, San Francisco, US, 1993, pp. 308–341.
Hickling, H 1965, *One Minute of Time: The Melbourne-Voyager Collision*, AandA Reed, Sydney, Australia.
Higgs, K 2015, *Collision Course: The Impossibility of Endless Growth on a Finite Planet*, MIT Press, Cambridge, US.
Hong, S, Candelone, J-P, Patterson, C and Boutron, C 1994, 'Greenland Ice Evidence of Hemispheric Lead Pollution Two Millennia Ago by Greek and Roman Civilisations', *Science*, vol. 265, no. 5180, 23 September, pp. 1841–1843.
Huesemann, M and Huesemann, J 2011, *Technofix: Why Technology Won't Save Us or the Environment*, New Society Publishers, British Colombia, Canada.
IPCC: Intergovernmental Panel on Climate Change 2014a, *Climate Change 2014, Mitigation of Climate Change*, WMO-UNEP, Geneva, Switzerland.
Jevons, W 1866, *The Coal Question*, second edition, Macmillan, London, UK.
Johnson, C 1990, 'The People Who Invented the Mechanical Nightingale', *Daedalus*, vol. 119, no. 3, Summer, pp. 71–90.

Jonas, H 1984, *The Imperative of Responsibility: In Search of an Ethics for the Technological Age*, University of Chicago Press.
Kondratieff, N 1926, 'The Long Waves in Economic Life', ('Die langen Wellen der Konjunktur' *Archiv fur Sozialwissenschaft und Sozialpolitik*, vol. 56, no. 3, pp. 573–609), Stolper, W (trans.) for *The Review of Economic Statistics*, vol. 46, no. 6, November 1935, reprinted in *Cycles* May/June 1982, pp. 107–111 and July 1982, pp. 151–154.
Kranzberg, M 1986, 'Technology and History: "Kranzberg's Laws"', *Technology and Culture*, vol. 27, no. 3, pp. 544–560.
Kurzweil, R 2013, 'Plant Bacteria Breakthrough Enables Crops Worldwide Take Nitrogen from the Air', *Kurzweil Accelerating Intelligence*, 1 August.
Kuznets, S 1971, 'Modern Economic Growth: Findings and Reflections', Nobel prize lecture, in Lindbeck, A (ed.), *Nobel Lectures, Economics 1969–1980*, World Scientific Publishing, Singapore.
Le Page, M 2014, 'World without Fossil Fuels', *New Scientist*, vol. 2991, 18 October, pp. 34–39.
Lovelock, J 2000 [1979], *Gaia: A New Look at Life on Earth*, Oxford University Press, Oxford, UK.
Marean, C 2010, 'When the Sea Saved Humanity', *Scientific American*, vol. 303, no. 2, August, New York, US.
Marlowe, C 2009 [1604], *The Tragical History of Dr Faustus* [eBook #779], Project Gutenberg, Salt Lake City, US.
Marx, K 1952 [1883], *Das Kapital – (Capital*, third edition, Encyclopedia Britannica, Chicago, US.
Matson, P, Clark, W and Andersson, K 2016, *Pursuing Sustainability: A Guide to the Science and Practice*, Princeton University Press, Princeton, US.
Meadows, D H 2008, *Thinking in Systems: A Primer*, Wright, D (ed), Sustainability Institute, Earthscan, London, UK.
Meadows, D H, Meadows D L, Randers, J and Behrens, W 1972, *The Limits to Growth: A Report for the Club of Rome's Project on the Predicament of Mankind*, Universe Books, New York, US.
Merriam-Webster 2014, online dictionary. www.merriam-webster.com/.
Moen, D 1997, 'The Japanese Organic Farming Movement: Consumers and Farmers United', *Bulletin of Concerned Asian Scholars*, vol. 29, no. 3, pp. 14–22.
Monbiot, G 2014a, 'The Impossibility of Growth: Why Collapse and Salvation Are Hard to Distinguish from Each Other', *The Guardian*, Manchester, UK, 28 May.
Mulongoy, K 1995, *Technical Paper 2: Biological Nitrogen Fixation*, ILRI Training Manual 2, Corporate Document Repository, Food and Agriculture Organization of the United Nations (FAO), Rome, Italy.
Narkus, S 2012, 'Kondratieff, N. and Schumpeter, Joseph A. Long-Waves Theory: Analysis of Long-Cycles Theory', Master's thesis for the degree of Master of Philosophy in Environmental and Development Economics, Universitetet I Oslo, Norway.
Newton, I 1952 [1686], *Principia: Mathematical Principles of Natural Philosophy*, University of Chicago/Encyclopaedia Britannica, Chicago, US.
Online Etymology Dictionary 2014. www.etymonline.com/.
Parkinson, G, and Gilding, P 2014, *Energy Futures: Reaching the Clean Energy Tipping Point*, Seminar, University of Tasmania, Hobart, Australia, 27 May.
Perez, R and Perez, M 2009, *A Fundamental Look At Energy Reserves For The Planet*, Draft for publication in the International Energy Agency/SHC Solar Update, Atmospheric Sciences Research Centre, Albany, US.

Phillimore, J 2001, 'Schumpeter, Schumacher and the Greening of Technology', *Technology Analysis and Strategic Management*, vol. 13, no. 1, pp. 23–37.

Plumwood, V 1993, *Feminism and the Mastery of Nature*, Routledge, London, UK.

Plumwood, V 2002, *Environmental Culture: The Ecological Crisis of Reason*, Routledge, London, UK, ISBN 0415178789.

Polanyi, K 2001 [1944], *The Great Transformation*, Beacon Press, Boston, US.

Polanyi, M 1968, 'Life's Irreducible Structure', *Science*, vol. 160, pp. 1308–1312.

Pyne, S 2015, 'The Fire Age', *Aeon*, 5 May.

Richardson, B 2013, 'Holarchies and Technology Evolvability', *Journal of Integral Theory and Practice*, vol. 8, no. 1–2, pp. 161–167.

Roden, D 2015, *Posthuman Life: Philosophy at the Edge of the Human*, Routledge, Oxford, UK.

Saunders, H 1992, 'The Khazzoom-Brookes Postulate and Neoclassical Growth', *The Energy Journal*, vol. 13, no. 4, pp. 131–148.

Schmidt, E and Cohen, J 2014, *The New Digital Age*, Vintage, New York, US.

Schrödinger, E 1944, *What Is Life? The Physical Aspect of the Living Cell*, Stanford, US.

Schumacher, E 1989 [1973], *Small Is Beautiful*, Harper and Row, London, UK.

Schumpeter, J 1939, *Business Cycles*, McGraw-Hill, New York, US.

Schumpeter, J 2011 [1942], *Capitalism, Socialism and Democracy*, Harper and Brothers, New York, US.

Sense-T 2014, How It Works, website. www.sense-t.org.au/about/how-it-works.

Shelley, M 1818, *Frankenstein; or, The Modern Prometheus*, Lackington, Hughes, Harding, Mavor and Jones, London, UK.

Shelley, M 1997 [1826], *The Last Man*, Jones, S (ed.), Romantic Circles, electronic edition.

Sgouridis, S, Bardi, U and Csala, D 2015, *A Net Energy-based Analysis for a Climate-constrained Sustainable Energy Transition*, Conference Paper, Club of Rome Conference, 15–16 October, Winterthur, Switzerland.

Simili, R 2014, *Erasmus Darwin, Galvanism, and the Principle of Life*, Inter-University Centre for Romantic Studies, University of Bologna, Italy.

Sloterdijk, P 2005, 'The Crystal Palace', *Im Weltinnenraum des Kapitals: Fur eine philosophische Theorie der Globalisierung* [*In the Global Inner Space of Capital: For a Philosophical Theory of Globalisation*], Suhrkamp, Frankfurt am Main, Germany, chapter 33, pp. 265–276.

Sloterdijk, P 2009, *Du mußt dein Leben ändern: Über Religion, Artistik und Anthropotechnik* [*You Must Change Your Life: Concerning Religion, Artistics and Anthropotechnics*], Suhrkamp Verlag, Frankfurt am Main, Germany.

Šmihula, D 2011, 'Long Waves of Technological Innovations', *Studia Politica Slovaca*, no. 2, pp. 50–68.

Snow, C 1959, *The Two Cultures: The Rede Lecture*, Cambridge University Press, Cambridge, UK.

Somers, J 2013, 'The Man Who Would Teach Machines to Think', *The Atlantic*, November.

Steffen, W, Persson, A, Deutsch, L, Zalasiewicz, J, Williams, M, Richardson, K, Crumley, C, Crutzen, P, Folke, C, Gordon, L, Molina, M, Ramanathan, V, Rockström, J, Scheffer, M, Schellnhuber, H and Svedin, U 2011, 'The Anthropocene: From Global Change to Planetary Stewardship', *Ambio*, vol. 40, no. 7, October, pp. 739–761.

Stiglitz, J 2014b, Interview by Jonathan Green, *Background Briefing*, Radio National, Australian Broadcasting Corporation, 29 June.

Stover, D 2011, 'The Myth of Renewable Energy', *Bulletin of the Atomic Scientists*, 22 November.
Tollefson, J 2014, 'Lockheed Martin's Fusion Goals Meet Scepticism', *Nature*, 17 October.
Truitt, E R 2015, 'Preternatural Machines', *Aeon*. Viewed 10 August 2015.
Van der Hout, S 2014, 'The Homeotechnological Turn: Sloterdijk's Response to the Ecological Crisis', *Environmental Values*, vol. 23, pp. 423–442.
Vein, C 2013, 'Interview with Leon Compton', *Statewide Mornings*, Australian Broadcasting Corporation (ABC) Radio, Hobart, Australia, 16 December.
Verne, J 1911 [1865], *From the Earth to the Moon*, in *Works of Jules Verne*, C Horne (trans.), V. Parke, New York, US.
Verne, J 1911 [1869], *Twenty Thousand Leagues Under the Sea*, in *Works of Jules Verne*, C Horne (trans.), V. Parke, New York, US.
Von Goethe, J 1870 [1828], *Faust: A Tragedy*, Bayard Taylor (trans.), Riverside Press, Cambridge, US.
WEF: World Economic Forum 2015, 'Deep Shift Technology Tipping Points and Societal Impact', Global Agenda Council on the Future of Software and Society, Survey Report, September.
Wells, H 1938, *World Brain*, Project Gutenberg of Australia eBook No.: 11003731h.html, June 2013, Methuen, London, UK.
Wheeler, M, 2013, 'Martin Heidegger', in Zalta, E (ed.), *The Stanford Encyclopedia of Philosophy*, Spring 2013 Edition.
Williams, E 1944, *Capitalism and Slavery*, University of North Carolina Press, Durham, US.
Wilson, E 2013, 'Beware the Age of Loneliness', *The Economist: The World in 2014*, 18 November, p. 143.
Wilson, G and Staffell, I 2018, "Winds of Change: Britain Now Generates Twice as Much Electricity from Wind as Coal", *The Conversation*, 6 January.
Zencey E 2013, 'Energy as Master Resource', *State of the World 2013: Is Sustainability still Possible?* Worldwatch Institute, blog, pp. 73–83.

7 Options

Consistent with Marshall McLuhan's (1967) remark that "there is absolutely no inevitability as long as there is a willingness to contemplate what is happening", this chapter considers the options that may lead to a sustainable future. Human impact on the biosphere approaches or has already reached unsustainable levels in several areas, including the extinctions of other species, fish harvesting, de-forestation, biochemical flows, and carbon pollution of the atmosphere. Options for sustainability are limited. Where to act: population, affluence or technology?

While fewer people on the planet is clearly most desirable for the entire biosphere, population measures are problematic. Direct population policy such as the one child policy in China and the sterilisation program in India have resulted in inequity, demographic distortion, resentment and abuse of human rights. Even if able to be applied in Africa or other high population growth areas, similar outcomes are inevitable. Better ways to check population are indirect – accessible contraception, upholding women's rights, education of girls, encouraging urbanisation, facilitating migration and increasing family incomes.

Nevertheless, neither approaches are at all likely to achieve global population reduction within coming decades such is the lag between action and result, that necessarily spans generations rather than the few years of a political cycle. Short of a super-pandemic, asteroid collision or nuclear war, our numbers are most likely to continue to increase until at least mid-century, and probably beyond that. There is no more lebensraum into which our society can expand. Those extra people will have to be accommodated, fed and educated within the bounds of one world.

If humanity has a future into the twenty-third century, then the sort of civilisation envisaged by Arthur Clarke becomes thinkable: hundreds of millions of us rather than billions; much of the Earth re-wilded and biologically diverse. That possibility depends on the many more people this century becoming not poorer, but more affluent – as Ricardo foresaw centuries ago, affluence is inversely related to population increase. That affluence, as the SDGs underline, includes access to modern sanitation, energy and medicines, and to education – especially in the use of technologies that advance sustainability.

It is neither feasible nor desirable to set out to diminish affluence as a political platform. It is challenging, but not impossible though, to reduce inequality and it is desirable from several perspectives. The inverse nexus between income and population growth underlines the absurdity of policies that aim to reduce average levels of affluence. It might possibly aid the biosphere that more billions of us are reduced to subsistence – there might be less resource use and chemical pollution – but those same billions would resent that decline; the politics of primitivism are clearly unsaleable. Alternatively, policies that tend to reduce inequality may benefit sustainability because the impetus for conspicuous consumption would diminish, luxury markets for private jets and swimming pools would shrivel, and especially because more people better off, better educated and better connected would have a stronger voice on the environmental quality that affects them.

Churchill's observation that capitalism tends to increase inequality has been borne out by Picketty, and considerable evidence is available that links inequality and environmental harm. But while policies that aim to reduce affluence are simply not viable, policies that tend to reduce inequality also can be difficult to make stick. The fate of the *Affordable Health Care for America Act* and its later variants is an example here. The 'Obamacare' Bill was diluted by the US Senate, yet passed in 2010 to the consternation of conservative Republicans. Although not wound back by 'Trumpcare' by mid-2018, this major step in the reduction of inequality would be undermined by continuing efforts to secure a congressional majority to do so (Sullivan 2018).

Nevertheless, elements of Kuznets' 'golden era' world of rising mass affluence and declining inequality might conceivably return on a global as well as a national scale, given determined public policies driven by the urgency of action to combat unsustainability. Such policies would include an approach to economics that measures impact on natural resources and emphasises non-material consumption rather than growth at all cost. Otherwise, unless the 'four horsemen of levelling'[1] revisit, inequality will inexorably rise, and the quest for sustainability will be even harder in a world where institutions pay only lip service to its pursuit. In any case, a rising affluence in the South implies both more resource use and waste in the shorter term. In the absence of significant technological development directed at these issues, sustainability will become a mirage, more elusive than an arctic fox in the snows of a Siberian winter.

Of the three main options implied by the IPAT identity, achieving a smaller global population may be feasible in the very long term, but the factor of population will continue to press probably at least until the end of the twenty-first century. The denial of greater human affluence may underlie some proposals for a return to 'simplicity', but there are very few poor people in the world who would prefer to remain so. This is shown not only by the dramatic increases in affluence in China and other countries of the East, but also by the increasing phenomenon of economic migration, whereby millions are motivated to seek the affluence of developed countries. Neither population nor affluence are factors that encompass practical hope for achieving a sustainable existence on this planet.

Only in technology is there opportunity to rein in unsustainable practices, while at the same time accommodating a still rising population, growing affluence and reducing inequality. But without a different approach to technology, the practices that have already wrought mass extinction and damaged the biosphere will continue. Those practices depend upon a form of passivity, an attitude of inevitability – a view that any innovation is good, industrial processes are progressive and that technologies give us mastery over the Earth.

Whilst humanity has been warned about technology throughout history, our relationship with it has continued to intensify. Commoner's (1972) point that technology most amplified environmental damage in the mid-twentieth century has been updated and detailed by Steffen et al. (2011) in the twenty-first, and the trends are arresting. Yet technology has transformed since Commoner's time. It increasingly encompasses wider alternatives, so much that it appears to be on the brink of a new era. In the critical energy sector, photovoltaics and other renewable sources are already viable, scalable and cheaper than established high emissions generation. In transport, electric and self-driving cars appear poised to predominate. Tesla is already worth more than Ford; India plans to ban petrol-powered cars by 2030; private urban car ownership is diminishing. High-speed rail has replaced much short-flight aviation in Europe, Japan and China. In agriculture, precision techniques and robotics minimise both inputs and waste, enabling an intensification that uses less land, and machine learning enables reliable prediction. In the chemical industry less toxic substitutes are being developed and deployed. Digital technologies enable communication networks without poles and wires, monitoring of the entire Earth, and the application of big data analytics to environmental harms.

On the demand side there are significant opportunities to cut domestic and industrial energy consumption, to reduce depletion of non-renewables and to increase demand for products from sustainable sources. This would amount to a redesign of developed economies to be more based on services rather than physical goods and to design objects for re-use rather than disposal as waste, as part of the circular economy concept.

For sustainability to be furthered, innovation is to be encouraged and channelled in that direction. And while innovation does not depend on capitalism, capitalism is quick to exploit its fruits. Its agility and practicality cannot therefore be discarded in the quest for sustainability. Yet capitalism's stripped down neo-liberal version tends to focus on short-term returns irrespective of other merits, and enables the massing of large fortunes and political power of the few. Because it is antipathetic to sustainability, it is evidently imperative that changes to the dominant technological paradigm within capitalism are effected if human civilisation is to have a long-term future. The next question is how best to do so.

Way forward

In 1976 Marshall McLuhan wrote a manuscript with his colleague Robert Logan titled 'The Future of the Library' that remained unpublished for nearly three decades, delayed by McLuhan's stroke in 1979 and death the following year. In it they noticed that:

> We live in a complex world of high technology and intractable social problems [where] we must confront the effects of exploding population, diminishing resources, information glut and over-specialisation. … We live in an era of information explosion. We are flooded by a plethora of data, yet seem unable to use our knowledge and understanding to come to terms with the difficulties facing us.
>
> (McLuhan and Logan 1976, p. 19)[2]

They observe that the tradition of Western thought implies a failure to consider the impact of technology on ourselves or our environment. Technocrats are "naïve about the effects of their technologies" as Henry Ford failed to foresee the carnage of the automobile alongside the mobility benefits it promised. In this "electric age", they say, "there can be no more monopolies of knowledge" (ibid., p. 26), a challenge yet to be met.

With the wisdom of hindsight, this manuscript identifies aspects of our relationship with technology that appear central to our present dilemma and suggests a way forward: to interrogate the vast amount of available data to help resolve sustainability; to pay much more attention to the effects of technologies; and in doing both to fuse the quantitative with the qualitative, the engineer with the artist – as was suggested in the same era by Robert Pirsig (1984 [1974]).

Alien overlords

So there is a need for stronger policy-making, especially about using technology to consume resources sustainably. Yet there remains an issue of values. The icons of contemporary new technology – algorithms, smartphones, virtual reality, artificial intelligence – associated with Silicon Valley may have vastly increased human connectivity and made fortunes for their creators, but physically and symbolically these creators and their clergy are isolated from normal life. These are our "alien overlords"– the "hoodied young software engineers who ride to work aboard luxury buses", oblivious to all but the music in their headphones and the screen of their mobile device (Solnit, cited in Beacock 2015). The term 'clergy' for these aliens suggests their hooded cassocks, their language of esoteric codes, and their overwhelmingly male numbers. They are also largely young and white, which tends to result in norms and values hostile to those who are not (Corbyn 2015). Yet while violent gaming, drone warfare and social trivia may reflect the values of this group, it is possible to imagine technologies more imaginative and creative – technologies driven by different values.

For those of us for whom the Google bus does not stop, our lives increasingly depend on technologies we do not understand, nor conceive of how we could change them. It is as if we are part of something we "cannot quite grasp, like there has been a phase shift from humans struggling to survive to humanity struggling to survive our success" (Bures 2016). Indeed, the present inevitability that we now use technologies different from each preceding generation only extends back in time about five generations. The struggle for survival for most people before the Industrial Revolution was waged with technologies unchanged for centuries – the plough, the terrace, the harness, the loom and domestic fire. Now with change compressing time and our adjustment to it, there is a renewed imperative to understand how technology might serve our highest priority – sustainability.

Resource productivity

There is evidence that sustainable prosperity by mid-century is possible through "mobilising technology" and delivering incentives to reduce environmental pressure. This would amount to a decoupling of economic growth from material impact through technology. It is claimed to be demonstrable through computer modelling by scientists at the Commonwealth Scientific and Industrial Research Organisation (CSIRO), Australia (Hatfield-Dodds et al. 2015). The Club of Rome has also undertaken extensive modelling of a new economy in which economic growth is decoupled from resource use in five European countries. Its report on the 'circular economy' links technology, policy reform and resource productivity:

> With a growing population, and in the developing countries a much-needed increase of per capita income (affluence), technology innovation, in combination with behaviour change – and underpinned by policy

> reforms – are the only options we have to bring down the environmental impacts. Luckily, there are many types of decoupling that could and should be achieved by improved technology, often complemented by behavioural change. … While the promotion of labour productivity has been a priority for economic policy-making in the past, resource productivity has been more or less neglected.
>
> (Wijkman and Skånberg 2015 p. 8)

Resolution

As Roden (2016) observes, technology is probably "not in control of anything, but is largely out of our control". If sustainability is to be progressed, then that control must be reasserted. But progressing sustainability is extremely complicated, involving judgments of "intergenerational ethics, science, economics, politics, and much else" to questions of "energy, agriculture, forestry, shelter, urban planning, health, livelihood, security, and the distribution of wealth within and between generations" (Orr 2006, p. 267). Ultimately, as historian Leo Marx in his *The Machine in the Garden* (1964) noticed, the resolution to the problem of technology and the environment is one not for historians or technocrats, whatever their values, but for politicians.

Notes

1 Mass warfare, revolution, state failure, and pandemics. See Chapter 5.
2 Quotation is from an edited excerpt published with the approval of Robert Logan.

References

Beacock, I 2015, 'Humanist among Machines', *Aeon*, 25 June.
Bures, F 2016, 'Dispatches from the Ruins', Essay, *Aeon*, 2 February. www.aeon.co.
Commoner, B 1972, *The Closing Circle,* Knopf, New York, US.
Corbyn, Z 2015, 'Silicon Valley Is Cool and Powerful: But Where Are the Women?', *The Guardian*, 8 March.
Friedman, D 2013, *The Puzzle of the Other Hockey Stick*, blog, 15 July.
Hatfield-Dodds, S, Schandl, H, Adams, P Baynes, T, Brinsmead, T, Bryan, B, Chiew, F, Graham, P, Grundy, M, Harwood, T, McCallum, R, McCrea, R, McKellar, L, Newth, D, Nolan, M, Prosser, I and Wonhas, A 2015, 'Australia Is 'Free to Choose' Economic Growth and Falling Environmental Pressures', *Nature*, vol. 527, 5 November, pp. 49–62.
Marx, L 1964, *The Machine in the Garden: Technology and the Pastoral Ideal in America*, Oxford University Press, New York, US.
McLuhan, M (and Fiore, Q illustrator) 1967, *The Medium Is the Massage*, Penguin, Middlesex, UK.
McLuhan, M and Logan, R 2015 [1976], 'The Future of the Library, Circa 1976', *Island*, vol. 140, pp. 14–28.

Orr, D 2006, 'Framing Sustainability', *Conservation Biology*, vol. 20, no. 2, pp. 265–268.
Pirsig, R 1984 [1974], *Zen and the Art of Motorcycle Maintenance*, Bantam, New York, US.
Roden, D 2016, 'New Substantivism in Philosophy of Technology', *Enemyindustry: Philosophy at the Edge of the Human*, blog, 23 July.
Steffen, W, Persson, A, Deutsch, L, Zalasiewicz, J, Williams, M, Richardson, K, Crumley, C, Crutzen, P, Folke, C, Gordon, L, Molina, M, Ramanathan, V, Rockström, J, Scheffer, M, Schellnhuber, H, and Svedin, U 2011, 'The Anthropocene: From Global Change to Planetary Stewardship', *Ambio*, vol. 40, no. 7, October, pp. 739–761.
Sullivan, P 2018, 'Conservative Groups Outline New ObamaCare Repeal Plan', *The Hill*, 19 June.
Wijkman, A and Skånberg, K 2015, *The Circular Economy and Benefits for Society: Jobs and Climate Clear Winners in an Economy Based on Renewable Energy and Resource Efficiency*, Study report for the Club of Rome, Pertaining to Finland, France, the Netherlands, Spain and Sweden.

8 Technology and governance

Humanity's relationship to technology is complicated; it is certainly one that involves more than just an extension of human faculties. Martin Heidegger (1976) once observed that "in its essence, technology is something that man does not control", rather that technology is an immensely powerful force in shaping human history. With this caution in mind, it is nevertheless critical to the future of this world that technology is consciously influenced in the direction of sustainability.

The 'technology as tiger' metaphor in the introduction highlights the need for control of its lethal power. Yet a different image, of an embrace, a dance with a duality, may be closer to our present relationship. As described in Chapter 2, some technologies can both advance and degrade sustainability, depending on their targets. This dual nature is intrinsic to our bond with its many forms. Humanity's relationship with technology is a tango with Janus without end. As technology compounds and advances, this embrace seems ever closer and the tempo ever quickens. Swept along together, humanity follows rather than leads, our partner switching between benevolence and malice. In this image, the issue is how to wrest control over the dance before calamity ensues.

A form of madness

While technology is key to sustainability, how it is governed is at issue. This is because there is an 'anything goes' approach to technology that has both negative and positive consequences for sustainability. Not only are we using technology to ransack our world to fuel our machines and to push consumption of the superfluous, technology remains largely free of constraint, direction or accountability for the loss of sustainability it causes. As Bertrand Russell (1946, p. 482) observed, the modern attitude to technology is "a form of madness", everything non-human is "mere raw material"; ends are not considered, only the skilfulness of the process. Yet, paradoxically, of the key factors affecting sustainability, only a changed relationship with technology holds practical hope of achieving it.

It has been said that, at least in the post-Enlightenment West, our time is always special, and much relates to technology:

> Temporal narcissism demands that we must always live in the most crucial, most urgent, most dangerous, yet most opportunity-rich time in human history. Technology must always be "moving faster than ever before." And law and policy, of course, must "not be keeping up with technology".
>
> (Vaidhyanathan 2018, p. 206)

It does seem that technology is accelerating and becoming harder to control through policy and regulation. This may be partly because the word 'technology' is increasingly assumed to mean the plethora of rapidly developing digital technologies rather than its wider meaning used here. 'Innovation' and 'disruption' are worshipped, yet much innovation is trivial and much disruption is economic; sustainability consequences are generally ignored. 'Innovation' has replaced the notion of 'progress', as perhaps it is more easily demonstrable; a diminished goal, free of annoying 'values'. If sustainable innovation is to be fostered and unsustainable innovation curtailed, new ways of governing technology are needed. Innovation may then contribute to genuine progress.

Humanity's relationship with technology is much deeper than is immediately apparent. Yet our entire past has been transformed by technology's cascade – from mastery of fire to the construction of hunting tools, to sewing and weaving, to the specialisation of Adam Smith, the industrial world of Karl Marx, to the telephone, electricity and radio, the car, the aeroplane and spacecraft, to the personal computer, the Internet and social media. Not only have we used technology to shape our world, as Engels (1940 [1876]) observed, equally humanity has been shaped by its creations. But not always in a manner obvious or benign:

> Everywhere we remain unfree and chained to technology, whether we passionately affirm or deny it. But we are delivered over to it in the worst possible way when we regard it as something neutral.
>
> (Heidegger 1953, p. 4)

A one-dimensional subject-object view of our relationship with technology is consistent with its ability to objectify our environment. For example, as geoengineering enters mainstream consciousness, such technologies can be seen as a threat based on objectification of the planet:

> by objectifying the world as a whole, geoengineering goes beyond the mere representation of nature as 'standing reserve'; it requires us … to see technology as a response to disorder breaking through.
>
> (Hamilton 2013)

In a similar way, Pope Francis (2015, s.53) believes there is a "techno-economic paradigm" that threatens to overwhelm ecosystems and politics. Russell, Engels, Heidegger, Hamilton and Francis recognise the significance of our bond with technology and its capacity to alienate us from the environment. The key to governing technology for sustainability then seems to be recognition of this relationship as a mutual interaction rather than a one-way street. The consequences of technology are increasingly hard to predict, but prediction is increasingly needed, and more and more the information is available on which to base such judgements.

It is also acknowledged that there is extensive critical Freudian and neo-Marxist literature concerning humanity's relationship with technology. This includes sociological works from the Frankfurt school, beginning in the 1920s, and involving Erich Fromm and Herbert Marcuse. Consistent with his own historical experience, the former was most concerned with the rise of totalitarianism and humanity's psychological "fear of freedom" (Fromm 1941), while the latter regarded technology as originally a liberating force due to its instrumental approach, but which "turns into a fetter of [human] liberation; the instrumentalisation of man" (Marcuse 1964, p. 159). He elaborates that:

> Nature, scientifically comprehended and mastered, reappears in the technical apparatus of production and destruction which sustains and improves the life of the individuals [while] subordinating them to the masters of the apparatus.
>
> (Ibid., p. 166)

Much of Marcuse's central assertions were based on those of the school's acknowledged leader, Max Horkheimer, who led the school in exile at Colombia University before reinstating it in Frankfurt after the war (Berendzen 2017). The essence of Horkheimer's view is that the expression of technology in industrialism and its associated consumption imperative creates a human existence that is a slave to rationality (such as efficiency), and which negates any moral dimension. He also criticises Science as overly specialised and divorced from society (ibid., s.2.5). Overall, contemporary critical theory aims to allow people to "transform themselves on their own terms" instead of demanding adaptation to existing conditions in order to survive (Schecter 2013, p. 133).

This book, however, goes further than the amorality of technology and its relationship with human society. All else in human aspiration is void if our species suffers the same fate as the extinction so many of our fellow creatures. And the key to that fate as well as to its avoidance is technology. Despite industrialism, efficiency imperatives and pressures to consume, humanity is self-aware. If that understanding is applied to technology and sustainability, then such forces can be blunted so that humanity may become sustainable over the long term.

Irrespective of its power to both magnify and shrink human impact, much of the present emphasis of governance and sustainability is not on technology, but rather on where it impacts – on particular parts of the biosphere and on the global 'earth systems' that are represented through computer modelling. Aspects of the biosphere are beginning to be governed as part of the global commons by international agreement – especially the atmosphere in terms of pollutants and the oceans in terms of sustainable catch. However, the terrestrial lithosphere is governed more or less directly by national and local governments while agriculture, resource extraction, industrialism and human populations intersect. Without diminishing this existing geographic governance, it would seem prudent to reinforce its web by incorporating a stronger weft – the governance of critical technologies to give priority to sustainability.

Even where technology assessments do take place in respect of policy, much of the assessment undertaken is focused not on environmental sustainability, but is rather diffused by social and economic impacts:

> based on normative filters such as notions of proportionality and precaution … various forms of impact analysis, such as sustainability impacts, cost-benefit analysis, environmental policy impact analysis etc., the application of particular consensual norms or prioritisation of norms … and the application of normative standards for product acceptability.
>
> (Von Schomberg 2011, p. 8)

Further, there is a lot of overlap between the methods of technology assessment, social impact assessment and environmental impact assessment. For technology assessment in the late twentieth century, the most common methods used were relatively simple: expert opinion, monitoring and trend extrapolation, as well as less frequently, scenario building, qualitative modelling, non-expert opinion, quantitative modelling, checklists, and matrices in that order (Porter 1995, p. 144). At issue here is that technology assessments were not focused on their contribution to environmental sustainability, and conversely environmental impact assessments did not concentrate on the impacts of technologies. This becomes self-perpetuating, as not only do "indicators arise from values (we measure what we care about), …[but] they create values (we care about what we measure)" (Meadows 1998, p. 2). As discussed in Chapter 5, the universal reliance on GDP as an indicator of economic well-being during the past century has meant that analysis and policy is directed to its growth at the expense of other possible measures, for example.

Two decades later, the field had become rather more complicated. Singh et al. (2012, pp. 296–297) list 41 major composite indices that are each themselves composed of dozens of different indicators used to measure "sustainability", or "sustainable development". They apply mainly to corporations, government, or processes rather than to technology directly and relate to social and economic as well as environmental sustainability. But even the "innovation, knowledge and technology" set of nine composite sustainability indices do not measure sustainability. They measure 'monetary investment in research and development', or the 'number of graduates or post-graduates in the sciences' for example (ibid., p. 288). A single indicator of a technology's contribution to sustainability remained elusive.

In the United States, the Congressional *Office of Technology Assessment* spanned the years from 1972 until 1996. Its purpose was to assess developing technologies so that members of Congress could make meaningful decisions about them. However, many of its several hundred background papers and reports concerned the social and industrial consequences of technology rather than their impact on sustainability. This included such reports as 'Demographic Trends and the Scientific and Engineering Workforce' (December 1985) and 'Displaced Homemakers: Programs and Policy' (October 1985), which provided no suggestions for sustainability governance and were only indirectly linked to technology. Further, some of its reports may have been true, but were unhelpful on the issue of technological sustainability. Its 1994 paper, *Studies on the environmental costs of electricity*, for example, concluded "that no clear consensus exists on quantitative estimates of environmental costs of electricity, or on methodologies for making those estimates" (US Congress Office of Technology Assessment 1994, p. 2), and the paper did not suggest any.

Some decades after the closure of the Office, its functions were picked up by the US Government Accountability Office (GAO), which is independent of Congress and has a much wider field of interest than technologies. While far fewer technology assessments have been undertaken by the GAO, some do directly relate to technological sustainability, such as its 2015 assessment on 'reducing freshwater use in hydraulic fracturing and thermoelectric power plant cooling' (USGAO 2015). But it seems telling that a 2017 keyword search of GAO assessments shows that the word 'sustainability' is rarely mentioned in titles and sub-titles. In the few cases where it is mentioned, it relates to fiscal sustainability and sustainability of nuclear weapons. Also, in 2012 the GAO won a satirical 'Ig Nobel' prize for literature for a report about reports that recommends a further report (USGAO 2012), which further tends to undermine confidence in its capacity to focus.

Nevertheless, the concept of a body to undertake technology assessments influenced the establishment of similar institutions in other countries, and especially in the EU, where it became the European Parliamentary Technological Assessment (EPTA) – to which, ironically, the US GAO technology assessment group later became affiliated. Again, EPTA reports do not always concern technology and sustainability, but at least some do. There are reports on technology and waste management, on energy systems, sustainable transport and genetically modified food, for example. But the basic notion of these assessment offices implies that it is only through *government* that technology *governance* is effected, whereas this may not be so in the practical 'real world'.

Governance

While the term 'governance' has been described as a 'weasel' word that is "slippery and elusive, used to obscure, not to shed light" (Bevir and Rhodes 2004, p. 133), in this context its means how we control technology in important conscious ways, rather than allowing its development unchecked by considerations of sustainability.

Crucially, 'governance' is not 'government'. Rather it is the "traditions, institutions and processes" that shape the exercise of power in making decisions of public concern (Graham, Amos and Plumptre 2003). Its emergence as a concept in the latter part of the last century, hinged on the "vast increase, in both numbers and influence, of non-state actors" as well as "the implications of technology in an age of globalisation" (Weiss 2000, p. 796). The World Bank, which is especially concerned with governance of developing states, says it is a process of interaction between state and non-state actors shaped by power. It involves government, international organisations, civil society organisations and business associations operating at various overlapping levels in "a complex network of actors and interests" (World Bank 2017b, p. 3). Importantly, governance – like technology – is systematic (Greer, Wismar and Figueras 2016, p. 3). It can apply to international society and to public and private organisations as much as to nation states and local communities. An important consideration is thus the boundaries in which it operates.

Old style command-and-control regulation was appropriate to simpler societies when governance was directly effected by government as its dominant agent. The key weakness in this style is of course that it not possible to command nor control fast changing technologies no matter how many administrators there are. Another is that controls tend to be implemented in case of rare events and exceptional cases, which unnecessarily degrades normal operations. If each step of a long process is under government control, then the process becomes ponderous; often accompanied by reams of paperwork and barriers to innovation. A more nuanced style of governance involving complex technologies is indicated to improve sustainability. On the other hand, there are fundamental sustainability issues that may not be abandoned by government. These are the issues of major, even existential risk as discussed in the case studies on nuclear power (Chapter 10) and artificial intelligence (Chapter 11).

As mentioned in Chapter 3, there is a contrast between attempting to control waste-producing technologies at the 'end-of-the-pipe' and encouraging more sustainable technologies in the first place (such as particulate filters for combustion engines compared with electric engines that do not produce particulates). Whatever the form of governance, it is important that non-polluting technologies are preferred, as well as those that sustainably make use of renewables or substitute for non-renewables (Daly 1990). Doing so, it would appear, tends to make governance simpler and easier. A clear trend in general regulation, however, is that it is ultimately governments that set standards for outcomes, rather than for technological and industrial processes. Ends are more important than means, and if the same ends can be achieved more efficiently, then this should not be impeded. Thus for technological sustainability, such regulation would specify standards of impact, in terms of yield, substitution and assimilation rather than how they would be achieved. Whilst outcome standard-setting is logically a government role, how it concludes and enforces them can vary considerably between industries and between jurisdictions. The glyphosate case study at Chapter 9 illustrates different approaches between the United States and the European Union, for example, where in the former there is a closer relationship with industry compared with the latter, while this is different again in the case of the nuclear power industry (Chapter 10) as the global French industry is particularly close to government. While the development of artificial intelligence is essentially left to Silicon Valley to determine (Chapter 11), the recent struggle over what should be counted for cost-benefit analyses of US rivers and wetlands illustrates how political ideology can drive and distort outcomes (Boyle, Kotchen and Smith 2017).

None of this denies that there is a relationship between regulation and innovation. But, rather, the relationship is not as straightforward as might be imagined. Strong regulation does not necessarily imply lack of innovation. It is probably more true that weak, confused and 'rule-book' regulation results in inertia and uncertainty. The key issue is the quality of the regulation. High standards appropriately enforced can enable competitive advantage built on that quality rather than commodity competition built on price. And the key issue in innovation is not its novelty, but ultimately its sustainability value. Implied political measures that would enable more sustainable technologies are discussed in 'alternative visions', the penultimate Chapter 13.

Most generally, governance is about the determination and implementation of collective decisions. 'Good governance' is therefore about the quality of that decision-making and the exercise of power. Whilst expressed in different detail, there is general consensus about what it entails across several areas. The UNDP, for example, lists nine good governance characteristics: participation, rule of law, transparency, responsiveness, consensus orientation, equity, effectiveness and efficiency, accountability, and strategic vision (UNDP 1997). The European Union condenses these to five broad categories more relevant to developed country programs: openness and transparency, participation of all stakeholders, accountability, effectiveness, and coherence (EU 2001, 2015). As to focus, Weiss (2000, p. 801) asserts that good governance centers on overcoming both "the unrepresentative character of governments and the inefficiency of non-market systems". Arguably, representation and efficiency remain important to its meaning.

The international Canadian *Institute on Governance* in reviewing environmental protection suggests five principles based on the UNDP formulation, which are similar to those of the European Union. These are: legitimacy and voice, direction, performance, accountability and transparency, and fairness (Graham, Amos and Plumptre 2003, p. 3). Pierre and Peters (2005, pp. 3–5) writing about state-related governance say that it involves (i) articulating goals and priorities; (ii) providing coherence and coordination; (iii) steering and implementing, directly or indirectly; and (iv) providing accountability to society. Biermann (2007) in discussing earth systems governance suggests that there are four related principles – credibility, stability, adaptiveness and inclusiveness – much due to the uncertainty and potentially extreme impacts of anthropogenic earth systems. Von Schomberg (2011, p. 16) of the European Commission advocates models of responsible innovation governance whereby actors become mutually responsive; companies identify the risks as well as the benefits of new technologies and NGOs consider the value of using new technologies as well as their potential hazards.

Ultimately, Greer, Wismar and Figueras (2016, pp. 28–29), whose focus is health policy within the EU, cite a comprehensive table linking 16 major authors writing between the years 1997 to 2013 against 23 dimensions of governance, in different contexts. Notably, 'sustainability' is mentioned only once as such a dimension – by the Council of Europe in 2012. However, from this and other information Greer, Wismar and Figueras distil a set of five broad dimensions, which they arrange into the acronym 'TAPIC': transparency, accountability, participation, integrity and capacity (ibid., pp. 116–120). These dimensions have three key advantages – an empirical rationale, an easily remembered acronym, and brevity. (There is also the serendipity of a near mirror of the IPAT acronym). Further, because they are universal categories and already apply to a field rife with technologies, these elements are applicable to the governance of technology. While the authors point out that none of these dimensions is unequivocally good (too much participation reduces efficiency, for example), they form a framework for gauging how technologies are and might be governed for sustainability. This structure is also used for the case study analyses in later chapters.

Transparency means that stakeholders and the public are informed of how and on what grounds decisions are made. Mechanisms include watchdog committees, inspectorates, and regular, clear and usable reports. Transparency mechanisms make it possible to understand an institution, identify corruption or incompetence, and ultimately rectify faults or adapt behaviours. Too much transparency can have contradictory results, however. For example, onerous freedom of information (FOI) rules can drive decision makers to conduct meetings without notes and in secret. At its best, transparency produces useful and accurate information that can be used by those who rely on or seek to influence the organisation. Ultimately, optimal transparency produces trust (Greer, Wismar and Figueras 2016, pp. 32–33).

Accountability involves both explanation and sanction. It is the relationship between an actor and a forum whereby the actor must inform of actions and explain decisions and can be sanctioned for infringements. Accountability mechanisms include contracts, rules that specify objectives and reporting mechanisms, conflict of interest policies, codes of conduct and regulation. But it can be difficult to implement and often focusses on the trivial at the expense of the important. Also, too much accountability can produce rigidity due to lack of discretion, and overly complex accountability systems produce inefficiency. Accountability goes wrong when regulators are captured (Friedman 1982, p. 128) or codes of conduct ignored, which implies that more than one accountability mechanism is desirable.

Participation means that affected parties have "access to decision-making and power to give them a meaningful stake in the work of the institution" (Greer, Wismar and Figueras 2016, p. 35). If decisions are made without the participation of affected groups, results tend to be sub-optimal. Conversely, Fung (2006, p. 66) argues that well-structured participation improves legitimacy and ownership, justice in outcomes, and policy effectiveness. Participatory mechanisms include stakeholder forums, consultations, elections or appointed representatives, choice mechanisms, advisory committees, partnerships, surveys and joint workings. However, participatory mechanisms can fail. One important issue is whether or not the purpose of participation is clear and justifiable. Another is whether the mechanism is effective in providing useful input. Creating many different participation mechanisms can create inefficiencies and delay important decisions.

Integrity counters corruption in that it creates a trustworthy and purposeful organisation. It involves ethics, predictability and rule of law, together with well-defined roles and responsibilities aligned with clearly specified processes of representation, decision-making and enforcement. Integrity mechanisms include internal career paths that obviate officials seeking profit outside government, audit processes, personnel practices that encourage talent and remove the unsound, legislative mandates, budgets, meeting procedures and documentation. On the downside it can also mean more bureaucracy, in that its procedures can proliferate at the cost of efficiency. Lastly, while the benefits of well-defined organisational roles can be overstated, clarity of mission and focus on key goals cannot (Greer, Wismar and Figueras 2016, pp. 38–39). Thus mission clarity might involve ensuring that current and emerging technologies contribute to sustainability. It would also involve identifying technologies that most jeopardise life within the biosphere. Goals might include the development of sustainability criteria for each cluster of technologies, the development of indices to measure progress, determining how best to substitute technologies found wanting.

Capacity is the capability of an organisation to undertake its mission. Whilst relevant to governance as a general concept, in the TAPIC framework 'capacity' has a more specific meaning. *Policy capacity* is the ability to develop policy in pursuit of goals aligned with resources. It is the "competencies, resources, and experience that governments and public agencies use to identify, formulate, implement, and evaluate solutions to public problems" (Forest et al. 2015, p. 265). In this sense, capacity means identifying opportunities to have an impact. It includes the ability to harness data, and to understand interest group and partisan politics. It also means undertaking activities that align with its mission, as well as having the technical and political skills to participate in discussions across sectors, and to create workable regulatory structures that produce desired policy outcomes (Greer et al. 2016, p. 120). Mechanisms that improve policy capacity include performance analyses to identify problems and gauge the effects of policies; process analysis to highlight legal, budgetary and systems issues; and procedures to incorporate specialist advice into policy formulation. While it is possible to have too little policy capacity, or one that lacks experience in particular areas, too much such capacity is rare, if not unheard of (Greer, Wismar and Figueras 2016, pp. 39–40).

Paradigms[1]

While any rigid form of direct global regulation over technologies would likely ossify progress towards sustainability, there are other ways to govern them. Three key paradigms are evident in which technologies may be governed. These are the *state-first, the market-based* and emergent *network* models, characterised by increasing numbers of actors respectively. These paradigms are outlined below and assessed against the TAPIC elements.

The state-led approach

The state-led paradigm is extensive over both time and geography. It involves significant input from government and has been especially prominent in times of war, both hot and cold. But as well as collaboration on military research and development, the state has also been especially involved in energy as well as medical technologies (Harvey 2014, p. 94). At the global level, states are its singular key actors, operating in order to gain military and industrial advantage over other states.

Apart from weapons, historically the governments of industrialised countries actively invested in new technologies. The first telegraph line was US government financed in 1842, for example (Stiglitz 2001). State-private partnerships resulted in railroads, aviation and new energy technologies (Jenkins, Swezey and Borofsky 2010). The greatest technology advances since World War I were derived from public investment in university research that became public goods, rather than via the private patent system. Penicillin was based on public research going back to Pasteur in the nineteenth century. The industrial techniques for its volume manufacture were developed at the US Drug Administration during the 1940s (ibid.).

Other state-developed technologies included radar and the digital computer during World War II, together with satellite communications, space travel and the Internet during the Cold War. As well as such benign examples, however, nuclear bombs, ballistic missiles, and biological weapons were also developed under this paradigm. The conflicts of both world wars and the post-war intense rivalry between the United States and the Soviet Union were central factors in these developments.

Strengthening the patent system protecting private research in biotechnology during the latter half of the twentieth century did not cause the bio-industrial revolution, but "was rather an outcome". The basis of the biotechnology industry had already been created by the state. Similarly, some of the superior technological achievements of the Soviet Union resulted from concentrating public investment on innovation in certain fields, rather than from patents and other protections provided under the neoliberal regime (Drahos and Braithwaite 2002, p. 213).

The 'Asian tiger' economies achieved growth through state-led industrialisation and technological innovation, including digital technologies and robotics (Japan) mobile computing and communications (South Korea), and computer hardware (Taiwan). Such regimes attempt to gain technological comparative advantage over other states to support their own development. In China, advances in photo-voltaics combined with the huge scale of their manufacturing have led to continuing price reductions in solar energy systems around the world. Much of this expansion results from the government's response to mounting concerns about air pollution due to coal-fired electricity generation.

However, over the past century it is in weapons technologies that the state-led approach has been most significant for sustainability governance. Biological[2] and chemical weapons such as the mustard gas of World War I and the Iran-Iraq war of 1980–88, and the smallpox and bubonic plague weapons developed during the Cold War are now prohibited. Cluster munitions and landmines are also prohibited, if not yet eliminated.[3] One reason for this is that all of these weapons have uncontrollable, unsustainable effects over the longer term that involve people and animals not part of the conflict, such as the mines laid down during the 1970s conflict in Cambodia and Laos. Notably, these governance measures are all simple bans. The technology is identified and it is prohibited. This suggests that state-led governance is a blunt instrument that may lack capacity in more complex situations.

But while nuclear weapons prohibition was the first item on the agenda of the UN General Assembly in 1946, nuclear weapons have since proliferated (Cochran 1995). This may be because nuclear weapons remain ultimate symbols of technological superiority. If so, this is in irrational contrast to other weapons, as Angela Kane, former UN High Representative for Disarmament, observed recently:

> How many states today boast that they are "biological-weapon states" or "chemical-weapon states"? Who is arguing now that bubonic plague or polio are legitimate to use as weapons under any circumstance, whether in an attack or in retaliation?
>
> (Kane 2014)

Especially because its ultimate role is destructive, it is odd that the military appears to have escaped scrutiny in relation to sustainability. Eisenhower's famous last US presidential speech warned of the conjunction of the military, industry and technology:

> we must guard against the acquisition of unwarranted influence, whether sought or unsought, by the military industrial complex ... largely responsible for the sweeping changes in our industrial-military posture, has been the technological revolution during recent decades.
>
> (Eisenhower 1961)

The destruction the US military has wrought since his speech has been immense, especially in South-East Asia and the Middle East.[4] But while military technologies have continued to be significant despite Eisenhower's warnings, nonetheless, US consumer technologies also developed in parallel. By contrast, the Soviet system of state planning over-emphasised military rather than consumer, energy or resource-efficient technologies. Non-military inventions were made but there was little incentive for them to be developed, until the failure of the system became obvious (Fukuyama 1989; Dresner 2008).

However, military-industrial technologies were not the only thing that Eisenhower's address warned about. Another caution anticipated the generational equity of Brundtland. It concerned sustainability, or eating the future:

> As we peer into society's future, we – you and I, and our government – must avoid the impulse to live only for today, plundering, for our own ease and convenience, the precious resources of tomorrow.
>
> (Eisenhower 1961)

Contemporary India offers a different take on the state-led approach to technology governance. India was a stimulus for early theorists on sustainability. The Ehrlichs wrote their book on overpopulation after a visit there in the 1960s and the other wife-husband team, the Meadows, "intuited" the basis of *The Limits to Growth* after visiting there shortly afterward (Dresner 2008, p. 28). Then, India was a country of widespread poverty, whereas today poverty remains, but India is regarded a 'middle income' rather than a 'poor' country. Since the 1960s its population has trebled, but affluence has risen considerably more (World Bank 2016).[5]

Rising affluence without loss of sustainability depends on access to low emissions energy. The election of the Bharatiya Janata Party (BJP) with Narendra Modi as Prime Minister in mid-2014 began to build on this principle. With cheap photo-voltaic panels sourced from China, Modi proposed that much of India's rural energy needs could be fulfilled by leapfrogging traditional large coal-powered electricity grids. Instead, rural households could use solar cells linked to battery storage, completely off-grid (Parkinson and Gilding 2014). This approach was demonstrated in his home state, Gujarat, where solar power also links to the established grid in urban areas.

Although also building new coal-fired and nuclear electricity plants where hundreds of millions are without access, India's energy path is different from China, which originally based its economic growth on a coal-fired grid. The development of both off and on-grid solar in India is due to several factors. First, the cost of coal in India from sources such as the Galilee field in Australia is rising (Buckley 2014), while the price of photovoltaics is continuing to fall. Further, India's climate would seem ideal for solar and where decentralised, would not require the construction of a grid. Lastly, the example of air pollution in Chinese cities is a major disincentive to coal fired generation. The outstanding issue for solar is how to resolve energy storage using either batteries or pumped hydro storage where it is feasible. The Indian renewable energy development agency is nevertheless clear about its goals to harness the "huge potential of renewable energy resources" to supply "24 x 7 power to all" with reduced emissions (IREDA 2016).

China's state capitalist economic development brought about a huge increase in consumption that is evidently unsustainable due to the emissions of its fossil-fuelled technologies, and its impact on fishing grounds in the South China Sea.[6] The challenges in shifting to a more sustainable economy "remain breathtaking" (Adams 2006 p. 8), but photovoltaic manufacturing is encouraged and closure of its dirtiest electricity plants is in train.

France's approach to industry is based on a strong centralised state grounded in Napoleonic institutions and an elite technocratic education system, favouring big state-supported enterprises that operate on a global scale. These include *Veolia*, which operates in waste, water and energy management, and *EDF* (Électricité de France), the world's largest electric company, which derives most of its capacity from nuclear power.[7] As well as energy-related institutions, the French state also developed high speed rail (*SNCF TGV*) and aero-spatial (*Air France-Industrie, Airbus* and *Dassault*) technologies. For these technologies, the French state-led approach indicates both opportunities and threats for sustainability governance. The opportunities centre on reach across continents, the standardisation of basic services and the capacity to determine priorities within the influence of a democratic state. Threats are the sheer size of its enterprises, possible over-reliance on nuclear power and a tendency to loosen control in favour of international market pressures.

Overall, it is typically the lack of participation in technological governance under this paradigm that stands out. The exercise of power that facilitates such large-scale technologies also depends on centralised control, whereas the capacity of the state to manage complex processes is questionable. Where the military is involved, transparency is especially deficient, allowing such programs as nuclear weaponry to proceed in secrecy. Participation outside of these two spheres is also negligible, while accountability to democratic institutions is mainly effected through budgetary measures, and is otherwise limited.

The market-based model

At the global level this paradigm involves both states and corporations acting within the world market to exploit new profitable technologies. This involves corporations funding their own research and development while the state provides subsidies and regulatory support to foster free trade and investment capital. Through international agencies such as the WTO and agreements such as TRIPS, 'mutual advantage' cooperation is encouraged, while 'winner-loser disadvantage' is rejected. Where specific technical issues risk incoherence, epistemic communities may be called on to broker technology regimes that are market compatible.

This dominant techno-economic paradigm is intended to promote innovation rather than sustainability. It is essentially a neoliberal approach that supports private sector innovation with tough global-scale laws on intellectual property. These laws probably originated in fifteenth-century Venice and were codified throughout Europe following the printing press (Sichelman and Veneri n.d.; Mgbeoji 2003). The current approach to intellectual property rights largely derives from nineteenth-century US and German industrial practice, internationalised with the 1883 Paris agreement (WIPO 2016) and globalised with the creation of the WTO in 1995. While the law increasingly lags technological innovation (Wadhwa 2014), in theory private firms can roll out technological innovations quickly and profits are protected in doing so. Government is supportive rather than a technology innovator itself.

This is the system that was exploited by Thomas Edison with his remarkable series of that defined the modern era. His industrial approach to invention led to more than a thousand separate patents for devices such as the electric light, the phonograph, the motion picture camera, the microphone, the electricity grid and the car battery, all of which depended on the platform innovation of generated electricity. The financier-industrialist, J P Morgan, was a prominent investor who helped promote and safeguard Edison's inventions and the New York attorney, Edwin Prindle, was a leading agent for patents to protect industrial profit. The dominant German chemical industry was similarly organised around protective patents and systematic innovation at around the same time (Drahos and Braithwaite 2002).

Yet while technological governance has focused on the protection of intellectual property, its weaknesses are that it may well discourage genuine innovation while encouraging outcomes that acknowledge no natural limits. This is the regime that so lacks sustainability governance that it gave us chlorofluorocarbons (CFCs), dichlorodiphenyltrichloroethane (DDT) and thalidomide. Each of these three examples would fulfil the fundamental requirements of current US patent law – that the product, design or process be genuinely new and "has a useful purpose" (USPTO 2016).

Measures enforcing intellectual property rights under this regime are part of global and regional trade agreements, including the WTO and the stillborn 12-nation Trans Pacific Partnership (TPP), which would have bound countries such as Japan, Vietnam and Australia to US-based regulation. The TPP followed years of negotiations over the application of patents in the different countries of the partnership, especially on the issue of how long each set of patents would apply. As at mid-2018, following US withdrawal under President Trump, the agreement may yet proceed with the remaining 11 members.

Under the 1995 WTO agreement on Trade-Related Aspects of Intellectual Property Rights (TRIPS), copyrights on software, designs, and patents are protected.[8] Patents protect "any inventions, whether products or processes, in all fields of technology, provided that they are new, involve an inventive step and are capable of industrial application" (article 27.1). The agreement also states that "intellectual property rights should contribute to the promotion of technological innovation and to the transfer and dissemination of technology" (article 7). Nonetheless, rather than from any desire to foster innovation and transfer, it seems reasonable to conclude that developed countries and their MNCs established TRIPS in order to profit from countries that lack Western technologies (Mercurio 2014).

The World Intellectual Property Organisation (WIPO), a UN agency, is closely related to TRIPS and the WTO. WIPO's role is to promote and protect intellectual property generally, not just in trade. Consistent with the TRIPS assertions, according to WIPO human progress depends on technological inventions,[9] and their legal protection encourages further innovation. WIPO also claims that intellectual property protection "spurs economic growth, creates new jobs and industries, and enhances the quality and enjoyment of life" (WIPO n.d., p. 3).

However, there is doubt that patent rights have substantially benefited innovation (Mgbeoji 2003), while others argue that intellectual property rights now actually constrain technological innovation. For example, a 'think piece' for the International Centre for Trade and Sustainable Development (ICTSD) and the World Economic Forum (WEF) argues there is a contradiction between innovation and patent protection, as patents impede diffusion and block potential gains from collaboration and competition (Mercurio 2014). It points out that patent-related legal costs amount to around US$20 billion each year and that overlapping patents result in incremental improvement of existing technologies rather than development of new technologies. 'Patent thickets'[10] and 'patent trolls'[11] are significant hazards for those trying to innovate, such that the social and economic costs of patent protection may now outweigh their benefits, as "firms divert resources away from RandD, and into lobbyists and lawsuits" (ibid., pp. 2–3).

Others, following their own research investigations at WIPO in Geneva, agree that intellectual property rights constrain innovation. Drahos and Braithwaite (2002, p. 2) make a case for this by outlining how innovation is stifled in the interests of rights holders. This occurs because rights holders increase the cost of information on which innovation depends.

In biotechnology especially, the patents market is a vehicle for commercial agglomeration rather than protection for start-ups, who often prefer "to be swallowed in one way or another by the very large fish... Patents act as a signal that they are worth swallowing" (ibid., p. 166). 'Swallowing' is also the business model in information technology. For example, WikiLeaks founder Julian Assange observes that Google "innovates through aggressive acquisition, then integrates what it has acquired" (Keane 2015, p. 32). But biotechnology has other ways of maintaining monopolies. During a 'data-exclusivity period' for biologic medicine,[12] manufacturers of generic copies cannot use original clinical trial data for approval to distribute their cheaper products. Many of these drugs are extremely expensive and are used for cancer treatment.[13] Under the TPP negotiations in 2015, the US originally sought 12 years exclusivity for its corporations, significantly longer than domestic laws in many other countries (Gleeson and Lopert 2015).

Thus the measures designed to enhance and encourage innovation under this paradigm, paradoxically tend to distort its most basic assumptions. Even if these measures were successful in promoting innovation, the sustainability of any innovation is irrelevant to the measures themselves. Sustainability is at best incidental to the corporate drive for possession, agglomeration and profit.

Within the neoliberal model, the 'knowledge economy' is an idea based on technological innovation. The notion emerged in the 1980s, relying on mathematical modelling by US economists, including Paul Krugman, Paul Romer and Robert Lucas, who worked in the new field of 'growth economics'. The central problem as they saw it was to design an economy that produced continuous growth to avoid the recessions of earlier eras. Technological progress is a product of economic activity, according to this theory, whereas earlier theories regarded technology as a product of non-market forces. Unlike physical objects, knowledge and technology are characterised by increasing returns, and it is these that drive growth (Cortright 2001).

An important concept in the knowledge economy is that of 'rival' versus 'non-rival' goods. Rival goods are like pizza: if I eat it, you cannot, whereas non-rival goods include ideas and information that can be shared without loss. Some kinds of ideas have great economic significance. These are the 'platform' ideas that are the basis of other innovations. For example, the Internet is a platform innovation whereas a router is not. Edison's electric generator is a platform whereas the microphone is a dependent device (Coan 2012).

The importance of the knowledge economy for sustainability is that it suggests that economic growth can be dematerialised, or 'de-coupled' from resource use, especially through digital technologies and the dissemination of ideas and innovation. There are two main issues with this proposition, however. First, it presumes the existence of a technologically advanced economy in which material needs are already largely satisfied, and also it supposes a digital infrastructure to operate. The knowledge economy cannot fulfil the unmet material demand that exists in much of the world; data is inedible, nor does it supply energy. Yet it can do more than just satisfy the ephemeral demands that emerge once basic needs are met. It can enhance and spread new knowledge. It can apply new knowledge to the global economy as much as to local ones. It can minimise human impact and help maximise resource intensity.

But such is the neoliberal paradigm that the very basis of the knowledge economy concept – the sharing of non-rival goods – can be constricted or made 'excludable' by the ownership of the data, the algorithms and the information on which it depends. For example, data from farm sensors on microclimates, soil and plant growth are uploaded in vast quantities by the US agricultural multinational John Deere and then resold as conglomerated information to others, such as seed and chemical companies, and in chunks back to the farmer. Farmers do not own the data from their own farms. Similarly, Facebook and Google sell their client data to advertisers, as does Microsoft, which also dominates the personal software market and cloud data repositories.

In the knowledge economy, data are as much a commodity as oil and wheat are to its industrial predecessor.

This is increasingly significant. Not only do intellectual property rights tend to constrain innovation and agglomerate market power, but as data proliferates and becomes potentially more valuable in the knowledge economy, it is owned, restricted and sold by the agglomerators of corporate capitalism. Governance over this aspect is weak and often years behind the data harvest and reuse techniques of major corporations. It demonstrably lacks both integrity and capacity.

The knowledge economy in its present form then, is not as attractive for sustainability as might first appear. It is rather its techniques that offer prospects to drive sustainability through technological governance. For example, the patent system itself can be trawled by algorithms that seek new combinations of ideas, new innovations that build upon existing inventions. This is a form of 'evolutionary mimicry' that designs progressively improved versions of devices or concepts that are tested in simulations and then the best of them selected for the next round of improvement (Marks 2015). There is no reason to suggest that improvements sought cannot be those that improve sustainability.

Likewise, the application of information technology to prosaic objects has already enhanced their sustainability. For example, the contemporary jet passenger aircraft visually resembles its 1960s ancestor, yet is an intelligent machine, first flown in cyberspace after its design, virtual manufacture and computer stress-testing optimised its construction. As it flies in reality, it transmits real-time information back to its manufacturers to enable further improvement and optimal maintenance (Mason 2015a). The Boeing 787–9 Dreamliner, which first flew commercially in 2014, looks like the 707–120 of 1958. Yet its body is made from one-piece composite material rather than riveted aluminium sheets and its in-flight engine performance is constantly monitored in real time by its engine manufacturer. Such intelligent machines are both more complex and more efficient than the visually similar but cruder technologies they replace.

In these ways, the techniques of the knowledge economy can enhance sustainability, but would be more valuable if data governance were inclusive, rather than the exclusive, low participation model that tends to be the case at present. Its capacity is lacking, and as most social media users know, integrity and transparency are governance qualities that are not prominent under this regime.

Market-based governance is also sometimes assisted by *'epistemic communities'* – associations of experts who help decision-makers define problems, identify policy solutions and assess outcomes, typically across national boundaries (Haas 1992). This aspect of governance is driven by increasing technical complexity that makes it harder for decision-makers to fully understand the issues they are dealing with. These experts typically may be academics, scientists or economists.

The UN Intergovernmental Panel on Climate Change (IPCC) is a high-level example of an epistemic community.

Expertise is not always associated with wise governance, however. An extreme case of the 'cult of the expert' is embodied by Alan Greenspan, chair of the US Federal Reserve 1987–2006, who oversaw the leveraging of sub-prime mortgages that resulted in the financial crash of 2008. Greenspan's reputation was such that when he spoke at international meetings of central bankers, "the distinguished figures at the table, titans in their own fields, took notes with the eagerness of undergraduates". Both politicians and markets regarded his words as infallible – which they turned out not to be (Mallaby 2016).

Yet one notable example of epistemic technology governance with global sustainability effects occurred in an earlier era. The US Nobel laureate Paul Berg helped organised the 1975 *Asilomar Conference on Recombinant DNA* to discuss the regulation of new forms of biotechnology involving the manipulation of DNA between species. This technology is potentially hazardous since it involves the creation of new life forms that might escape the laboratory and multiply with unpredictable effects. The conference was to decide whether to lift a voluntary moratorium on such research and determine guidelines so that it could safely proceed (Berg 2008).

His account is intriguing because he says it was scientists rather than politicians who were concerned about the issue. While the call for restraint was "hotly debated", the public "seemed comforted" because the cautionary call was made by the science community, and the public were kept informed of discussions. As well as scientists, conference participants included lawyers, journalists and government officials (ibid.). *Asilomar* was probably inspired by the international *Biological Weapons Convention* (BWC), which opened for signature in 1972. Berg does say, however, that the threat of government legislation helped overcome disagreement. Resolution of the issue was based on assessments of experiments and safety measures that varied according to the degree of risk.[14] The conference recommendations were ultimately adopted as official US guidelines in 1976 and subsequently influenced global practice. At least according to Berg, "they have proved remarkably effective".

Now though, Berg says securing public trust for experimental biotechnologies is "much more difficult". In the 1970s such research was mainly within public institutions, whereas scientists now tend to work for corporations where commercial pressures preclude open discussion. Despite such pessimism, a similar conference met in 2015 to consider the emerging technology of genome editing, which enables heritable alteration of nuclear DNA. The conference and a subsequent study committee ultimately recommended that the technology could be used on human embryos under stringent conditions, including an 'only option' to prevent serious disease. Significantly, the study committee consulted with "patient and disability rights advocates, clinicians, scientists, ethicists, and public engagement specialists", a list considerably more controlled than in 1975. Resolution was found despite apparent tensions between democracy and corporate influence (Alta Charo and Hynes 2017).

Overall, while these examples concerned two very particular and controversial technologies, their process and results indicate that the TAPIC elements were largely fulfilled. In this paradigm, participation is at risk due to its highly technical nature. But while scientists led the process, in both cases there was significant participation of other interests. Notably, both cases were about regulating a narrow and emerging field, and were a 'one off' event. This does not mean that the same results would occur in ongoing wider technology governance where transparency is obscure and participation can be exclusive.

Overall, this market-based model, tends to stress technology ownership at the expense of sustainability. Nevertheless, some digital and geospatial technologies can enable not only resource efficiency and evolutionary improvement, but also the transparency and participation that governance of complex systems demands. At the same time, its tendency to allow exclusive data ownership threatens integrity and precludes capacity development. The market-based approach will probably increasingly rely on epistemic groups as technologies become more intricate and bring greater risk, while transparency is threatened by the same dynamic.

Network governance paradigm

Network governance has much greater participation than the other two models. Due to the many organisations involved in contemporary society, combined with shifting boundaries, complex coordination and different levels of authority, the concept of network governance emerged fairly recently in several disciplines and much of its theoretical underpinning remains unclear (Sørenson and Torfing 2007). Emphasising cooperation and partnership rather than laissez-faire or exclusion, the approach still involves the state, often combined with international institutions, but more in orchestration roles than as direct regulators in the traditional sense (Greer, Wismar and Figueras 2016). Corporations are also key actors as are civil society organisation, and often the interested public as well.

Its fundamental structure is informed groups of different stakeholders that share governance responsibilities within a defined sphere. Two of its prominent features are the central role of private actors combined with a voluntary, or 'soft law', rather than 'hard law' treaty approach (Abbott and Snidal 2009). While not yet applied to any particular technology, some of its characteristics are apparent in the governance of technical processes such as manufacturing supply chains and agricultural practices.

These network characteristics overlap with the *corporate social responsibility* (CSR) approach to business ethics. Both have arisen in parallel with the rise of civil society organisations, and both depend on digital communications. CSR began as a defensive response to accusations of exploitation against multinational firms that use third world workers to produce first world consumer goods. It describes situations where the firm goes beyond regulatory compliance and engages in "actions that appear to further some social good" (McWilliams, Siegel and Wright 2005, p. 3). Although derided by an ascendant Milton Friedman in the early 1980s (Friedman 1982), beginning with labour conditions and notions of 'fair trade', CSR has since expanded to especially include the environment in the business 'triple bottom line' of economic, social and environmental sustainability. Yet there is much massaging of what CSR means. Even within industries there is a proliferation of codes and the maintenance of standards beyond legal minima is voluntary and often less than transparent. But ultimately business' compliance with CSR measures depends on consumer knowledge and consumer pressure. Digital technologies can facilitate both.

The joint ILO-World Bank program, *Better Work,* is an example of network governance combined with CSR. Under this program, the compliance of apparel factories in developing countries with hundreds of labour and environmental standards[15] is accessible through its multi-lingual global database. This program involves diverse stakeholders in its governance including nine national governments, employer groups, labour unions, national and international NGOs, the UN's ILO and the Bank's International Finance Corporation, as well as the global apparel brands such as *Nike*, *Adidas* and *The Gap.* Results of factory monitoring are entered directly onto handheld tablet computers onsite and are automatically uploaded to the database for reporting and decision making. Consistent with the 'soft law' approach, though, much of the focus is on remediation rather than sanction (Bolwell 2015). Because command-and-control bureaucracies are incapable of governing complex societies (Offer 2006), this network approach of varied stakeholder involvement combined with digital technology is a way to make CSR and its sustainability aspirations realisable across global supply chains.

In the Amazon, two networked governance measures have helped reduce deforestation. They were both aided by the "political sea change" after the 2003 election of Lula's[16] Workers' Party and the consequent loss of influence of big landowners. The *Soy Moratorium* (*SoyM*) cut soy-related deforestation from nearly 30 percent to only one percent between 2006 and 2013, due to a supply-chain governance that involves farmers, buyers, NGOs and government agencies (Gibbs et al. 2015, pp. 377–378). Similarly, the Amazon beef industry has been governed along these lines since 2008. A new code reduced to 50 percent the previous unenforceable requirement that 80 percent of land should remain forest, and allowed landowners to trade forestation credits. Public prosecutors stopped beef wholesalers buying from illegally deforested areas and a bilateral financial arrangement with Norway bolstered political

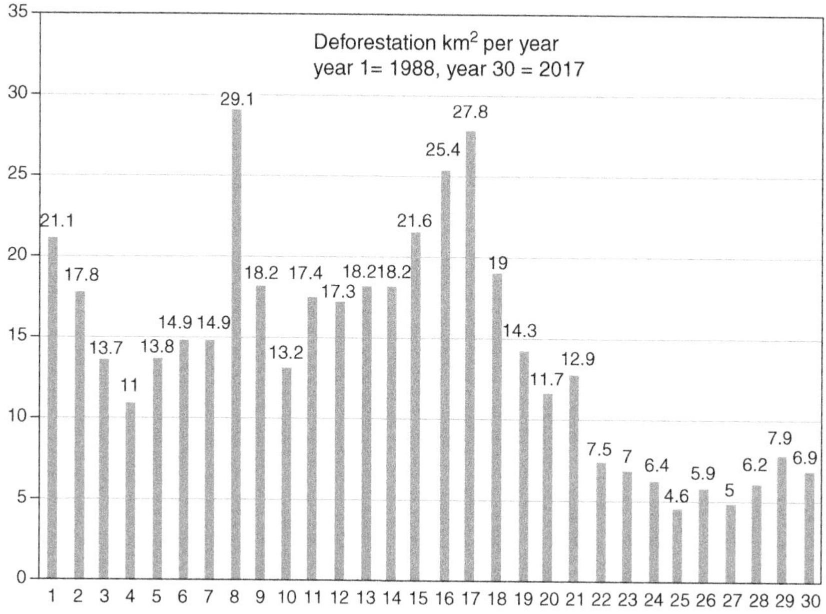

Figure 8.1 Annual deforestation rate in the Legal Amazon, 1988–2017.

Source: based on data from INPE Instituto Nacional de Pesquisas Espaciais (National Institute for Space Research), Brazil 2018.

will. Technologies include satellite imagery, GPS tracking and remote sensing. Combined with a digital rural cadastre registry to track land ownership, the rate of deforestation has significantly slowed (Chomitz 2015). Figure 8.1 shows the significant decline since 2004 (year 17) in the Legal Amazon,[17] when 27,800 square kilometres were lost. However, there continues a net loss of forest cover such that 6,900 square kilometres were lost in 2017 (year 30).

These governance examples depend on geo-spatial and digital technologies. Yet the same enthusiasm is not apparent in applying the concept to technologies themselves. Agricultural technologies used in the Amazon for example, include synthetic fertilizers, herbicides and pesticides, as in most farming areas. Despite risks to wildlife, workers and food consumers, they are largely uncontrolled, unsustainable and based on access to cheap petroleum (Jordan 2001, pp. 161–163).

As its name suggests, one of the characteristics of this form of governance is participation – a particular virtue in areas of complexity and potential controversy. The network structure also provides both a resilience and redundancy safeguard in that capacity weaknesses in some organisations or individuals are more likely to be compensated by the capacities of other stakeholders.

Transparency is also a characteristic because network governance, which necessarily involves many participants, cannot function without it. Integrity, too, is reinforced because such an approach tends to reflect the interests and views of the many, rather than the few and its actions are more visible to stakeholders. The remaining *TAPIC* characteristic, accountability, may be diluted by the numbers involved; when everyone is accountable, no-one is accountable. At the global level, accountability may also be weakened by distance from stakeholder constituencies. However, that risk is not unique to this form of governance. But while networks may not always be appropriate, they often work better "when there is some kind of authority in the background to give them legal force or oblige them to do their jobs well" (Greer et al. 2016, p. 15).

These three governance paradigms (state-led, market-based and network) help clarify issues of how technology can be regulated and their strengths and weaknesses against the TAPIC criteria. The state-led approach has several weaknesses, especially in transparency, accountability and capacity. Whereas the capacity of the market-led is improved through use of epistemic communities, its tendency to restrict data ownership weakens accountability and participation. Network approaches are yet undeveloped, but indicate strengths due to its wider participation and capacity reserves.

Fabric of global governance

While there are over 77,000 multinational firms operating in the global economy (Ruggie 2007, p. 823) and over 900 international environmental agreements in force (Biermann 2007, p. 335), there is little focus on technology and sustainability.

The WTO and the World Bank are part of the institutional framework of global environmental governance, as are several UN agencies. These especially include the UNEP, UNDP, UNESCO and the IAEA, as well as the IEA.[18] Only the two latter energy agencies relate directly to technology, if not to sustainability, whereas UNESCO mentions its carriage of technology transfer policy for development. The other two UN institutions mention 'sustainable development' or 'sustainability' as one of their objectives, but not technology at all. The Bank mentions a 'clean technology fund' among its functions, while WTO decisions can affect the environment indirectly.

Many global environmental interventions are modelled on the 1992 UN Framework Convention on Climate Change (UNFCCC), a pact between almost all countries that accepts commitments that are the responsibilities of individual states. The UNFCCC aims to prevent dangerous human interference with the climate system. It was developed by the International Panel on Climate Change (IPCC), itself formed by the World Meteorological Organisation (WMO) and the UNEP in 1988. The IPCC has an annual plenary of individual governments, a secretariat located within the WMO in Geneva, and several working groups with technical support on different aspects of the issue. The UNFCCC has three sister Conventions – on biological diversity, on desertification, and on wetlands (UNFCCC website 2017).

While global summits beginning with Stockholm in 1972 and especially including Rio in 1992 have aimed to drive consensus action for the environment, results have been mixed. Nevertheless, other governance initiatives have also been undertaken with more specific aims that involve technology. Born from frustration with the 2002 Johannesburg Earth Summit, the OECD set up the High Seas Task Force[19] to make practical proposals on the governance of illegal, unreported and unregulated (IUU) fishing. Its 2006 report, *Closing the Net,* recommended several important governance measures that involve technologies, such as strengthening the international monitoring, control and surveillance (MCS) network; establishing a global information system on high seas fishing vessel; and developing an electronic database of port state measures. The report further recommended measures aimed at wider stakeholder participation and developing model regulation.

One proposal prompted by the study is to close the high seas to fishing altogether supported by satellite surveillance technologies and an expanded automatic identification system (AIS) for shipping. Modelling indicates that this would not reduce the overall catch but would would improve equity between countries (Sumaila et al. 2015). But while all these measures involve various technologies to improve regulation, none of them address the regulation of fishing-related technologies themselves. For example, relatively simple measures to reduce by-catch could be mandated in the interests of biodiversity, yet are not.[20]

In 2001, all UN member states agreed to eight Millennium Development Goals (MDGs) for the period 2001–2015, including the last minute goal seven: *ensure environmental sustainability*. The 21 targets included 'reversing losses of environmental resources and biodiversity', but the MDGs tended to concentrate on measures such as poverty reduction, health and education.

However sustainability measures took more priority for the following years to 2030, when a set of 17 Sustainable Development Goals (SDGs) and 169 targets (Clark 2014) were adopted – more than twice as many goals, and eight times as many targets than the MDGs. Consistent with an analysis that helped shape the process, the much wider consultations undertaken for the SDGs were probably a factor in this conglomeration of objectives (Rippin 2013). However, another recommendation from the analysis was the need for 'focus' (ibid.), which seems to have been ignored.[21] To overcome this lack, Secretary-General Ban Ki-moon's 2014 report to the General Assembly framed the SDGs in six themes: dignity, people, prosperity, planet, justice, and partnership for sustainable development (UNGA 2014).

Themes #4 and #6 are particularly relevant to sustainability as defined in this book. But while this thematic overlay was not entirely successful at making the package clear, the same report did make important points about sustainability and technology:

- Access to new technologies will be crucial to a sustainable world (s.118).
- Environmentally sound technologies are unevenly spread, both within and between countries. The poor in many developing countries are "locked out" from access (s.119).
- We must phase out unsustainable technologies (s.120).
- The UN will build an online platform to map technology needs and initiatives in sustainable development areas such as agriculture, cities and health, to promote networking and technical assistance and to scale up the application of clean technologies (s.125).
- Member states should build a "technology bank" for least developed countries (LDCs); global intellectual property regimes including TRIPS should contribute to sustainable development; public resources should shift from harmful technologies towards the sustainable development goals; and the innovation-to-market-to-public good cycle of clean technologies should be accelerated (s.126).

While these points do address technologies directly, much so far has proven more aspirational than concrete. Progress has been slow in the construction of the technology platform, for example. While the MDGs were developed by a handful of people in a basement, the SDGs by contrast appear to have suffered from too much participation.

The 2015 Paris Agreement, however, does have clear focus. The *COP21 UN Climate Change Conference*[22] aimed to achieve a legally binding and universal agreement on climate changing emissions from all nations. Negotiations were led by French Foreign Minister, Laurent Fabius, whose experience helped make for a positive outcome.[23] The equivalent conference in Copenhagen in 2009 had foundered on the same reef that Brundtland had tried so deftly to avoid thirty years before – the issue of current versus future equity, or "progress versus the planet" (Charlton 2011). In Paris, not only was consensus reached on the draft agreement, but most countries committed to specific measures before the meeting. The biggest emitters all pledged substantial emissions cuts,[24] although the US pledge was since signalled to be revoked under the Trump administration in 2017.

Commitments under the agreement aim to limit emissions so that global warming will be no more than two degrees Celsius above pre-industrial levels and desirably less than 1.5 degrees. Targets are to be reviewed every five years. Developed countries agreed to fund $100 billion a year for developing country investment in clean technologies and adapt infrastructure to minimise damage from climate change. A record 168 countries signed on day one, and 14 ratified immediately.[25] The agreement entered into force in November 2016 (Jeyaratnam, Whitmore and Hopkin 2016). Much of these emissions reductions depend on emissions reductions in electricity generation, transport and agriculture. Fossil-fuelled technologies are under increasing pressure as a result. Emissions trading schemes, carbon taxes and renewable energy certificates are some of the methods now used to ensure that more sustainable technologies are deployed.

In summary, the many environmental institutions and treaties tend not to address technology and sustainability as linked issues. Even the High Seas Task Force, which is essentially all about using technologies to limit unsustainable catch, fails to address technologies directly in the same aim. For the MDGs, sustainability was an afterthought, while the SDGs are a form of global governance that still lack a genuine focus on sustainability, with results yet unclear.

Against the *TAPIC* criteria, the High Seas Task Force recommendations are consistent with each element of the framework, although much depends on the capacity of states to implement and its accountabilities are vague. The MDGs were relatively successful despite the extremely narrow participation in their formulation, while the SDGs effectively fudge transparency due to their complexity, possibly due to too much participation. In a related way, they also lack accountability, because responsibility for their implementation is split over several UN agencies. They do, however, at least attempt to grapple with technologies, with an effectiveness yet to be adjudged.[26] The Paris Agreement is a focused global pledge that puts the onus on individual states to progressively cut emissions by using more sustainable technologies. The major governance issue for Paris is singularly 'capacity', especially the political capacity of states to deliver.

Three case studies

The arrangements outlined in this chapter are all part of humanity's tango with technology. But to advance human control over the dance, it can help to know in some detail how key technologies were developed and are governed. While the first part of this book approached the notion of sustainability through the IPAT structure, this second part centres on a comparative case study approach. The following three chapters outline each case study separately: the herbicide glyphosate, electricity generation by nuclear reactor, and the development of robotics and artificial intelligence.

These particular case studies were chosen because they are all important applications of major technologies in key industries, including chemicals, agriculture, electricity generation, manufacturing, and information technology together with its wider applications across other sectors.

Glyphosate governance relates to potentially dangerous chemicals in general. The global chemical industry, built by German, British and US multinational corporations such as Bayer, Imperial Chemical Industries (ICI) and Dow is pervasive across the 'effortless industrialism' of the West and its technologies are part of almost all the processes of modernity. While the case study relates particularly to agriculture, glyphosate also pervades our homes, the streets upon which we walk and the parks in which our children play. It has done so only since the past few decades. It is the centre of considerable controversy and relates directly to sustainability, corporate behaviour and governance.

Nuclear power is a core example of managing the risks of hazardous waste as well as ultimate disaster. It is associated with the possibility of nuclear proliferation. It has been lauded as a reliable source of 'baseload power' that has few emissions compared with other means of power generation. Yet it also brings with it issues of location – as it needs geological stability, large amounts of water and somewhere to store waste, as well as, preferably, distance from major population centres. Nuclear plants are also expensive and take a long time to construct. Nonetheless, as the consumption of electricity increases, there are temptations for policy-makers to turn to nuclear to assume a larger share of generation technologies. This is despite forms of governance that have demonstrable flaws.

Much of robotics and artificial intelligence governance – or lack of it – relates to the oversight of all new technologies, especially those with potential for drastic societal and environmental impacts. Yet this set of technologies is at the brink of a new industrial era, a time long-promised, where inexorable logic, data and systems are combined to release humanity from drudgery so that people can benefit from the affluence of time and creativity. Or, perhaps, a darker future looms where dull machines roam the world beyond human control, or where artificial intelligences dwarf human capacity and resist the finger on their 'off' switch.

Many other technologies such as the steam engine, the Internet, or the compact fluorescent lightbulb might also have served as case studies. However, the steam engine is well-documented as a nineteenth-century platform and its relationship to sustainability is obvious and – except for electricity generation – uncontentious. The Internet is a profound and recent technology. However, its governance is largely focussed on issues of privacy rather than anything connected with sustainability, although it has considerable value in its capacity for collaboration and the dissemination of information and ideas. The compact fluorescent lightbulb produces more light from less electricity than the earlier incandescent technology, yet is a mid-step towards the still more sustainable light emitting diode (LED) form. It is interesting as an illustration of how it superseded incandescents yet itself was superseded so quickly by LEDs. Its governance is unremarkable and essentially relies on pure market forces.

The three chosen technologies – glyphosate, nuclear power and robotics and artificial intelligence – are all quite different in their underlying platforms, but all were commercialised after World War II. All are associated with the 'great acceleration' of human impact since. All three are contentious. All have profound environmental impacts, as well as influences on human society.

In assessing the qualities of their governance, in each case reference is made to the *TAPIC* dimensions, which cut across these examples. Other lessons are elaborated in the case study summary at Chapter 12 and support a preferred governance model in the 'alternative visions' at Chapter 13.

Notes

1 In the interest of variety and readability, synonyms such as 'regime', 'model' and 'form' are also used instead of 'paradigm' in this discussion.
2 US President Nixon ordered an end to the US biological weapons program in 1969.
3 Under the 1998 *Ottawa Treaty*, signatory countries promise not to manufacture, stockpile or use anti-personnel mines. As of 2015, it has been signed by 162 countries, notably not including China, Russia or the United States.
4 Including carpet bombing Vietnam, Laos and Cambodia during the 1960s and 1970s, and the invasions of Iraq and Afghanistan from 2003.
5 Despite trebling her population from about 400 million in 1961 to more than 1300 million in 2015, India's GDP per person rose five-fold in the same period (World Bank 2016).
6 China's assertion of territorial rights over the reconstructed Spratly Islands may be motivated by the decline of fishing resources closer to its coast.
7 France has 56 nuclear power plants due to a state development program originating in 1945 and accelerated in 1973 due to the OPEC crisis. *EDF* controls plants throughout Europe and the Americas.
8 TRIPS also protects other copyrights, trademarks, regional labels (such as wine origins), industrial designs (such as on textiles), and undisclosed commercial information.
9 As well as new cultural works.
10 Thickets: a series of patents surrounding related inventions that can trigger litigation and act to prevent their development.
11 Trolls: Those who derive profit from litigious action on patents rather than from the invention patented.
12 Biologic medicines are made from living organisms.
13 Such drugs can cost more than US$100,000 per patient per year.
14 Safety measures varied from open benches for no risk, to flow hoods, to airlocks and to negative pressure, as well as to special facilities for high risk containment.
15 There are over 500 standards based on ILO Conventions and national labour law.
16 'Lula': Luiz Inácio Lula da Silva, President of Brazil 2003–2011.
17 *Amazônia Legal* or 'Legal Amazon' is Brazil's largest socio-geographic division, containing all nine states of the Amazon basin. The region was created in 1948 and has a population of about 24 million.
18 The IEA is an agency of the OECD.

19 The *Task Force* was a ministerial-level group of several maritime nations including Australia, Canada, Chile, Namibia, New Zealand and the United Kingdom, with the WWF, the IUCN and the Earth Institute.

20 These include bright lights mounted at the front of prawn nets, devices that emit audible signals to warn dolphins and whales away from nets (GWA 2011), and line weighting and streamer lines to exclude seabirds from longlines. The report does refer to the UN Fish Stocks Agreement, which encourages measures to mitigate by-catch, although they are "far from fully realised" (Balton and Koehler 2006, pp. 8–9).

21 Although the UN's central agency responsible for the SDGs, the UNDP, did at the same time produce its Strategic plan 2014–2017 that centred on seven sets of outcomes, compared to the previous 35 (UNDP 2014a).

22 This was the 21st yearly session of the Conference of the Parties (COP21) to the United Nations Framework Convention on Climate Change (UNFCCC) and the 11th session of the Meeting of the Parties (CMP11) to the 1997 Kyoto Protocol.

23 Fabius was France's youngest ever Prime Minister under Socialist President Mitterand (1984–1986) and Foreign Minister under Socialist President Hollande (2012–2016).

24 The European Union pledged to cut emissions by 40 percent compared with 1990 levels by 2030. China vowed its emissions would peak by 2030. The United States pledged to cut its emissions by 26–28 percent compared with 2005 levels by 2025.

25 Of the 14 countries that ratified immediately, 12 were small island states vulnerable to sea level rise. Belize, which has the world's second longest barrier reef was one, and the other was landlocked Switzerland, which was concerned for its glaciers and snowfields, and for its own financial security.

26 SDG goal 14, 'Life below water', does pick up some of the High Seas Task Force concerns, as well as regulation of destructive fishing practices, at target 4.

References

Abbott, K and Snidal, D 2009, 'Strengthening International Regulation through Transnational New Governance: Overcoming the Orchestration Deficit', *Vanderbilt Journal of Transnational Law*, vol. 42, pp. 501–577.

Adams, W 2006, *The Future of Sustainability: Re-Thinking Environment and Development in the Twenty-First Century*, International Union for the Conservation of Nature (IUCN), Gland, Switzerland.

Alta Charo, R and Hynes, R 2017, 'Evolving Policy with Science', *Science*, vol. 355, no. 6328, 3 March, p. 889. doi: 10.1126/science.aan0509.

Berendzen, J, 2017, 'Max Horkheimer', in Zalta, E N (ed.), *The Stanford Encyclopedia of Philosophy,* Fall 2017 edition.

Berg, P 2008, 'Meetings that Changed the World: Asilomar 1975: DNA Modification Secured', *Nature*, vol. 455, 18 September, pp. 290–291. doi: 10.1038/455290a.

Bevir, M and Rhodes, R 2004 'Interpreting British Governance', *British Journal of Politics and International Relations*, vol. 6, pp. 130–136.

Biermann, F 2007, 'Earth System Governance as a Crosscutting Theme of Global Change Research', *Global Environmental Change*, vol. 17, pp. 326–337.

Biswas, S 2014, 'Will India's Narendra Modi Be a Reformer?', *BBC News India*, 26 May.

Bolwell, D 2015, *To the Lighthouse: Towards a Global Minimum Wage – Lessons from the Better Work Program*, Research Gate, GmbH, Berlin, Germany.

Boyle, K, Kotchen, M and Smith, V 2017, 'Deciphering Dueling Analyses of Clean Water Regulations', *Science*, vol. 358, no. 6359, pp. 49–50. doi: 10.1126/science.aap8023.

Buckley, T 2014, 'No Market for Galilee Coal, Not Even in Energy Poor India', *Renew Economy*, 6 May.

Charlton, A 2011, 'Man-Made World: Choosing between Progress and Planet', *Quarterly Essay*, vol. 44, pp. 1–72.

Chomitz, K 2015, *Can Tradeable Forest Conservation Obligations Enhance Economic, Biodiversity, and Carbon Outcomes?* Seminar, 15 October, Institute for the Study of Social Change and School of Social Sciences, University of Tasmania, Hobart, Australia.

Clark, H 2014, *Lecture on The Future We Want – Can We Make it a Reality?* At the Dag Hammarskjold Foundation, Uppsala, Sweden, 4 November.

Coan R 2012, 'What's the Theory Behind Innovation and the Knowledge-Based Economy', *Journal of Applied Research in Economic Development*, vol. 9, August.

Cochran, T 1995, *Reducing and Controlling the U.S. and Russian Nuclear Weapon Arsenals*, Discussion paper for Panel Discussion at the National Assembly on the United States and the United Nations, 31 August, Washington, DC, US.

Cortright, J 2001, New Growth Theory, Technology and Learning: A Practitioners Guide, Reviews of Economic Development Literature and Practice: No. 4, Impresa, Portland, US.

Daly, H 1990, 'Toward Some Operational Principles of Sustainable Development', *Ecological Economics*, vol. 2, pp. 1–6.

Drahos, P and Braithwaite, J 2002, *Information Feudalism*, Earthscan, London, UK.

Dresner, S 2008, *The Principles of Sustainability*, second edition, Earthscan, London, UK.

Eisenhower, D 1961, *Military-Industrial Complex Speech*, Public Papers of the Presidents, Dwight D. Eisenhower, 1960, pp. 1035–1040.

Engels, F 1940 [1876], *The Dialectics of Nature*, International Publishers, New York, US.

EU: European Union 2001, 'European Governance – A White Paper', *Communication from the Commission of 25 July*, [COM (2001) 428 final – Official Journal C 287 of 12.10.2001].

EU: European Union 2015, *A Guide to Multi-Level Governance for Local and Regional Public Authorities*, Coopenergy Consortium.

Forest, P-G, Denis, J-L, Brown, L and Helms, D 2015, 'Health Reform Requires Policy Capacity', *International Journal of Health Policy and Management*, vol. 4, no. 5, pp. 265–266.

Friedman, M 1982 [1962], *Capitalism and Freedom*, University of Chicago Press, Chicago, US and London, UK.

Fromm, E 1941, *Escape from Freedom*, iBook, Open Road Integrated Media, New York, US (also published elsewhere as *The Fear of Freedom*).

Fukuyama, F 1989, 'The End of History?' *The National Interest*, Summer.

Fung, A 2006, 'Varieties of Participation in Complex Governance', *Public Administration Review*, vol. 66, pp. 66–75.

Gibbs, H, Rausch, L, Munger, J, Schelly, I, Morton, D, Noojipady, P, Soares-Filho, B, Barretto, P, Micol, L and Walker, N 2015, 'Brazil's Soy Moratorium', *Science*, vol. 347, no. 6220, 23 January, pp. 377–378.

Gleeson, D and Lopert, R 2015, 'How the Battle over Biologics Helped Stall the Trans Pacific Partnership', *The Conversation*, 6 August.

Graham, J, Amos, B and Plumptre, T 2003, *Principles for Good Governance in the 21st Century*, Policy brief No.15, Institute on Governance, Ottawa, Canada.

Greer, G 1985, *Sex and Destiny: The Politics of Human Fertility*, Picador, London, UK.

Greer, S, Wismar, M and Figueras, J (eds) 2016, *Strengthening Health System Governance: Better Policies, Stronger Performance*, European Observatory on Health Systems and Policies, McGraw Hill Open University Press, Maidenhead, UK.

Greer, S, Wismar, M, Figueras, J and Vasev, N 2016, 'Policy Lessons for Health Governance', in Greer, Wismar and Figueras (eds) 2016, *Strengthening Health System Governance: Better Policies, Stronger Performance,* European Observatory on Health Systems and Policies, McGraw Hill Open University Press, Maidenhead, UK, pp. 105–125.

GWA: Government of Western Australia 2011, *Bycatch*, Fisheries Fact Sheet, June, Department of Fisheries, Perth, Australia.

Haas, P 1992, 'Epistemic Communities and International Policy Coordination', *International Organization*, vol. 46, no. 1, Winter, pp. 1–35.

Hamilton, C 2013, *What Would Heidegger Say About Geo-Engineering?*, 11 September.

Harvey, D 2014, *Seventeen Contradictions and the End of Capitalism*, Oxford University Press, New York, US.

Hegel, G 2001 [1820], *Philosophy of Right*, Dyde, S W (trans.), Batoche Books, Kitchener, Canada.

Heidegger, M 1953, 'The Question Concerning Technology', in Krell, D (ed.), *Martin Heidegger: Basic Writings*, Harper, San Francisco, US, pp. 308–341.

Heidegger, M 1976, 'Nur noch ein Gott kann uns retten' (Only a God Can Save Us), interview of 23 September 1966, Richardson, W (trans.), *Der Spiegel*, vol. 30, Mai, pp. 193–219.

High Seas Task Force 2006, *Closing the Net: Stopping Illegal Fishing on the High Seas*, Summary recommendations, Governments of Australia, Canada, Chile, Namibia, New Zealand, and the United Kingdom, WWF, IUCN and the Earth Institute at Columbia University, published by the Department for Environment, Food and Rural Affairs and the Department for International Development, London, UK.

IREDA: Indian Renewable Energy Development Agency Ltd, 2016 *Compendium of State Government Policies on Renewable Energy Sector in India*, Issued up to June 2016.

Jenkins, J, Swezey, D and Borofsky, Y (Eds) 2010, *Where Good Technologies Come From: Case Studies in American Innovation*, Breakthrough Institute, Oakland, US.

Jeyaratnam, E, Whitmore, J, and Hopkin, M 2016, 'The Paris Agreement Signing Ceremony at a Glance', *The Conversation*, 22 April.

Jordan, C 2001, 'The Interface between Economics and Nutrient Cycling in Amazon Land Development, in McClain, M, Victoria, R and Richey, J (eds), *The Biogeochemistry of the Amazon Basin*, Oxford University Press, New York, US, pp. 156–164.

Kane, A 2014, *Interview for the first International Day for the Total Elimination of Nuclear Weapons*, 13 September, UNFOLD Zero.

Keane, J 2015, 'Why Google Is a Political Matter', *The Monthly*, vol. 112, June, pp. 24–33.

Mallaby, S 2016, 'The Cult of the Expert – and How It Collapsed', *The Guardian*, 20 October.

Marcuse, H 1964, *One-Dimensional Man: Studies in the Ideology of Advanced Industrial Society*, Beacon Press, Boston, US.

Marks, P 2015, 'Eureka Machines', *New Scientist*, vol. 227, no. 2036, 29 August, pp. 32–35.

Mason, P 2015a, 'The End of Capitalism Has Begun', *The Guardian*, 17 July.

McWilliams, A, Siegel, D, and Wright, P 2005, *Corporate Social Responsibility: Strategic Implications*, Working Paper in Economics no. 0506, Department of Economics, Rensselaer Polytechnic Institute, New York, US.

Meadows, D H, 1998, *Indicators and Information Systems for Sustainable Development*, Report to the Balaton Group, The Sustainability Institute, Hartland, US.

Mercurio, B 2014, *TRIPs, Patents and Innovation: A Necessary Re-appraisal?* E15 Initiative, International Centre for Trade and Sustainable Development (ICTSD) and World Economic Forum, Geneva, Switzerland.

Mgbeoji, I 2003, 'The Juridical Origins of the International Patent System: Towards a Historiography of the Role of Patents in Industrialisation', *Journal of the History of International Law*, vol. 5, pp. 403–422.

Offer, A 2006, *The Challenge of Affluence: Self-Control and Well-Being in the United States and Britain since 1950*, Oxford University Press, Oxford, UK.

Parkinson, G, and Gilding, P 2014, *Energy Futures: Reaching the Clean Energy Tipping Point*, Seminar, University of Tasmania, Hobart, Australia, 27 May.

Peon, B 1995, 'Is Hinchcliffe's Rule True?', *New Scientist*, 16 August 2014.

Pierre, J and Peters, B 2005, *Governing Complex Societies: Ttrajectories and Scenarios*, Palgrave Macmillan, Hampshire, UK.

Pope Francis 2015, *Laudato si'*, Libreria Editrice Vaticana, 24 May.

Porter, A 1995, 'Technology Assessment', *Impact Assessment*, vol. 13, no. 2, pp. 135–151. doi: 10.1080/07349165.1995.9726087.

Rippin, N 2013, *Progress, Prospects and Lessons from the MDGs*, Background research paper, submitted to the [UN] High Level Panel on the Post-2015 Development Agenda, May.

Ruggie, J 2007, 'Business and Human Rights: The Evolving International Agenda', *American Journal of International Law*, vol. 101, no.4, pp. 819–840.

Russell, B 1979 [1946], *History of Western Philosophy*, Allen and Unwin, London, UK.

Schecter, D 2013, *Critical Theory in the Twenty-First Century*, Bloomsbury, New York, US.

Sichelman T and Veneri T n.d., *A Proper Interpretation of the Venetian Patent Act of 1474*, abstract only.

Singh, R, Murty, H, Gupta, S and Dikshit, A 2012, 'An Overview of Sustainability Assessment Methodologies', *Ecological Indicators*, vol. 15, no. 1, April, pp. 281–299.

Sørenson, E and Torfing, J (eds) 2007, *Theories of Democratic Network Governance*, Palgrave Macmillan, Basingstoke, UK.

Stiglitz, J 2001, 'Foreword to Polanyi K 2001 [1944]', in *The Great Transformation*, Beacon Press, Boston, US, pp. vii–xvii.

Sumaila, U, Lam, V, Miller, D, Teh, L, Watson, R, Zeller, D, Cheung, W, Côté, I, Rogers, A, Roberts, C, Sala, E and Pauly, D 2015, 'Winners and Losers in a World Where the High Seas Is Closed to Fishing', *Nature, Scientific Reports* vol. 5, 8481. www.nature.com/articles/srep08481. doi:10.1038/srep08481.

UNDP: United Nations Development Program 1997, 'Characteristics of Good Governance', *Governance for Sustainable Human Development*, New York, US.

UNDP: United Nations Development Program 2009, *Supporting Capacity Development: The UNDP Approach*, Bureau for Development Policy, New York, US.

UNDP: United Nations Development Program 2014a, *Changing with the World: UNDP Strategic Plan 2014-2017*, New York, US.

UNFCCC: United Nations Framework Convention on Climate Change 2017, website.

UNGA: United Nations General Assembly 2014, *The Road to Dignity by 2030: Ending Poverty, Transforming All Lives and Protecting The Planet*, Synthesis report of the Secretary-General on the post-2015 sustainable development agenda, Sixty-ninth session, Agenda items 13 (a) and 115, 4 December, New York, US.

US Congress Office of Technology Assessment 1994, *Studies of the Environmental Costs of Electricity*, OTA–ETI–134, September, Government Printing Office, Washington, DC, US.

USGAO: United States Government Accountability Office 2012, *Actions Needed to Evaluate the Impact of Efforts to Estimate Costs of Reports and Studies,* GAO-12-480R, 10 May.

USGAO: United States Government Accountability Office 2015, *Technology Assessment, Water in the Energy Sector: Reducing Freshwater Use in Hydraulic Fracturing and Thermoelectric Power Plant Cooling*, GAO-15–545, August, Center for Science, Technology, and Engineering, Report to Congressional Requesters.

USPTO: US Patent and Trademarks Office 2016, website, August.

Vaidhyanathan, S 2018, *Anti-Social Media: How Facebook Disconnects Us and Undermines Democracy*, Oxford University Press, New York, US.

Von Schomberg, R 2011, 'Prospects for Technology Assessment in a Framework of Responsible Research and Innovation' in Dusseldorp, M. and Beecroft, R. (eds), *Technikfolgen abschätzen lehren: Bildungspotenziale transdisziplinärer Methoden*, Vs Verlag, Wiesbaden, Germany.

Wadhwa, V 2014 'Laws and Ethics Can't Keep Pace with Technology', *MIT Technology Review*, 15 April.

Weiss, T 2000, 'Governance, Good Governance and Global Governance: Conceptual and Actual Challenges', *Third World Quarterly*, vol. 21, no. 5, pp. 795–814. doi: 10.1080/713701075.

WIPO: World Intellectual Property Organisation n.d., *What Is Intellectual Property?*, Publication No. 450(E), Geneva, Switzerland.

WIPO: World Intellectual Property Organisation 2016, *Summary of the Paris Convention for the Protection of Industrial Property* (1883), Geneva, Switzerland.

World Bank 2016, *Poverty and Shared Prosperity 2016: Taking on Inequality,* World Bank, Washington, DC, US.

World Bank 2017b, *The Sunken Billions Revisited: Progress and Challenges in Global Marine Fisheries*, World Bank Environment and Sustainable Development Series, Washington, DC, US.. doi: 10.1596/978-1-4648-0919-4.

9 Case study: Glyphosate

All substances are poisons; there is none which is not a poison. The right dose differentiates a poison from a remedy

— Paracelsus 1567

Glyphosate is the world's most used herbicide, designed and used as a poison. As such it warrants close scrutiny and effective governance. But it is also associated with powerful multinationals, with the military and with persistent controversy. Its governance involves more than just matters of fact, and it illustrates how chemicals in general are governed.

Paracelsus' observation that 'the dose makes the poison' remains fundamental to the science of toxicology. And the dose of glyphosate required to produce toxic effects in mammals is extremely high, considerably higher than substances such as caffeine and common table salt, for example. Therefore, in line with numerous research findings, it is safe to use in agriculture, especially at safety margins of many times less than toxic levels. Yet despite these reports there persists a counterbalance of investigations that identify reasons for caution, especially in its effects on non-mammals. The development of this substance illustrates how the neoliberal paradigm of patent protection, takeover and commercialisation evolves. It also shows how its governance is entangled with science, economics and politics. As a result, its status is contested along several axes.

The military

The close relationship between institutions of the military, chemical corporations and regulatory bodies is illustrated in the development of herbicides during the twentieth century. The broadleaf herbicide, 2,4 dichlorophenoxy-acetic acid (2,4-D) was developed independently by both British and US researchers during World War II. Its purpose was to starve enemy populations through destruction of their staple crops – potatoes in Germany and rice in Japan – as well as the destruction of "food supplies of dissident tribes" within the British empire (Perera and Thomas 1985, pp. 34–35). Whilst fortunately 2,4-D was not then cost-effective, according to an early research report it was remarkably effective in killing broadleaf plants:

> [2,4-D] at a concentration of 1,000 ppm in water was applied as a spray to two 100-foot rows of apple nursery stock infested with bindweed. ... The sprayed plants showed change within a few hours following application.
>
> (Hamner and Tukey 1944, p. 155)

Before this herbicide was developed, the control of weeds throughout history was almost entirely a matter of mechanics. Weeds were covered, pulled, ploughed, hoed or otherwise removed from the ground, largely by hand or with the animal-powered technologies.[1] Their removal by chemical spray therefore was a major leap for technology and agricultural productivity.

A similar compound, 2,4,5 trichlorophenoxy-acetic acid (2,4,5-T), was also found to be likely "effective as an herbicide" at around the same time (ibid.). Manufactured by US multinationals such as Dow and Monsanto, these chemicals were widely used as weed killers from the 1950s. Their military applications were also tested during this decade — by the British in East Africa and in the 'Malayan emergency' as a defoliant. As the mixture *Agent Orange,* they were sprayed by US aircraft over crops and more than 17 percent of Vietnam's forests in order to reduce food and cover for the North Vietnamese Army during the Vietnam War. UN resolutions that sought to outlaw the practice were vetoed by the United States and the United Kingdom, on the technical grounds that they were not direct military weapons as such. The UN General Assembly condemned the practice in 1969, but without apparent effect (Bach 2015).

Both these compounds kill broadleaf plants by overstimulating their growth processes. But the production of both substances is associated with dioxins, '*persistent organic pollutant*' (POPs), that contaminate the herbicides and accumulate in the bodies of animals. As top predators, humans are most at risk of suffering from their effects (WHO 2014a). These include skin lesions, as well as developmental and carcinogenic effects associated with prolonged exposure.

While both Dow and Monsanto benefited from US military contracts during the Vietnam War,[2] Monsanto in particular would have assumed military support for its herbicide development efforts during the 1940s and 1950s. Monsanto was directly involved in the nuclear bomb Manhattan project during the World War II through its research director Charles Thomas,[3] who declined an offer to co-direct the Manhattan project with Robert Oppenheimer in New Mexico, instead preferring to stay in Ohio, working on the linked 'Dayton project' to extract polonium used in the atomic bombing of Japan (Shook and Williams 1983). According to an Army report, the nuclear program continued with Monsanto at Dayton until the mid 1980s (USDoA 2004).

Restrictions

But as doubts grew about their effects on both people and ecosystems, these herbicides became subject to restriction and outright bans (especially on 2,4,5-T), including in the United States, Canada, Australia and parts of Europe and Asia from the 1980s onward (Cohen and Michelmore 2013). Whilst pursuing legal action in Canada against bans on 2,4-D as late as 2008 (McKenna 2011), Dow chemical finally gave up trying to defend 2,4,5-T in the 1980s after spending more than $10 million on US hearings, stating that the ban was due to "public concern and public misinformation" rather than scientific reasons (Holusha 1983). Public anxiety certainly was increased by the 1976 dioxin explosion in Seveso, Italy that resulted in a toxic cloud (Bertazzi et al. 1998), cases brought by Vietnam War veterans against manufacturers in the early 1980s citing serious health effects and the 1982 evacuation and permanent closure of the town of Times Beach, Missouri[4] due to dioxin contamination (Holusha 1983).

All this created an opportunity for a new, hopefully less dangerous herbicide to be developed and rolled out. Enter *glyphosate*, or '*Roundup*', as it is known commercially. A chemist who worked for the Swiss pharmaceutical company, Cilag, discovered glyphosate (N-phosphonomethyl glycine) in 1950. The discovery was not patented or reported in professional journals as it did not find any pharmaceutical application. The company was taken over by the US Johnson and Johnson corporation in 1959, which sold research samples of the compound to Stauffer chemicals and to Monsanto, amongst others. Stauffer first patented[5] glyphosate as a 'chelator', which removes minerals from solutions, in 1964. A Monsanto chemist identified the plant killing properties of glyphosate in 1970 and patented it as a herbicide in 1973.[6] It was first commercialised in 1974 (Dill et al. 2010).

Glyphosate works differently from the components of Agent Orange. While Agent Orange selectively destroys only broadleaf plants, glyphosate destroys all plants with which it comes into contact – except for plants that have been specifically engineered to resist it, or have become resistant to it as a result of natural mutation.

For the first few decades after commercial release, glyphosate spraying was limited to places where all vegetation was to be destroyed, such as between rows in orchards and vineyards, and along train tracks and power lines. But in 1996 Monsanto introduced *Roundup Ready* genetically modified crops into the United States and subsequently elsewhere. These crops were specially engineered to resist the effects of glyphosate and have since transformed the practice of agriculture throughout the world. Now, many genetically modified crops[7] are sprayed with glyphosate as a matter of course, negating the need for mechanical weed control, without the limitations of earlier broadleaf herbicides. Global agricultural use of glyphosate "mushroomed" following introduction of these herbicide tolerant crops. The total amount applied by farmers rose nearly 15-fold, from 51 million kilograms in 1995 to 747 million kilograms in 2014 (Benbrook 2016) and was used in more than 130 countries by 2008 (Dill et al. 2010). As of 2015, on average more than half a kilogram of the salt was sprayed over each hectare of cropland throughout the world (Benbrook 2016). Its normal application rate, however, is about one kilogram per hectare, and the increase in total use is mainly due to the increased area treated, so this is not especially concerning according to the eminent toxicologist Professor Keith Solomon.[8]

But not all of this vast tonnage of glyphosate represents profit for the company that patented it. Monsanto patents for glyphosate as a herbicide began to expire during the 1990s and many other companies began to produce it as a generic substance. Now less than a third of Monsanto's $15 billion yearly revenue comes from glyphosate herbicide products. Most revenue comes from the patented genetically modified crops that work in synergy with glyphosate's properties as a universal weed killer (Gillam 2016). As the patents on these crop seeds were to expire beginning in 2014, Monsanto since applied for a third patent on glyphosate as a treatment for microbial parasitic infections such as malaria. The patent was granted in 2010.[9]

Risks

This most recent 2010 patent has particular significance because much of the environmental risk of glyphosate turns on two concepts. One is the notion of its inhibition of an enzyme through the *shikimate pathway*,[10] found in plants, fungi and bacteria, but not found in animals, whereby it achieves its effect. Thus, in theory, glyphosate cannot be harmful to animals, including humans – except that all animals possess large numbers of gut bacteria that do have such enzymes and pathways. Indirect adverse effects on animals are therefore possible, based on the very property relied on for its latest patent.

The other concept is the 'surfactants' in the weed killing spray, which are not an active ingredient themselves, but are rather intended to enhance glyphosate's efficiency by helping it more easily penetrate foliage (Dill et al. 2010). While glyphosate itself may or may not be safe for animals and the environment, the presence of surfactant in some preparations "has been found to have greater toxicity to aquatic organisms than glyphosate (Folmar et al. 1979, cited in Dill et al. 2010, p. 10), although in field conditions such as the Ontario wetlands it fell "well below" toxic concentrations observed in laboratory conditions (ibid., p. 13). Although others warn that not only are the surfactants themselves toxic, but they also enhance the toxicity of the active ingredient (Antoniou et al. 2012).

While chemicals may be patented, in the US all pesticides such as glyphosate also must be registered by the Environmental Protection Agency (EPA) under the '*Federal Insecticide, Fungicide, and Rodenticide Act*' (US Congress 2012). Pesticides are registered providing there are no "unreasonable adverse effects on the environment" (including people) "taking into account the economic, social, and environmental costs and benefits" of its use. To support registration, applicants must provide research data in a form prescribed by the agency. If the Agency is satisfied following its own assessments, then the product may be distributed. However, the EPA assessment process is based on mathematical computer modelling, rather than its own direct investigation (EPA email of 15 September 2016), as the following rather arcane extract from one of its guides illustrates:

> The Surface Water Concentration Calculator (SWCC) estimates pesticide concentrations in water bodies that result from pesticide applications to land. The SWCC is designed to simulate the environmental concentration of a pesticide in the water column and sediment and is used for regulatory purposes by the USEPA Office of Pesticide Programs (OPP). The SWCC uses PRZM version 5.0+ (PRZM5) and the Variable Volume Water Body Model (VVWM), replacing the older PE5 shell (last updated November 2006), which used PRZM3 (Carousel et al., 2005) and EXAMS (Burns, 2003). This updated model will improve users' interactions with the program and facilitate maintenance and operation of the software.
>
> (Fry et al. 2014, p. 2)

As this example also shows, both applicant and EPA assessments are quantitative and decidedly technical, which tends to diminish transparency. It also opens the question of whether the system might be able to be gamed, as Volkswagen were able to evade EPA standards in the emissions scandal by using software that reacted to the conditions of testing (WVU 2015; Merkel 2015). Further, for 'commercial reasons', many of the Monsanto studies in support of glyphosate are not available for scrutiny by the broader scientific community or the public.

Agencies

The EPA was born from related areas of several agencies in 1970, and continues in "the extended shadow of Rachel Carson" (Lewis 1985). As the largest regulatory agency in the United States, the EPA is evidently a global leader in establishing chemical and other environmental impact standards, in cooperation with equivalents such as the European Environment Agency and global agencies such as the WHO and the FAO. The EPA can, however, be hoodwinked. In getting approval to sell a mixture of glyphosate and 2,4-D in 2011, Dow chemical told the EPA that the concoction was no more toxic than the two chemicals considered separately. It was later shown in court that Dow had told the Patent Office the opposite – that the combination of herbicides had a "synergistic effect" that increased its toxicity (Charles 2015).

Also, there have been different findings from different regulators as to tolerance. As it did with DDT under its founder head, William Ruckelshaus, the EPA has the power to ban undesirable chemicals. That it has not done so with glyphosate indicates that as yet it is satisfied that the substance is, on balance, beneficial. The EU, however, is less certain: its daily chronic reference dose for human exposure is set at 3.5 times lower than the EPA rate, and a level 17 times lower has been recommended (Benbrook 2016). China officially regards the substance as "harmless" with no tolerance level set (Chen 2014).

The differing tolerance standards of the United States and the European Union may be due to culture and economics rather than science. The United States, like Australia and Canada, has a broad-acre efficient approach to the farming of commodities. Whereas Europe tends to smaller scale agriculture based on tradition and the desirability of maintaining a village-scale demographic. Hence industrialised agriculture in North America is more tolerant of toxicity in the name of efficiency, while in Europe, where small-scale farming is directly subsidised, farmers and consumers are more sensitive to how food is produced. China's position is probably due to an authoritarian government most concerned with economic development. While its official position has been challenged by several hundred "Beijing food safety volunteers", greater transparency or tighter regulation seems unlikely as it would jeopardise China's position as the world's largest producer of the substance. Recently Chinese authorities refused a petition to release the research on which it based glyphosate registration, made in 1988.[11] In a familiar story, Monsanto provided the original (1985) toxicology for the registration process, but it could not be released because it contained "commercial secrets". The Chinese government said it would "seek permission" from Monsanto to release the research, so far without apparent result (ibid.).

Of the multi-lateral agencies, the WHO is concerned only with human health. In 2015 its cancer research agency found that glyphosate is "probably carcinogenic to humans" (Guyton et al. 2015, p. 491). However this was contested by the European Food Safety Authority meeting under German leadership a few months later, which drew the opposite conclusion, despite several "missing" pieces of information (EFSA 2015, pp. 1–3).

In its own response to the WHO verdict, Monsanto paid consultants to convene an "expert panel" similar in size to the WHO group that with some minor caveats found "the data do not support" the WHO finding. Monsanto then paid for the expert panel's conclusions to be published in a 'special supplement' of a toxicology journal (Williams et al. 2016, p. 16) and publicised the article. All but four of the Monsanto panel were former consultants or employees of the corporation (Monsanto 2015; Williams et al. 2016). The lead author, a former employee, had earlier published a review of glyphosate, which found that glyphosate is "practically non-toxic" (Williams, Kroes and Munro 2000, p. 117) and so was probably a safe choice for appointment to the panel. Another safe choice had publicly attacked the WHO finding before his appointment (Arnason 2015). According to email records, "Monsanto officials discussed and debated scientists who should be considered, and shaped the project" (Hakim 2017). Further, while all of the panel appear to be experts in their fields, most are consequently at a career stage where they are probably less sensitive about reputation – aged in their 60s and 70s with the word "emeritus" often appearing in their titles. While the article was peer-reviewed, Monsanto's involvement in the process appears less than exemplary.

Also in 2015, Monsanto filed a lawsuit against California's health hazard assessment agency for listing the substance as a carcinogen following the WHO finding (Plume 2016), which at the time of writing is unresolved. In a further contradiction, the EPA had subsequently found glyphosate to be not likely carcinogenic on preliminary review, with a final assessment scheduled for 2017 (Huffstutter 2016). But as part of the legal process, the federal court considering glyphosate's health effects unsealed internal Monsanto emails and other emails between Monsanto and the EPA. The Monsanto emails showed that an EPA deputy director had vowed "to beat back" an attempt by the Department of Health and Human Services to conduct its own glyphosate review. In another email a Monsanto executive told other company officials that they could hire academics to put their names on papers about glyphosate that were actually ghost written by Monsanto. The company has denied the practice (Hakim 2017).

With the FAO, the WHO has also published regulatory guidelines for states, especially developing countries, on the management of pesticides, which are not substance-specific (FAO-WHO 2014). However, the FAO-WHO *Codex Alimentarius*, does list individual pesticides and shows maximum limits for glyphosate residues in 31 different foodstuffs and fodder. One of its most important functions according to its own website is to limit tighter standards by importing countries that would form a restraint on trade: the *Codex* is an instrument under the WTO's free trade provisions. In effect, it puts a limit on standards higher than its own assessments. The range of food residues for glyphosate covered by the *Codex* is far narrower than the EPA (31 against 158), the classification descriptions are not consistent between the two and, as with dosage tolerances, food residue limits are different in several cases. In the *Codex*, meat from both mammals and poultry have a maximum glyphosate residue of 0.05 parts per million, whereas the equivalent EPA levels are 5.0 and 0.10.

Use

How glyphosate is used also differs. Toxicology studies typically refer to the mixing and use of glyphosate compounds "in accordance with the label" instructions. But as the compound is used in over 130 countries, many people involved may not even be able to read the instructions, let alone take care in its handling. Further, in industrialised countries, glyphosate is typically sprayed from a tractor towed boom. But North American farmers tend to be enclosed in air conditioned cabs, unlike farmers in South America, for example, who are typically open to the elements and are therefore exposed to glyphosate drift. These risks have been ignored in some regulatory studies, as have other adverse findings in Ecuador and Argentina (Antoniou 2012).

There are some further concerns. There is evidence that any research critical of glyphosate is subject to a coordinated academic censure: many of the critics are in the employ of Monsanto or benefit from its funding and therefore cannot be free of apprehended bias (Séralini et al. 2014). Also, one public relations counter to negative publicity is 'GMO Answers', a website that responds to individual concerns about genetically modified organisms and glyphosate. According to the site itself, "GMO Answers is funded by the members of the 'Council for Biotechnology Information'", which includes BASF, Bayer, Dow AgroSciences, DuPont, Monsanto and Syngenta. A random sample of answers to concerns in August 2016 indicated that most of the "qualified experts" answering were in fact employees of Monsanto or the Council.

Beyond the Anglosphere, where there is less leverage from such corporations, more critical studies of the effects of glyphosate have emerged. In Sweden in 2008, it was reported that glyphosate exposure is a risk factor for non-Hodgkin lymphoma. In Argentina there was controversy in 2009 after a scientist reported a high incidence of human birth defects and cancers in those living near crop-spraying areas, as well as genetic malformations in amphibians (Gammon 2009).

A particular example of how the science surrounding glyphosate is contested is the French molecular biologist, Gilles-Éric Séralini cited earlier, whose 2012 study of the effects of genetically modified corn that had been sprayed with glyphosate drew a wave of criticism, including from reviewers who were associated with Monsanto. Séralini's paper, which found that rats fed on the corn tended to develop tumours, was subsequently withdrawn by its original 2012 publisher following criticism of its methods. However, a revised version was re-published in different open-access journals two years later and reported in the prestigious *Nature*, which tended to increase publicity of its findings and the controversy it stirred (Casassus 2014).

Later biological research at the same institution[12] found major discrepancies between the official view of glyphosate and the reality of its formulation:

> Despite its reputation, Roundup was by far the most toxic among the herbicides and insecticides tested. This inconsistency between scientific fact and industrial claim may be attributed to huge economic interests, which

> have been found to falsify health risk assessments and delay health policy decisions.
>
> (Mesnage et al. 2014)

The Anglo-Brazilian paper cited above on the effect of glyphosate on embryos (Antoniou et al. 2012) found that German and EU approval of glyphosate levels was partly based on "unpublished industry-sponsored studies", and "minimised" earlier findings of malformed fetuses allowing potentially unsafe daily intakes of the substance. They further cite several studies showing a link between birth defects and miscarriage in Canada and Argentina from glyphosate spray exposure. Based on these sorts of studies, glyphosate is now banned in several countries — Norway, Denmark, Sweden, Sri Lanka, El Salvador, Brazil, and India. Unlike the US EPA registration system, in these countries there is significant distance between industry experts and government regulators who decide what is or is not toxic (Davis 2016).

By contrast, the EPA released a report on 29 June 2015 that concluded that glyphosate was not an endocrine disruptor (USEPA 2015). However, a popular critical review points out that much of the evidence considered was studies funded by Monsanto and other companies. Of the five independent studies considered, three were cautionary, but were outweighed by industry provided evidence (Strauss 2015). Interestingly, the EPA report does refer to one study that raised concerns about the adverse effects of a surfactant commonly used with glyphosate (USEPA 2015, p. 60), but this is not mentioned in the summary findings.

Overall, one of the most comprehensive reviews of the substance, Székács and Darvas' (2012) *'Forty Years with Glyphosate'*, which refers to almost 200 earlier studies, found that glyphosate formulations can increase fungal infection of soy crops, are toxic to amphibians and can produce tumours in both amphibians and birds at relatively low concentrations. According to the review, glyphosate can also cause malformations and developmental disorders in a range of species (ibid., pp. 262–266). Importantly, the article further puts a figure on Monsanto's unpublished studies: "of 180 research reports of Monsanto, 150 are not public, or have never been presented to the scientific community" (ibid., p. 264).

Cuhra, Bøhn and Cuhra's (2016) Norwegian review study concludes that glyphosate regulation should be tightened, not relaxed as has happened in the US, because the industry is too close to regulators, there is evidence of fraudulent practice and many research studies do not assess its effects in its commercial formulations. They found that more than 62,000 articles[13] on "glyphosate" were published between 1965 and 2014, with about 20,000 concerning "safety" or "risk'.[14]

Similar to the US, European corporations involved with glyphosate have a website (www.glyphosate.eu) that publishes positive articles and 'fact sheets' about the substance in English, German, Spanish and French. One such fact sheet[15] does admit that spray surfactants "could be harmful to aquatic organisms by impairing the integrity of cellular membranes". Reassuringly, it says that risks are "mild or negligible if glyphosate is used in accordance with label instructions and good agricultural practices". Less comforting is the label instruction for Roundup that it should not be used in or near freshwater to protect amphibians and other wildlife. Confusingly though, the Australian 11-page guide to Roundup for example, similarly says it should not be used near freshwater (Nufarm 2011) but also refers to its use "for weeds in aquatic situations" as long as "the entry of spray into water" is minimised (ibid., p. 7). On the one hand, it is not to be used in or near freshwater; on the other, it can be used to kill aquatic weeds. Both cannot be true.

Influence

It has been further suggested that the glyphosate industry influences government policy and safety regulations in favour of its products. Because glyphosate has been applied in increasing amounts only days before harvest especially in the North America, there are now much higher residues in the harvested crops than previously. After Monsanto and others requested substantial increases in glyphosate tolerance levels for these crops, typically they were granted. In the most extreme examples, alfalfa hay and silage tolerances have increased by 2000 times between 1993 and 2015, from 0.2 to 400 parts per million.

Large tolerance increases were also granted in Europe on application by Monsanto in 2012, after it was found that imported Canadian lentils showed high residual levels of glyphosate. The existing residue limit for glyphosate in lentils was then 0.1 mg/kg and the application recommended raising it to at least 10 mg/kg — a factor of at least 100. The EU acceded to the 10mg/kg limit in May of the same year (EU 2012, L 135/15; Cuhra, Bøhn and Cuhra 2016).

Despite increasing concerns, as late as 2008 one paper by leading plant biologists described glyphosate as "very toxicologically and environmentally safe" and a "virtually ideal herbicide" (Duke and Powles 2008, p. 319). Although eight years later the lead author was wary when asked if he still held the same views, instead referring the question to a toxicologist.[16] But irrespective of its toxicity, these authors caution that there are sustainability challenges in the very success of the world's number one herbicide because "glyphosate-resistant weeds will emerge" so that there needs to be less reliance on glyphosate and more on new chemical, "mechanical and precision application technologies" (ibid., p. 324). Professor Duke advised in 2016 that weed resistance "has grown much worse since our paper was published."

But as far as its toxic effects are concerned, he observes that "papers more critical of glyphosate than Séralini's … [are] mostly in predatory ('pay to play') journals", whereas there are many studies showing 'no effect' – despite the difficulty of getting a journal to publish a 'no effect' paper. Duke further comments that "unfortunately, this topic is politically charged, with many unqualified people taking sides. That is not the way science should be done".[17] This issue is one that lingers across its governance. For example, a 2015 article in a Canadian agricultural newspaper 'Toxicologist pans UN glyphosate report' drew more than 30 pages of online comments reflecting divergent views of the substance and its effects (Arnason 2015). Indeed, this is a matter of politics as much as science.

Bearing in mind the old joke about Russian roulette whereby five out of six scientists declare the practice safe, there are patterns apparent in the politics. On the basis of the information reviewed for this assessment, industry tends to support more tolerant glyphosate standards because it is good for profit, whereas civil society represented by NGOs is more concerned about health and the environment. Toxicologists, who concentrate on "active ingredients" and by-the-label instructions (and who are often employed by chemical companies) tend to support industry, whereas more independent entomologists tend to be concerned about its effects on insects, especially pollinators. Anglophones, who commercialised the substance and have large scale agriculture, tend to be more tolerant than say francophones, who are more suspicious of genetically modified crops and favour smaller scale intensive agriculture. The large producer states – the United States and China – tend to be more tolerant than the consumers of Europe and South America, while national environment agencies tend to be more tolerant of glyphosate than multilateral agencies of the UN and EU that are more removed from commercial pressures.

Agreements

Apart from the WHO and FAO institutional involvement, there are several related multilateral agreements relevant to the sustainability of chemical technologies. These include the Montreal Protocol (adopted 1987) banning ozone-depleting substances, the Basel Convention (1989) restricting the transboundary disposal of hazardous waste, the Rotterdam Convention (1998) restricting trade in hazardous chemicals and the Stockholm Convention (2001) that limits persistent organic pollutants, such as dioxins. There is also the Minamata Convention (2013) on the control of mercury. These Conventions typically involve a global secretariat, a committee of scientific experts and regular meetings of the signatories, each known as a 'conference of the parties'. Signatories can be states as well as regional groupings such as the European Union.

All of these measures resulted from major disasters, such as the emergence of an 'ozone hole' in the early 1980s, the dumping of thousands of tons of US industrial waste on a beach in Haiti in 1986 (Avril 2002), and thousands of deaths in Japan and Iraq between the 1950s and 1970s due to mercury poisoning. None, however, capture glyphosate within their net. Any future Convention involving glyphosate would only arise due to a major international calamity, which appears unlikely. It is therefore more probable that the present mixed voluntarist regime plus industrialised state regulation will continue.

'Responsible Care' is a form of voluntary private governance in this industry, which began in Canada in 1985 and now claims most of the world's chemical production. As such, it does at least theoretically encompass glyphosate. It is essentially a commitment to a program of continuous improvement in both workplace safety and environmental impact. Its leading safety indicators are the number of worker fatalities (about 40 per year globally) and the 'lost workday case rate' (about four days per year per 'case'). Its environmental indicators are emissions of various gases (NOx, SO_2, CO_2) and 'chemical oxygen demand', energy and water consumption, as well as 'distribution incidents', or spillage of dangerous chemicals (around 1800 cases per year).

Its origins were not entirely altruistic: "there is no question that the threat of regulation served as a major incentive for the development of the program" (Moffet, Bregha and Middelkoop 2004, p. 204). Also its indicators relate to only a few impacts of its technologies – the ones easy to measure; indicators of environmental accumulation and toxicity, for example, are missing. It lacks consumer involvement and its coverage is not universal. Yet it does link with other elements of global governance such as the UN Global Compact and the International Standard Organisation's 14000 series on environmental impact management.[18] *Responsible Care* also acknowledges the UN Sustainable Development Goals (2015–2030), especially goal 6 (clean water and sanitation) and goal 12 (responsible production and consumption) that target waste reduction and chemical pollution in general, especially as they affect water supplies – an area pertinent to glyphosate and its effects on amphibians.

The *Strategic Approach to International Chemicals Management* (SAICM) is the intergovernmental response to increasing chemical risk. It is a 'policy framework' that focuses directly on the chemical technologies concerned. Developed in 2006, it concedes one of the challenges is assessing the risks (and opportunities) "associated with more than 100,000 different chemicals". While participation is again voluntary, it encourages better regulation from its stakeholders, which include nearly 300 governments, NGOs and industry bodies (UNEP 2015, p. 5). Much of it is focussed on the developing world and it remains relatively underfunded.[19] The SAICM also continues to face serious obstacles. At its recent meeting in Brasilia, some stakeholders, argued for concentrating on the development of a solid chemical regulatory regime and building technical capacity. A proposal for a science-policy interface for chemicals and waste similar to the IPCC was rejected. Yet to be agreed too, is how to practically implement the strategy (ENB 2017).

At the US national level though, the professional body for the chemical industry has adopted a set of principles called 'green chemistry'. These principles focus on eliminating hazardous effects of chemicals by, for example, ensuring that all materials used in chemical processes are incorporated into the final product, by reducing energy requirements of chemical processes, and by ensuring that processes minimise waste (Anastas and Warner 1998; ACS 2017). The principles can be 'drilled down' for greater detail on the society's site. Their adoption followed several major chemical disasters and the resultant introduction of the US Pollution Prevent Act of 1990. Europe and the OECD were involved in complementary initiatives during the same era (ACS 2017). If implemented, such principles would mark a big step towards a sustainable chemical industry.

Weaknesses

On balance then, there are several weaknesses evident in the governance of glyphosate, all of which militate against sustainability. They also tend to undermine the credibility of the wider chemical industry, as do the dangers of substances including persistent organic pollutants (POPs) such as DDT, dieldrin and dioxins that are banned in many jurisdictions (Ritter et al. 1997).

First, the military-industrial relationship involved in its development undermines confidence because it is associated with secrecy, destruction and the use of force. Also US multinationals such as Monsanto have opposed unfavourable research and regulation concerning the herbicide. Monsanto has been particularly aggressive in attacking adverse findings and there is evidence that even pays academic toxicologists to publically support its cause. While its human toxicology nevertheless still appears relatively benign, evidence for glyphosate's effects on other life forms, especially aquatic animals, appears increasingly less favourable. Further, a technical approach to its control has resulted in acceptance of research based on a questionable adherence to instructions on the label. In any case, its instructions are contradictory in relation to its use around water. There is thus a participation deficit in its governance.

The continuing controversy over glyphosate and its increasing use world-wide are linked issues. Much of the research that supports it is secret, and mostly focuses on the active ingredient rather than its hundreds of commercial formulations. Those commercial applications vary in mixtures with other chemicals, and many such combinations remain untested. Its transparency is lacking.

Affecting both participation and transparency, with the exception of the SAICM, the breadth of stakeholder involvement in its governance is narrow, which must weaken both understanding and acceptance of the science involved. As with other chemicals, this can be exacerbated by the silo-like technical expertise about the substance.

There is no single program, process or institution that has global carriage of its regulation. Globally, its sustainability risks are dealt with through voluntarist strategies at state level, together with regulation in industrial countries. Leading regulators take different positions on the substance and its toxicity. While some of this may be due to cultural and economic factors, there is evidence that the rapport between industry and government may be rather too cosy in the United States and China. And both the WHO and FAO are focused on human health rather than biodiversity, but they also act to limit tighter safety standards in the cause of free trade in food. Despite claims of higher yield and less pesticide use through glyphosate and genetically modified crops, evidence is to the contrary (Hakim 2016). Sustainability is better where there is less glyphosate and no genetic modification. This issue is central yet still to be addressed by many regulators. Regulation of technologies in this industry is a patchwork rather than a coherent fabric. The problem of how these measures can be integrated remains largely unsolved. At this level, there is an absence of authority and thus a lack of accountability.

An internationally consistent glyphosate regulation would support better governance. A recent letter to stakeholders from the EPA director of pesticide programs points to a new 'twenty-first century approach' to pesticide testing and assessment (Housenger 2016), based on digital technologies. It is aimed at faster evaluations of more pesticides, using fewer research animals with lower costs. The approach emphasises international cooperation on data sets that relies more on models of established toxicity relationships rather than direct testing. As such it promises a more coherent international approach to regulation. Developing this sort of approach is discussed further in Chapter 13.

However, glyphosate governance now faces another challenge. The US$66 billion takeover of Monsanto by Bayer announced in September 2016 (Roumeliotis and Burger 2016) and completed less than two years later with the demise of the very name 'Monsanto' (Chow 2018) created a globally dominant chemical and seed conglomerate across agriculture as well as pharmaceuticals. Governance must now grapple with the immense leverage of a global enterprise that stands astride the traditional twin centers of the world chemical industry, in Germany and the United States. Integrity is further threatened.

Yet the October 2016 finding of the international Monsanto Tribunal indicates that the law can be used against any unsustainability threat posed by glyphosate as much as it has been to negate adverse findings. Five judges meeting in The Hague using the legal principles of the International Criminal Court, brought down a verdict that Monsanto is guilty of crimes against the planet and humanity due to the effects of glyphosate and genetically modified crops on the ecosystem. While ecocide is yet to be established as an international crime,[20] and Monsanto denounced the tribunal as a "kangaroo court", the presiding judge said that civil society could help that development in international law (Bamforth 2017).

Glyphosate is an example of a market-led approach to governance in which the state is supportive of private development and market access through trade; its support strengthened by strong military links to herbicide and nuclear weapons development. Its governance issues can apply to other chemical products. These issues include association with the military, commercialisation secrecy, incoherent global regulation, litigation that menaces dissenting research, and the political leverage of big corporate producers. Against the TAPIC governance criteria, transparency, accountability and participation are all sub-optimal due to secrecy and regulatory capture. Integrity is jeopardised by corporate influence. Capacity is sapped by incoherent standards, different institutional focus and lack of stakeholder engagement.

Counterweights that favour sustainability include the creation of legal categories that would thwart environmental damage and the use of digital technologies to speed more coherent testing through data analysis.

Finally, irrespective of whether it is possible to find Paracelsus' dosage 'sweet spot', glyphosate's sales success as a weedkiller points to the return of mechanical methods due to the problem of evolved resistance. Robotics look to be one precision alternative that has no harmful environmental effects and thus a sustainability advantage. This is further discussed in Chapter 11.

Notes

1 Although there were limited attempts at steam and even sulphur control in the 19th Century.
2 In fact, by 1967 supply to the US military accounted for the entire commercial production of 2,4-D and 2,4,5-T (Perera and Thomas 1985, p. 36).
3 Later Monsanto President between 1951 and 1960 and Board Chairman from 1960 to 1965.
4 Involving 800 families, or more than 2,000 residents. The area is now a ghost town, ironically used for high-temperature disposal of dangerous chemicals.
5 US patent no. 3,160,362.
6 US patent no. 3977860 A.
7 Such as wheat, soybeans, barley, legumes, corn, sunflower, kiwifruit, grapes, raspberries, apples, alfalfa and sugar cane.
8 Personal email of 22 August 2016.
9 US patent no. 7771736 B2.
10 Although exactly "how glyphosate-induced inhibition of the shikimate pathway actually kills plants is not entirely clear" (Duke and Powles 2008, p. 320).
11 Ministry of Agriculture registration number PD73-88.
12 The University of Caen, Normandy.
13 These were peer-reviewed scientific articles and related posts such as technical reports and patent documents found on Google scholar.
14 Similar results were obtained by the author of this book in October 2016.
15 *Glyphosate the environment and wildlife faqs*, p. 3.
16 Personal email of 20 August 2016.

17 Ibid.
18 The 14000 series are essentially sets of checkboxes that guide and evidence an enterprise's conformity with a standard of environmental management, rather than a set of environmental standards that must be met. It appears to be a more extensive approach than Responsible Care, is complementary to it and is likewise voluntary.
19 SAICM has "supported projects worth more than US$110 million in more than 100 developing countries" (UNEP 2015, p. 2).
20 Ecocide is nevertheless a crime in ten national jurisdictions, including Vietnam and Ukraine.

References

ACS: American Chemical Society 2017, *History of Green Chemistry*, webpage.

Anastas, P, and Warner, J 1998, *Green Chemistry: Theory and Practice*, Oxford University Press, New York, US.

Antoniou, M, Habib, M, Howard, C, Jennings, R, Leifert, C, Nodari R, Robinson C and Fagan, J 2012, 'Teratogenic Effects of Glyphosate-Based Herbicides: Divergence of Regulatory Decisions from Scientific Evidence', *Journal of Environmental and Analytical Toxicology*, S4. doi: 10.4172/2161-052.S4-0061.

Arnason, R 2015, 'Toxicologist Pans UN Glyphosate Report', The Western Producer, 27 March.

Avril, T 2002, 'Traveling Trash: Years Later, Long-Fought Ash Returning', *The Philadelphia Inquirer*, 15 June.

Bach, W 2015, *Agent Orange in the Indochina Wars*, Veterans for Peace UK, London, UK.

Bamforth, A 2017, 'Monsanto Is Probably Guilty of Ecocide and War Crimes Rules Global Tribunal', *Signs of the Times*, 23 April.

Bertazzi, P, Bernucci, I, Brambilla, G, Consonni, D and Pesatori, A 1998, 'The Seveso Studies on Early and Long-Term Effects of Dioxin Exposure: A Review', *Environmental Health Perspectives*, vol. 106, Supplement 2, April, pp. 625–633.

Benbrook, C 2016, 'Trends in Glyphosate Herbicide Use in the United States and Globally', *Environmental Sciences Europe*, vol. 28, no.3, December. doi: 10.1186/s12302-016-0070-0.

Casassus, B 2014, 'Paper Claiming GM Link with Tumours Republished', *Nature*, 24 June.

Charles, D 2015, 'Busted: EPA Discovers Dow Weedkiller Claim, Wants It Off the Market', *The Salt*, 25 November, National Public Radio, Washington DC, US.

Chen, A 2015 'Seeing into the Future: Does Philip Tetlock Hold the Key to Accurate Predictions?', *The Chronicle Review*, 5 October.

Chen, I-wan 2014, 'Chinese People Fight Back on Monsanto against Glyphosate-based Roundup', blog report, cheniwan@cei.gov.cn. http://blog.sina.com.cn/s/blog_4bb17e9d0102edk0.html.

Chow, L 2018, 'Bayer to Drop Monsanto Name After $63 Billion Takeover', *EcoWatch*, 4 June.

Cohen, J and Michelmore, K 2013, 'Chemical Time Bomb', *Four Corners*, Australian Broadcasting Corporation, 23 July.

Cuhra, M, Bøhn, T and Cuhra, P 2016, 'Glyphosate: Too Much of a Good Thing?', *Frontiers of Environmental Science*, 28 April.

Davis, D 2016, 'Potential Dangers of Glyphosate Weed Killers', Oxford University Press, blog, 19 March.

Dill, G, Sammons, R, Feng, P, Kohn, F, Kretzmer, K, Mehrsheikh, A, Bleeke, M, Honegger, J, Farmer, D, Wright, D and Haupfear, E 2010, 'Glyphosate: Discovery, Development, Applications, and Properties' in Nandula, V (ed.), *Glyphosate Resistance in Crops and Weeds: History, Development, and Management*, Wiley, Hoboken, US.
Duke, S and Powles, S 2008, 'Glyphosate: A Once-in-a-Century Herbicide', *Pest Management Science*, vol. 64, pp. 319–325.
EFSA: European Food Safety Authority 2015, 'Conclusion on the Peer Review of the Pesticide Risk Assessment of the Active Substance Glyphosate', *EFSA Journal*, vol. 13, no. 11, pp. 107. doi: 10.2903/j.efsa.2015.4302.
ENB: Earth Negotiations Bulletin 2017, 'First Meeting of the Intersessional Process for Considering SAICM and the Sound Management of Chemicals and Waste beyond 2020', vol. 15, no. 241, 7–9 February.
EU: Official Journal of the European Union 2012, *Commission Regulation (EU) No. 441/2012* of 24 May.
FAO: Food and Agriculture Organisation of the United Nations 2016, *International Code of Conduct on Pesticide Management Guidelines on Highly Hazardous Pesticides*, Rome, Italy.
FAO-WHO: Food and Agriculture Organization – World Health Organization of the United Nations 2014, *The International Code of Conduct on Pesticide Management*, Rome, Italy.
Fry, M, Milians, K, Young, D and Zhong, H 2014, *Surface Water Concentration Calculator User Manual*, USEPA/OPP 734F14001, Environmental Fate and Effects Division, Office of Pesticides, United States Environmental Protection Agency, Washington, DC.
Gammon, C 2009, 'Sustainability: Weed-Whacking Herbicide Proves Deadly to Human Cells', Environmental Health News, *Scientific American*, 23 June.
Gillam, C 2016, 'Conflict of Interest Concerns Cloud Glyphosate Review', *US Right to Know*, web post, 12 May.
Guyton, K, Loomis, D, Grosse, Y, El Ghissassi, F, Benbrahim-Tallaa, L, Guha, N, Chiara Scoccianti, C, Mattock, H and Straif, K 2015, 'Carcinogenicity of Tetrachlorvinphos, Parathion, Malathion, Diazinon, and Glyphosate, *Lancet Oncology*, vol. 16, pp. 490–491.
Hakim, D 2016, 'Doubts about the Promised Bounty of Genetically Modified Crops', *The New York Times*, 29 October.
Hakim, D 2017, 'Monsanto Weed Killer Roundup Faces New Doubts on Safety in Unsealed Documents', *New York Times*, 14 March.
Hamner, C and Tukey, H 1944, 'The Herbicidal Action of 2,4 Dichlorophenoxyacetic and 2,4,5 Trichlorophenoxyacetic Acid on Bindweed' *Science*, vol. 100, no. 2590, pp. 154–155.
Holusha, J 1983, 'Dow Halts Fight to Sell Herbicide', *New York Times*, 15 October.
Housenger, J 2016, *Letter to Stakeholders on EPA Office of Pesticide Program's Goal to Reduce Animal Testing from Jack E. Housenger, Director Office of Pesticide Programs*, US EPA document EPA-HQ-OPP-2016-0093-0003, 16 March.
Huffstutter, P 2016, 'EPA Says Glyphosate, Used in Monsanto Herbicide, Likely Not Carcinogenic', *Reuters*, 16 September.
Lewis, J 1985, 'The Birth of EPA', *EPA Journal*, November.
McKenna, B 2011, 'Controversial Pesticide 2,4-D Deemed Not 'Dangerous', But Still Banned in Quebec', *The Globe and Mail*, 26 May.
Merkel, R 2015, 'Where Were the Whistleblowers in the Volkswagen Emissions Scandal?', *The Conversation*, 30 September.
Mesnage, R, Defarge, N, Spiroux de Vendômois, J, and Séralini, G-E 2014, 'Major Pesticides Are More Toxic to Human Cells Than Their Declared Active Principles', *BioMed Research International*, vol. 2014, 179691. doi: 10.1155/2014/179691.

Moffet, J, Bregha, F and Middelkoop, M 2004, 'Responsible Care: A Case Study of a Voluntary Environmental Initiative', in Webb, K (ed.), *Voluntary Codes: Private Governance, the Public Interest and Innovation*, ch. 6, Carleton University, Ottawa, Canada, pp. 177–208.

Monsanto 2015, *2015 Glyphosate Expert Panel*, webpage, Monsanto.com.

Nufarm 2011, *Roundup Herbicide by Monsanto*, guide to use, pamphlet. www.herbiguide.com.au/Labels/GLY36_31393-54214.PDF.

Perera, J and Thomas, A 1985, 'This Horrible Natural Experiment', *New Scientist*, 18 April, pp. 34–36.

Plume, K 2016, 'Monsanto Sues to Keep Herbicide off California List of Carcinogens', *Reuters Business News*, 21 January. www.reuters.com/article/us-usa-monsanto-glyphosate-idUSKCN0UZ2RN.

Ritter, L, Solomon, K, Forget, J, Stemeroff, M and O'Leary, C 1997, *Persistent Organic Pollutants: An Assessment Report on DDT-Aldrin-Dieldrin-Endrin-Chlordane, Heptachlor-Hexachlorobenzene, Mirex-Toxaphene, Polychlorinated Biphenyls, Dioxins and Furans*, International Programme on Chemical Safety (IPCS) within the framework of the Inter-Organization Programme for the Sound Management of Chemicals (IOMC), Guelph, Canada.

Roumeliotis, G and Burger, L 2016, 'Bayer Clinches Monsanto with Improved $66 Billion Bid', *Reuters*, 15 September.

Séralini, G-E, Mesnage, R, Defarge, N and Spiroux de Vendômois, J 2014, 'Conflicts of Interests, Confidentiality and Censorship in Health Risk Assessment: The Example of an Herbicide and a GMO', *Environmental Sciences Europe*, vol. 26, no. 13.

Shook, H and Williams, J 1983, 'Building the Bomb in Oakwood', *Dayton Daily News*, 18 September.

Strauss, M 2015, 'New Documents Expose the EPA's Relationship with Monsanto', *Inverse*, 4 November.

Székács, A and Darvas, B 2012, 'Forty Years with Glyphosate', in Hasaneen, M (ed.), *Herbicides – Properties and Control of Weeds*, Intech Europe, Croatia and Intech, China, pp. 247–286.

UNEP: United Nations Environment Program 2015, *Strategic Approach to International Chemicals Management: Good Chemistry Together*, Geneva, Switzerland.

US Congress 2012, *Federal Insecticide, Fungicide, and Rodenticide Act*, 28 September, Washington, DC, US.

USDoA: United States Department of the Army 2004, *Combined Preliminary Assessment/Site Inspection Report Dayton Unit*, Contract No. DACW49-01-D-0001, Delivery order no. 0001, September, Prepared for Department of the Army Buffalo District, Corps of Engineers, New York, US, pp. 14207–3199.

USEPA: United States Environmental Protection Agency, Office of Chemical Safety and Pollution Prevention 2015, *EDSP Weight of Evidence Conclusions on the Tier 1 Screening Assays for the List 1 Chemicals: Analysis of Potential Interaction with the Estrogen, Androgen or Thyroid Pathways, Chemical: Glyphosate*, Memorandum 29 June, Washington, DC, US.

WHO: World Health Organisation 2014a, *Dioxins and Their Effects on Human Health*, Fact sheet N°225, Geneva, Switzerland, updated June.

Williams, G, Aardema, M, Acquavella, J, Berry, C, Brusick, D, Burns, M, Joao Viana de Camargo, J, Garabrant, D, Greim, H, Kier, L, Kirkland, D, Marsh, G, Solomon, K, Sorahan, T, Roberts A and Weed, D 2016, 'A Review of the Carcinogenic Potential of Glyphosate by Four Independent Expert Panels and Comparison to the IARC

Assessment', *Critical Reviews in Toxicology*, vol. 46, issue sup, 28 September, pp. 3–20. doi: 10.1080/10408444.2016.1214677.

Williams, G, Kroes, R and Munro, I 2000, 'Safety Evaluation and Risk Assessment of the Herbicide Roundup and Its Active Ingredient, Glyphosate, for Humans', *Regulatory Toxicology and Pharmacology*, vol. 31, pp. 117–165.

WVU: West Virginia University 2015, 'WVU Study Found Elevated Levels of Emissions from Volkswagen Vehicles', *WVU Today*, 24 September.

10 Case study: Nuclear-electric power

Generating electricity involves technologies that form a key platform of modernity, and which are linked to societal affluence (Garrett 2014). They combine significant benefits with big risks. The energy sector generates about two-thirds of global greenhouse gas emissions (IEA 2013), sourced mainly from fossil fuels. And about 41 percent of the energy sector's emissions come from electrical power generation (ibid.). While greenhouse gas emissions are by no means the only issue in electricity generation, they are the major cause of climate change. Yet, despite the 2015 Paris climate agreement that limits emissions, there remains a political-technological divide between those who are reluctant to replace fossil-fuelled electricity generation and others who want electricity generation from renewable sources progressed urgently. Both views might rely on Margaret Thatcher's famous dictum, TINA: *There is no alternative.*

One group argues that only coal-fired plants can provide viable base load power, while the other says that only renewables will prevent the overheating of the biosphere. But contrary to Thatcher's dictum, there is an alternative to these extremes. It is neither fossil-fuelled, nor does it use renewable energy. It provides continuous base-load power and it emits no greenhouse gases during operation. As do fossil-fuelled power plants, it uses turbines spun by steam pressure to turn the electric generators. The key difference is that the heat is supplied not by coal but by nuclear fission.

Nuclear-electric power is favoured by '*ecomodernists*', who have been described as the environmental 'centre-left' of the political spectrum (Lynas 2015). They believe that decoupling production from nature using energy-dense technologies will benefit humanity, while at the same time leaving nature more space and less interference. Ecomodernists argue that nuclear power has a demonstrated potential "to reduce human demands on the environment, allowing more room for non-human species" and that nuclear fission is "the only present-day zero-carbon technology with the demonstrated ability to meet most, if not all, of the energy demands of a modern economy". But due to social, economic and institutional "challenges", large scale deployment of the technology is unlikely in the shorter term. New generation fission and fusion technologies are, however, viable in the longer term future. According to the *Ecomodernist Manifesto*: "Human civilisation can flourish for centuries and millennia on energy delivered from a closed uranium or thorium fuel cycle, or from hydrogen-deuterium fusion" (Asafu-Adjaye et al. 2015).

Many of these "challenges" to nuclear power arise from major accidents that involve contamination of large areas and extensive publicity about nuclear catastrophe. The downside also concerns the proliferation of nuclear weapons material, the high cost of the power plants including construction, operation and decommissioning, the risk of nuclear terrorism and the difficulty of radioactive waste disposal. Further, nuclear plants need long lead times for planning and construction. The inevitability of accidents means that the electricity generated cannot always be relied on as other nuclear plants in the same grid are shut down when catastrophes occur, as happened in Japan after Fukushima in 2011. Given these challenges it is unsurprising that the nuclear industry has struggled to present itself as the answer to future energy needs, despite millions of dollars spent on major public relations efforts (Farsetta 2007).

Background

While the ancient Greek, Democritus, had proposed the existence of fundamental particles or atoms, it was only in 1896 that radioactive substances were discovered,[1] and in the early 1930s Rutherford at Cambridge found that splitting lithium atoms released vast amounts of energy. The first self-sustaining atomic pile was built in Chicago in 1941. The first (experimental) reactor producing electricity was built in 1951, and the first commercial nuclear power plant was opened at Windscale in the United Kingdom in 1956. In the United States, the 1957 Price-Anderson Act limited the liability of firms for nuclear accidents and helped them secure capital with federal loan guarantees, such that more than 100 nuclear plants were built by the early 1970s (Jenkins et al. 2010). Construction of nuclear electric plants elsewhere continued in a stuttering way until by 2016 there were 449 plants in operation in the world, 60 under construction, 2 in long-term shutdown and over 30 permanently closed (IAEA-PRIS 2016). All use uranium as a fuel. Apart from the four major incidents that received extensive publicity, Windscale (United Kingdom 1957), Three Mile Island (United States 1979), Chernobyl (Ukraine 1986), and Fukushima (Japan 2011), there were about 100 "significant accidents"[2] at nuclear power plants between 1952 and 2009, involving more than US$20 billion in damages (Sovacool 2010).

Connections

There are people connections between the regulation of pesticides and the generation of nuclear power. In the United States, part of the nuclear-electric public relations effort has involved setting up industry funded pro-nuclear front groups. People with 'green' associations are highly valued recruits. Former EPA chiefs, Christine Whitman and Carol Browner have taken prominent positions with pro-nuclear organisations since they left the Agency. Whitman is now co-chair of the 'Clean and Safe Energy Coalition' (*CASEnergy*), which is funded by the industry *Nuclear Energy Institute* (Farsetta 2007). Browner is a leader of the advocacy group *'Nuclear Matters'*. Another former EPA chief, William Reilly, is also a prominent campaigner (Nuclear Matters 2014; Manjunatha 2016). In these cases, two were appointed to the EPA by Republican presidents, one by a Democrat,[3] so it is a bipartisan phenomenon that evidences close ties between government regulators and the nuclear power industry in the United States.

Also, both herbicides and nuclear power grew from military applications fostered by government, especially in the United States and United Kingdom. Nuclear military applications were also developed in France from 1945 and its first nuclear electric plant opened in 1962. By 2016 France had the highest proportion of electricity generated from nuclear power of any country – around 75 percent. Due to the 1973 oil crisis, the 'Messner Plan' was implemented aiming to produce all France's electricity from nuclear power on the popular rationale of: "no oil, no gas, no coal, no choice" (Palfreman 1997). Despite major accidents elsewhere, nuclear power remains popular in France for several reasons: a relatively good safety record, lack of alternatives, a national regard for elite technocrats, strong government ownership, and a low retail price of electricity (ibid.). These factors have also made nuclear-electric power an important export industry, as discussed in Chapter 8.

Dosage is yet another nexus. Both industries are critically concerned with the 'safe dose' for risk estimates. The nuclear risk profile has relatively low probability of happening combined with high impact should an accident occur, while pesticide contamination may be more likely but with lower impact. A common assumption of nuclear radiation risk is that no dosage is safe and there is a linear relationship between exposure and harm (IAEA 2013). Exposure at significantly more than the level of natural background radiation is therefore cautionary. Table 10.1 lists different radiation exposures, measured in millisieverts (mSv),[4] that relate to human biological impact (Chandler 2011). It shows that the recommended limit for workers exposed to nuclear radiation is 20 millisieverts per year against typical natural radiation of 2 millisieverts per year – a tenfold difference.

The increased dosage limit (250mSv) for emergency workers at Fukushima is an example of the regulatory 'bracket creep' that has also occurred with

Table 10.1 Radiation exposure comparisons

2 mSv/year	Typical background radiation
1.5 to 2.0 mSv/year	Average extra dose to Australian uranium miners, and medical staff
2.4 mSv/year	Average extra dose to US nuclear industry employees
up to 5 mSv/year	Typical incremental dose for aircrew in middle latitudes
9 mSv/year	Exposure by airline crew flying polar routes
10 mSv/year	Maximum dose for Australian uranium miners
20 mSv/year	Current average limit for nuclear industry employees and uranium miners
50 mSv/year	Former routine limit for nuclear industry employees
100 mSv/year	Lowest level where increase in cancer clearly evident
250 mSv	Limit for emergency workers at Fukushima 2011
350 mSv/lifetime	Criterion for relocating people after Chernobyl accident
500 mSv	Symptoms of radiation poisoning evident
1,000 mSv/cumulative	Probably cause fatal cancer in five of every 100 people exposed
1,000 mSv/single dose	Radiation sickness (nausea and decreased white blood cell count)
5,000 mSv/single dose	Death for c.half those exposed within a month (Hiroshima 1945)
10,000 mSv/single dose	Fatal within a few weeks.

Sources: Based on World Nuclear Association 2004; Chandler 2011.

glyphosate tolerances. It suggests that regulations were adjusted to accommodate hazards, as distinct from measures developed to avert them. This compares with above background dosage limits according to three international organisations, which are consistent and considerably lower than dosages experienced at both Chernobyl and Fukushima. The IAEA, the EU and the International Commission on Radiological Protection (ICRP), an international NGO, all recommend limits of 1 mSv/year for the general public and 20 mSv/year for licensed workers over 18 years.

They also concur on a limit of less than 6mSv/year for "workers 16–18 years" (IAEA 2011; ICRP 2018; EU 2013). However, given that young people are particularly susceptible to radiation, the employment of 16-year-olds at nuclear plants is a probable breach of ILO Conventions on child labour including C138: "*The minimum age for admission to any type of employment or work which by its nature or the circumstances in which it is carried out is likely to jeopardise the health, safety or morals of young persons shall not be less than 18 years*". Further, C182 prohibits "*work which, by its nature or the circumstances in which it is carried out, is likely to harm the health, safety or morals of children*".

But this kind of information can be difficult to interpret. It is not immediately obvious that the rate of exposure is as important as the total amount. Further, there is a difference between the *external* exposures reflected in most of these figures and the *internal* exposures that result from longer term inhalation, absorption and ingestion of radioactive substances due to fallout. In contaminated areas those high in the food chain are at greater risk because contamination is accumulated through diet. These internal exposures produce cumulative genetic damage leading to cancers – later and separate from the effects of the external exposures described in most of the dosage limits above (Caldicott 2011). Along this fault line, a typical pro-nuclear argument is that there were no deaths directly due to radiation exposure at Fukushima. The contrary view is that there will be many indirect deaths due to fallout and ingestion of contaminated food over an area shaped by the prevailing winds originating from the disaster site. There will also be many cancers, especially of the thyroid, that may not register as fatalities (Tsuda et al. 2016) and to which children are particularly susceptible (WHO 2016).

Regulation

Regulations about relatively safe exposure levels can be confusing as well as contested. For example, current standards are based on the conservative assumption that risk is directly proportional to the dose, called the 'linear no-threshold (LNT) hypothesis', although other bases are proposed. These other (higher) bases, include the '*as low as reasonably achievable*' (ALARA), the '*as high as naturally existent*' (AHANE), and '*as high as relatively safe*' (AHARS) – about 1000mSv per year. All are mentioned as possible approaches to the issue of limits by the industry-based World Nuclear Association, which argues that there is probably a threshold below which there is no risk and that exposure to radiation is much less harmful than is usually assumed (WNA 2016).

As far as patents are concerned, nuclear technology is "like Gaul … divided into three parts: a zone of non-patentability, a zone of government patents, and a zone of private patent rights" (Riesenfeld 1958, p. 51). Only some nuclear technologies may be patented. The seminal US 1954 *Atomic Energy Act* has been supplemented several times but remains largely intact. It followed a 1946 law that was primarily to ensure government control over its military applications. By contrast, a central purpose of the 1954 Act is "to make available to cooperating nations the benefits of peaceful applications of atomic energy as widely as expanding technology and considerations of the common defence and security will permit" (s. 3e). Following considerable political pressure, the Act excludes patent rights for inventions that can be used solely in atomic weapons (USPTO 2016), but provides for patents in non-military applications such as electricity generation. There is also an 'intermediate zone' whereby the patent office must report all nuclear inventions (to the Department of Energy) and the government may subsume patent rights with appropriate compensation, irrespective of their application. As such it is an attempt to balance the need for security in military matters against commercial incentives for innovation.

The Act also establishes the *Atomic Energy Commission*, and asserts a licensing requirement for all activities concerning nuclear material (s. 101). However it mentions nuclear safety only fleetingly – but twice in its 276 pages – in relation to cooperation with the states on a federal radiation council (s. 274) and in relation to standards for uranium mill tailings (s. 275) (!).

As to international regulators, the independent ICRP[5] was formed in 1928 to provide scientific advice on the effects of all forms of radiation, well before nuclear power plants were conceived. This NGO is accredited with several UN agencies including the WHO, FAO and the ILO, as well as the International Atomic Energy Agency (IAEA). Its recommendations on policy and standards are widely followed by national health authorities.

The IAEA is a semi-independent UN body with statutory responsibilities for nuclear safety that develops non-binding standards[6] based on ICRP recommendations. Following Eisenhower's 1953 *"Atoms for Peace"* speech to the UN General Assembly and the expediting 1954 US *Atomic Energy Act*, the IAEA was set up in Vienna in 1957 to "to accelerate and enlarge the contribution of atomic energy to peace, health and prosperity through the world" as well as, in consultation with other agencies, "to establish or adopt ... standards of safety for protection of health and minimisation of danger to life and property". The IAEA is also responsible for measures to contain the spread of nuclear weapons, a task at which it has not been entirely successful.[7]

The UN Scientific Committee on the Effects of Atomic Radiation (UNSCEAR), set up in 1955, is an authoritative source of information specifically on the effects of the ionising radiation that result from nuclear materials (as distinct from the wider radiation concerns of the ICRP), based on assessments of the scientific literature. It provides appraisals to both the ICRP and the IAEA. National governments set radiation protection standards generally in line with ICRP recommendations, keeping exposure as low as reasonably achievable (ALARA).

Also based on ICRP recommendations, the European Commission (EC) develops binding directives that its member states must transpose into national law. The OECD *Nuclear Energy Agency* (NEA) investigates emerging issues in radiological protection when invited by member countries. It also tests draft recommendations of the ICRP before they are finalised (Lazo 2009).

It is in this global regulation of nuclear power that there is an intriguing asymmetry involving the IAEA and the WHO. The IAEA was primarily created to *promote* non-military nuclear energy. Its *regulation* of that energy, however, is a secondary function that involves consultation with other agencies and member states. Yet that consultation works both ways and the agency's overriding promotional objective tends to overshadow its relationships. For example, before the IAEA was established, the WHO issued strong public warnings about the health risks of the nuclear industry, as in 1956 when it said "the health of future generations is threatened by increasing development of the atomic industry and sources of radiation". However, in 1959, after the IAEA was established, the WHO became muted due to a formal pact with the IAEA that required them to consult on relevant activities "with a view to adjusting the matter by mutual agreement" (clause 12.40, cited in Caldicott 2011). Current WHO statements consist of factual information on the health effects of radiation rather than dire warnings of nuclear threats to posterity (WHO 2016).

This clubby arrangement of dispersed responsibility appeared lax when the scale of the 1986 Chernobyl disaster became apparent. In response, the IAEA facilitated a binding *Convention on Nuclear Safety* (CNS), which all countries with nuclear power plants (except Iran) agreed to – albeit a full eight years after the disaster, in 1994. The CNS commits countries (and the EU) to specific nuclear safety measures that include licensing (article 7) principles of oversight (article 8) and systematic safety assessment (article 14), although not to any technical benchmarks, which remain voluntary. Every three years under the Convention there is a peer-review system of country reports on the management of nuclear safety (article 5) aimed at identifying good practice as well as flaws in national supervision.

However the Convention itself proved inadequate when Japan reported in 2010 that it was in compliance, whereas in fact it was not – especially concerning regulatory independence (article 8.2) and safe siting of nuclear power plants (article 17). The Fukushima disaster occurred the following year, exposing "certain weaknesses in Japan's regulatory framework" (Dahl 2015). According to the IAEA Director General, Yukiya Amano – himself Japanese – "responsibilities were divided among a number of bodies and it was not always clear where authority lay". Underlying this,

> a major factor that contributed to the emergency was a widespread assumption in Japan that its nuclear power plants were so safe that an accident of such a magnitude was simply unthinkable. This assumption was accepted by nuclear plant operators and was not challenged by regulators or by the government.
>
> (Ibid.)

Russia, Switzerland and the IAEA itself then proposed that IAEA safety standards and peer reviews be compulsory. The proposition failed. Switzerland and the EU then further proposed that the CNS include compulsory safety targets for all reactors, but this was again opposed by countries that feared extra costs and domestic interference. It is probable that the United States especially was wary of cost implications since its reactors vie in intensely competitive electricity markets, whereas European reactors tend to operate in more regulated electricity monopolies where costs are less critical (Hibbs 2015).

Beyond these inadequate governance arrangements, in response to Chernobyl the industry itself formed an association that focused uniquely on safety. The World Association of Nuclear Operators (WANO) was formed in 1989 with headquarters in London, representing almost all operators of individual nuclear power plants. It undertakes external peer reviews and provides operational and technical support and professional development. Immediately after Fukushima, WANO convened a commission to report on needed changes to the organisation. Its recommendations were:

1. To extend the scope of WANO to include design and accident management
2. To set up an event response strategy
3. To increase WANO's credibility through stronger internal control
4. To increase WANO's transparency by making WANO regular reports accessible to the public
5. To increase internal consistency between its four regional centres. (Chudakov 2014)

In effect, these measures accept that nuclear power plant accidents are inevitable. Beyond maximising safety through prevention, they recognise that there must be an effective response when accidents occur. In respect of the transparency recommendation #4, however, as of mid-2017 very few reports were available to the public on the WANO website, apart from some brochures on global performance indicators and a list of safety measures. Detailed reports appear to be restricted to members.

Human mortality

Despite the setbacks due to nuclear accidents, pro-nuclear public relations agents[8] and academic research have both tackled the nuclear power safety issue directly. Articles comparing nuclear-electric mortality with other means of electricity generation have become more common. For example, an article published in the influential US *Forbes* business magazine (Conca 2012) argued that nuclear is the least harmful form of electricity generation to humans. The piece, which asserted that nuclear power causes only 90 deaths compared with global coal's 100,000 deaths per trillion kilowatt hours, since prompted hundreds of online comments over a period of four years. However, the figures were dubious: they were unsourced, undated, and all the energy sources totalled up to 112 percent of global electricity. These figures were subsequently updated in 2016 so that they did total 100 percent, but still lacked source and year.

The International Energy Agency (IEA) figures for 2014 differ. For example, 66.7 percent of world electricity was generated from fossil fuelled plants according to the IEA, (IEA 2016, p. 30) compared with 71 percent according to Forbes article. This suggests that, as with glyphosate, nuclear-electric technology is a crucible of controversy. In fact, the conflict over nuclear power has been described as having "an intensity unprecedented in the history of technological controversies" (Kitschelt 1986, p. 57), visible in mass demonstrations in many countries, especially those following the four major and most public nuclear accidents.

Another study, still favourable to nuclear power, came from NASA's Goddard centre. According to the study:

> nuclear power prevented an average of over 1.8 million net deaths worldwide between 1971–2009 [over 47,000 per year]. ... This amounts to at least hundreds and more likely thousands of times more deaths than it caused. An average of 76,000 deaths per year were avoided annually between 2000–2009 ... with a range of 19,000–300,000 per year.
>
> (Kharecha and Hansen 2013)

Dr Kharecha confirms that the higher rate of avoided deaths in the later period is attributed simply because there were more nuclear plants then operating.[9] But accepting that these estimates are valid, the logic is constrained. The numbers are based on estimated deaths mainly from the inhalation of particulates from fossil-fuelled plants. The article assumes that "nuclear energy is cancelled and replaced entirely by energy from either coal or natural gas" (ibid.). Yet this does not have to be the case. Renewable sources can also substitute for nuclear; they are much more quickly constructed than nuclear plants and have more public acceptance. In fact, the *Lancet* article on which Kharecha and Hansen rely for their mortality estimates, indicates that a decision to replace current nuclear power plants:

> would be welcome in health terms if the nuclear plants were replaced by capacity in renewable production additional to the level of renewable production that would otherwise occur.
>
> (Markandya and Wilkinson 2007, p. 988)

Also, the wide range of 'deaths avoided' in the study — between 19,000 and 300,000 per year —indicates uncertainty about the actual figure. That figure must be high as there are an estimated seven million early deaths each year from all sources of air pollution (WHO 2014b), many of which must be from coal and gas-fired electricity, especially in China and India. But on the nuclear 'deaths caused' side of the ledger, estimates also vary widely. For example, the Belarussian scientist Malko's 2006 study on Chernobyl estimated more than 90,000 excess cancer deaths, and his later study showed more than 115,000. This contrasts with a WHO prediction of 9,000 additional deaths due to

Chernobyl (Dawe et al. 2016). Peplow (2011) reporting in *Nature* says study estimates "range from a few thousand to hundreds of thousands" of deaths due to the disaster. Further, contamination continues over many decades: Cesium 137, which forms much of the radioactive fallout from both Chernobyl and Fukushima has a half-life of 30 years (Schneider 2011).

In China, the Fukushima effect was more economic than physical, probably because the prevailing winds are westerly, blowing fallout across the Pacific rather than towards the Asian continent. Land prices within 40 kilometres of a nuclear power plant in China fell by about 18 percent a month after the disaster, eventually returning to normal some years afterwards (Zhua et al. 2016). China's State Council did suspend nuclear plant development (Schneider 2011), but construction since resumed.

Comparisons

Apart from these accident risks, the emissions and costs of nuclear power can be compared with other sources of electricity, both key factors in any political consideration. Figure 10.1 shows emissions and costs from existing and possible future sources, based on IPCC data. Importantly, calculations are for the life cycle of each source, including materials, construction, and ultimate remediation.[10] Nuclear emissions and costs include the mining, processing and transport of uranium, construction of reactors and insurance costs against the risks of disaster. Hydro-electric generation includes the emissions and running costs of the construction machinery used, as well as for the dams and turbines. Social and direct environmental costs such as the loss of habitat due to dams and coal mines are not included. Notably, the data assume there is a distribution grid, which skews cost advantage away from rooftop solar in developing countries, where nuclear and all other forms of generation would require a grid to be built.

If cost were the only consideration, several sources are all more viable than nuclear. However, when emissions and cost are combined as in this comparison, nuclear is the third most viable form of generation, after hydro and onshore wind. The data show that the range of costs (from 35 to 220 USD/mWh) is much narrower than the range of emissions (from 11 to 820 kgCO_2/mWh). Whilst it has been argued to the contrary, the cost of fossil fuels is unlikely to rise much in the medium term. Supplies are not yet near exhaustion due to new extraction methods and there are still considerable reserves (Helm 2013). Yet largely due to economies of scale in manufacturing, the cost of solar photovoltaic has been continually falling in real terms. Increasingly, renewable sources (that use free fuel) are among the cheapest and becoming cheaper, rivalling the cost-emissions advantage of nuclear. It is also a considerable disadvantage that nuclear has very high upfront costs as well as long lead times (Findlay 2010).

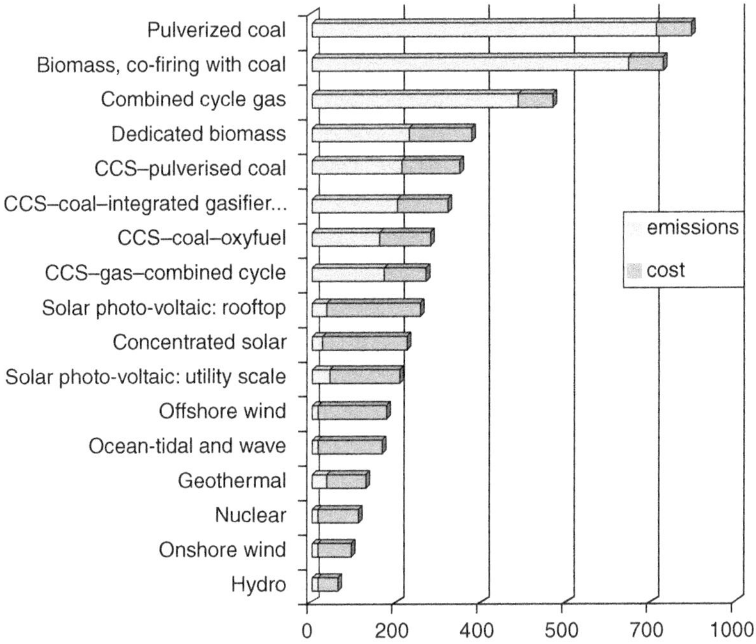

Figure 10.1 Total greenhouse gas emissions plus total cost of electricity by source per megawatt-hour. Zero carbon price. CCS: carbon capture and storage (not yet commercial*). Emissions: kgCO_2eq/mWh. Cost: US$_{2010}$/mWh.

* Although according to Service (2017) it is close to being so in Texas, US.

Source: Author, based on data from IPCC 2014c, pp. 1333–1335.

However, these two categories are necessary but not sufficient to determine a viable energy policy and the place of nuclear within it. Not only is energy desirably clean and cheap, it is also an advantage that it is continuous, whether provided directly or as a result of storage, and desirably inexhaustible. There are also locational factors: it is obviously not possible to locate offshore wind farms in landlocked countries, for example. Nor is hydro power feasible without rivers. Coal and nuclear power stations have locational problems associated with pollution and safety, and nuclear carries additional risks of major catastrophe, nuclear proliferation and waste disposal.

The prolific energy policy analyst, Benjamin Sovacool (2010) judges nuclear and renewable technologies against six criteria: cost, fuel availability, land degradation, water use, safety and security, as well as 'climate change' – the equivalent of greenhouse gas 'emissions'. Writing before Fukushima, Sovacool finds that for nuclear, the costs of construction, fuel, and decommissioning will probably increase. Further, nuclear reactors are prone to accidents and failures, imminent shortages of quality uranium ore, the degradation of cooling water and the issue of nuclear waste storage. Writing after Fukushima, Tickell (2012) is even more critical. He points out that in a world that relied on nuclear energy:

> Serious accidents, such as those at Windscale, Three Mile Island, Chernobyl and Fukushima – the last of which came very close to making Tokyo uninhabitable for decades to come – would become commonplace events.

Disasters

Another leading energy academic is cited as predicting the total demise of nuclear electricity if there is one more major disaster within the next five or ten years (Beale 2016). Yet there appears to be a disconnect between this reality and energy policy research, which tends to be technical, narrow and masculine. Sovacool's (2014) meta study of 4,444 full-length articles published between 1999 and 2013 showed that the social dimensions of energy use were undervalued; studies were biased towards science, engineering and economics over other disciplines; there was little interdisciplinary collaboration; and women and minority authors were under-represented. Layperson viewpoints were ignored and quantitative research dominated over qualitative.

These findings resonate for nuclear electricity, because the societal and ecological impacts of disaster are its greatest challenge. The sheer scale of disaster haunts imaginings over both space and time. For example, 100,000 people were evacuated in 1986 from the Chernobyl exclusion zone within 30 kilometres of the accident site. Later a further 200,000 people were evacuated from the three countries most contaminated by radioactive fallout – Belarus, Russia and the Ukraine (IAEA 2006). Release of radioactive particles into the atmosphere continued for ten days, so that different wind directions produced three main plumes over the period. Many countries including Germany, Switzerland and Greece were also affected by fallout, as shown in Figure 10.2. In Scandinavia, the native Saami people were unable to eat contaminated reindeer meat (ibid.). In large areas closer to the site children drank contaminated milk unaware of the risk of thyroid cancer.

While some fallout was relatively short-lived with half lives a matter of days, cesium 137 continues to cause disquiet as it formed the bulk of the fallout and has a 30-year half-life. In these contaminated areas, the highest concentrations of cesium 137 are found in woodland organisms because it is recycled in forest ecosystems. The level is still above limits in many countries and "can be expected to continue for several decades to come" (IAEA 2006, p. 3). For example, Norway's reindeer remain radioactive as they feed on contaminated lichen and mushrooms. Despite another example of bracket creep in which the Norwegian food contamination limit is five times the EU limit, by 2014 reindeer were found still too radioactive to slaughter for food (Taylor 2016).

Figure 10.2 Europe and USSR: contamination from Chernobyl. Surface density of cesium 137. kBq/m2: kiloBecquerels per square metre; Ci/km2: Curies per square kilometre. NPP: nuclear power plant.

Source: Reprinted from Steinhauser, G, Brandl, A and Johnson, T 2014, 'Comparison of the Chernobyl and Fukushima nuclear accidents: A review of the environmental impacts', *Science of the Total Environment,* Vols 470–471, I February, pp. 800–817, Copyright 2014, with permission from Elsevier.

The Fukushima accident led to more than 150,000 people evacuated from towns within 30 kilometres of the plant, as shown in Figure 10.3. It had "a massive impact on the atmospheric and natural environment, the economic and political situation, and human psychology and health" (Barletta et al. 2016). Most evacuees remained in temporary accommodation five years after the

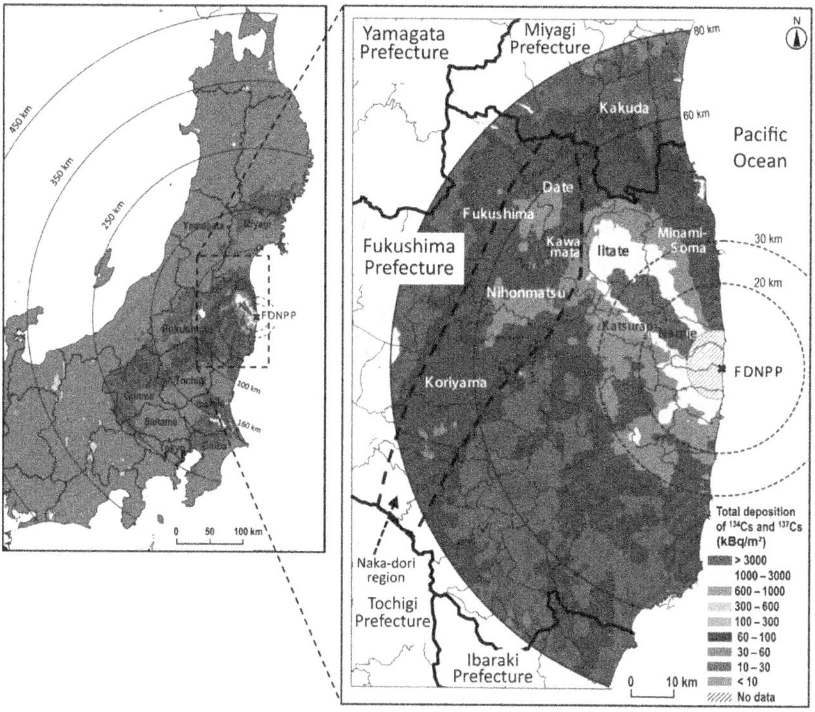

Figure 10.3 Honshu: contamination from Fukushima. Surface density of cesium 134 and 137. kBq/m2: kiloBecquerels per square metre. FDNPP: Fukushima Daiichi nuclear power plant.

Source: Reprinted from Steinhauser, G, Brandl, A and Johnson, T 2014, 'Comparison of the Chernobyl and Fukushima nuclear accidents: A review of the environmental impacts', *Science of the Total Environment,* vols 470–471, I February, pp. 800–817, Copyright 2014, with permission from Elsevier.

explosion. Contamination was much less than Chernobyl, however, as can be seen by comparing Figures 10.2 and 10.3.

This was because, despite the evident flare to the North-West, much of the Fukushima fallout was swept over the ocean and the volume of radioactive particles released was only about a tenth of the Chernobyl explosion (ibid.). Health effects were further reduced because Fukushima happened on 11 March, before the main agricultural season, and fallout tended to avoid agricultural areas. The Chernobyl accident was on 26 April, when the growing season had begun and fallout settled directly on crops and pasture (Steinhauser, Brandl and Johnson 2014).

Different regulators made divergent efforts concerning their own citizens living in or visiting Japan at the time of the disaster. The US Nuclear Regulatory Commission, for example, advised Americans in the region to evacuate to 80 kilometres from the site (Eisler 2012). So did Sweden. Others, such as South Korea and Germany facilitated evacuation from Tokyo and other cities, as did many other countries (*Philippine Daily Inquirer*, 18 March 2011).[11] Although the logistics would have been prodigious, the Japanese government also considered such a measure for its own people in Tokyo.

Despite greater population density, far fewer will die due to the lower fallout. One modelling study suggests radiation exposure will result in an extra 130 (range: 15–1300) deaths from cancer and 180 (range: 24–2500) cancer-related deaths (Ten Hoeve and Jacobson 2012). Others estimate around 1000 fatal cancers due to the disaster, which is within the above range (Beyea, Lyman and von Hippel 2013; von Hippel 2011). This last figure is also reasonably consistent with Malko's estimates for Chernobyl (90,000–115,000 deaths) that was ten times more severe and lasted much longer. Non-fatal cancers are ignored in these estimates. Even when in future decades the human impact can be better ascertained, it is still likely that different methods and researchers will produce different results.

Waste

Yet while the impact of nuclear disasters may haunt us over decades, the time periods involved in nuclear waste disposal are very much greater – on a such a scale that the intervals of all other terrestrial technologies are indiscernible:

> Since the only way radioactive waste finally becomes harmless is through decay, which for high-level wastes can take hundreds of thousands of years, the wastes must be stored and finally disposed of in a way that provides adequate protection of the public for a very long time.
>
> (USNRC 2016)

The regulation of nuclear waste storage is again nationally determined, hopefully consistent with IAEA guidelines. But one overwhelming problem is that as yet there is no permanent repository for high-level waste. At present it is contained on site, usually within the boundaries of nuclear power plants in temporary containers. It is universally agreed that 'deep geological burial' is the long-term solution to the problem, yet not one such shaft has been opened since the first nuclear power plant began operating in 1956 (Findlay 2010), although Finland is preparing a such a facility due for commercial use in 2020 (ABC 2016).

Understandably there is extreme political sensitivity about the establishment of a nuclear waste dump. Japan once attempted to encourage its more than 3,000 municipalities to volunteer a site in 2002. Several years later not one willing candidate had emerged (*Japan Times*, 21 January 2014). Regional repository schemes have been proposed, but do not overcome the issue of local reluctance. A proposed referendum in the state of South Australia on the issue appears doomed (Wills 2016). Even twenty years after the event, a comprehensive program for waste management had not been established at the Chernobyl site (IAEA 2006). Politically it has taken a long time to develop facilities for waste disposal, although nowhere near as long as needed for the waste to become harmless.

It does seem that the effective international regulation of the nuclear power industry is just too much of a challenge for current human institutions, many of which have overlapping and unenforceable responsibilities, as much as its sustainability founders on the twin issues of disaster and waste. Even if, despite the doubts of energy analysts (Brutoco 2014; Johnstone et al. 2016), safer nuclear technologies are developed over the medium term as ecomodernists hope, it may be too late for the future of the industry.

This is primarily a state-led example of technology governance, due to its implications for national security and evidenced by the heavy involvement of the five veto powers – the United States, the United Kingdom, Russia, France and China – in its development. Despite its major risks, governance seems ineffective. On the TAPIC measures, the transparency promise of WANO appears yet to eventuate. Accountability is largely national and weak at the international level, its military origins notable. Participation is limited to the industry. Integrity is moderate and responsibilities generally well-defined, but is undermined by pro-nuclear front groups, especially in the United States. Capacity is affected by the tension between the different aims of the IAEA and its influence over the WHO.

Over the last few years the forests of northern Ukraine have become part of a valuable export industry. Locals pick thousands of tons of wild berries from the contaminated area despite official warnings of danger. The berries are exported to the EU via Poland. Buyers use Geiger counters to check radioactivity levels, not to reject 'hot' berries, but rather to pay lower prices for them. They then mix contamination levels so that each quantity is below the generous post-Chernobyl EU limit of 600 becquerels per kilogram. The berries are marketed in Europe as 'organic', which they are. In this way, the hazards of ingesting nuclear fallout are shared throughout Europe (Brown and Martynyuk 2016).

Visiting Chernobyl and its dormitory town, Pripyat, thirty years after the nuclear accident – safely by Google Earth and YouTube drone — two images stand out. One is the rapid re-vegetation of what was once a town of 50,000 people. The other is the remains of a playground where stands a rusting Ferris wheel; yellow gondolas, still and silent in an eerie landscape, long abandoned by workers and their children.

Notes

1 By Henri Becquerel (France).
2 Incidents resulting in loss of human life or more than US$50,000 of damage, based on US mandatory reporting criteria.
3 Whitman was appointed by Bush the elder, Reilly by Bush the younger and Browner by Clinton.
4 This is essentially a measure of the amount of potential damage to the body from a given amount of radiation.
5 The Swede, Rolf Sievert, was the first president of the Association and is commemorated in the unit of ionising radiation that bears his name.
6 Although any state that accepts help from the IAEA must also accept its safety standards.
7 Given that there are nine nuclear weapons states possessing a total of more than 15,000 nuclear warheads, according to the Arms Control Association (2016).
8 According to the European energy consultant, Mycle Schneider, "the international nuclear lobby has pursued a ten-year long massive propaganda strategy aimed at convincing decision-makers that atomic technology has a bright future as a low-carbon option" (Schneider 2011).
9 Personal email of 2 September 2016.
10 The only significant remediation cost is for nuclear; others are assumed to be negligible.
11 'Foreigners Stream out of Tokyo'.

References

ABC: Australian Broadcasting Corporation 2016, *Finland to Bury Nuclear Waste for 100,000 Years in World's Costliest Tomb*, 8 June.

Asafu-Adjaye, J, Blomqvist, L, Brand, S, Brook, B, Defries, R Ellis, E, Foreman, C, Keith, D, Lewis, M, Lynas, M, Nordhaus, T, Pielke, R Jr, Pritzker, R, Roy, J, Sagoff, M, Shellenberger, M, Stone, R and Teague, P 2015, *An EcoModernist Manifesto*, April 2015.

Barletta, W, Stoop, J, O'Hara, C, Bailiff, I, Hermanspahn, N, Hugtenburg, R, Alcock, R, Covaci, A, Barceló, D, Gan, J, and Karlsson, A 2016, 'Five Years After Fukushima – Insights From Current Research', *Elsevier Connect*, 10 March.

Beale, C 2016, 'Has the Chernobyl Disaster Affected the Number of Nuclear Plants Built?', *The Guardian*, 30 April.

Beyea, J, Lyman, E, and von Hippel, F 2013, 'Accounting for Long-Term Doses in "Worldwide Health Effects of the Fukushima Daiichi Nuclear Accident"', *Energy and Environmental Science*, vol. 6, pp. 1042–1045.

Brown, K and Martynyuk, O 2016, 'The Harvests of Chernobyl', *Aeon*, 29 November.

Brutoco, R 2014, *Nuclear Power: Totally Unqualified to Combat Climate Change, World Business Academy*, 14 September.

Caldicott, H 2011, 'How Nuclear Apologists Mislead the World Over Radiation', *The Guardian*, 11 April.

Chandler, D 2011, *Explained: Rad, Rem, Sieverts, Becquerels: A Guide to Terminology about Radiation Exposure*, MIT News Office, 28 March.

Chudakov, M 2014, *Post-Fukushima WANO Changes*, WANO Moscow Centre, Presentation for the 9th INPRO Dialogue Forum International Collaboration on Innovations to Support Globally Sustainable Nuclear Energy Systems, IAEA Headquarters, Vienna, Austria, 18–21 November.

Conca, J 2012, 'How Deadly Is Your Kilowatt? We Rank the Killer Energy Sources', *Forbes*, 10 June.

Dahl, F 2015, *IAEA Delivers Major Report on Fukushima Accident to Member States*, 14 May, IAEA Office of Public Information and Communication, Vienna, Austria.

Dawe, A, McKeating, J, Labunska, I, Schulz, N, Stensil, S-P and Teule, R 2016, *Nuclear Scars: The Lasting Legacies of Chernobyl and Fukushima*, Greenpeace, Amsterdam, the Netherlands.

Eisenhower, D 1953, *Atoms for Peace Speech*, Address by Mr. Dwight D. Eisenhower, President of the United States of America, to the 470th Plenary Meeting of the United Nations General Assembly Tuesday, 8 December 1953, 2:45 pm, International Atomic Energy Agency website 2018.

Eisler, R 2012, *The Fukushima 2011 Disaster*, CRC Press, Taylor and Francis, Boca Raton, US.

EU: European Union 2013, 'Council Directive 2013/59/Euratom of 5 December Laying Down Basic Safety Standards for Protection against the Dangers Arising from Exposure to Ionising Radiation, and Repealing Directives 89/618/Euratom, 90/641/Euratom, 96/29/Euratom, 97/43/Euratom and 2003/122/Euratom', *Official Journal of the European Union*, 17 January.

Farsetta, D 2007, 'Moore Spin: Or, How Reporters Learned to Stop Worrying and Love Nuclear Front Groups', *PR Watch*, 14 March.

Findlay, T 2010, *The Future of Nuclear Energy to 2030 and Its Implications for Safety, Security and Nonproliferation*, Nuclear Energy Futures Project, Centre for International Governance Innovation, Ontario, Canada.

Garrett, T 2014, *The Physics of Long-Run Global Economic Growth*, University of Utah, US.

Helm, D 2013, *The Carbon Crunch: How We're Getting Climate Change Wrong – and How To Fix It*, Yale University Press, London, UK.

Hibbs, M 2015, *A Failed Effort to Toughen Nuclear Safety Standards*, Carnegie Endowment for International Peace, 18 February.

IAEA: International Atomic Energy Agency 2006, *Environmental Consequences of the Chernobyl Accident and Their Remediation: Twenty Years of Experience*, Report of the Chernobyl Forum Expert Group 'Environment', Radiological Assessment Reports Series, Vienna, Austria.

IAEA: International Atomic Energy Agency 2011, *IAEA Safety Standards Radiation Protection and Safety of Radiation Sources: International Basic Safety Standards for Protecting People and the Environment*, interim edition, No. GSR Part 3 (Interim), General Safety Requirements Part 3 ISBN 978–92–0–120910–8, ISSN 1020–525X, Vienna, Austria.

IAEA: International Atomic Energy Agency 2013, *Framework for Assessing Dynamic Nuclear Energy Systems for Sustainability: Final Report of the INPRO Collaborative Project GAINS*, Nuclear Energy Series No. NP-T-1.14, Vienna, Austria.

IAEA-PRIS: International Atomic Energy Agency – Power Reactor Information System 2016, website.

ICRP: International Commission on Radiological Protection 2018, *ICRPaedia: Dose Limits*, website.

IEA: International Energy Agency 2013, *World Energy Outlook Special Report: Redrawing the Energy-Climate Map*, 10 June, Paris, France.

IEA: International Energy Agency 2014, World Energy Investment Outlook – Special Report, IEA, Paris, France. Viewed 30 July 2015.

IEA: International Energy Agency 2015, *Energy and Climate Change: World Energy Outlook Special Report*, Paris, France.

IEA: International Energy Agency 2016, *Key Electricity Trends: Excerpt from Electricity Information*, Paris, France.

Johnstone, P, Sovacool, B, MacKerron, G and Stirling, A 2016, 'Nuclear Power: Serious Risks', *Science*, letters, vol. 354, no. 6316, 02 December, p. 1112. doi: 10.1126/science.aal1777.

Kharecha, P and Hansen, J 2013, *Coal and Gas Are Far More Harmful than Nuclear Power*, Earth Sciences Division, Goddard Institute for Space Studies, National Aeronautics and Space Administration, New York, US.

Kitschelt, H 1986, 'Political Opportunity Structures and Political Protest: Anti-Nuclear Movements in Four Democracies', *British Journal of Political Science*, vol. 16, no. 1, pp. 57–85.

Lazo, E 2009, *The International Systems of Radiological Protection: Key Structures and Current Challenges*, Nuclear Energy Agency (NEA), Organisation for Economic Cooperation and Development, pp. 49–63.

Lynas, M 2015, 'Ecomodernism Launch Was a Screw-Up of Impressive Proportions', *The Guardian*, 30 September.

Manjunatha, Y 2016, 'Former EPA Administrator Advocates Role of Nuclear Energy in Reducing Carbon Emissions', *The Chronicle*, 8 September.

Markandya, A and Wilkinson, P 2007, 'Electricity Generation and Health', *Lancet*, vol. 370, no. 9591, 15 September, pp. 979–990.

Nuclear Matters 2014, *Carol M. Browner, Longest Serving EPA Administrator, Joins Nuclear Matters*, Press release, 22 April, Washington, DC, US.

Palfreman, J 1997, 'Why the French Like Nuclear Energy', *Frontline*, Public Broadcasting Service, Washington, DC, US.

Peplow, M 2011, 'Chernobyl's Legacy', *Nature*, vol. 471, 28 March, pp. 562–565. doi: 10.1038/471562a.

Riesenfeld, S 1958, 'Patent Protection and Atomic Energy Legislation', *California Law Review*, vol. 46, no. 1, pp. 40–68.

Schneider, M 2011, 'Fukushima Crisis: Can Japan Be at the Forefront of an Authentic Paradigm Shift?', *Bulletin of the Atomic Scientists*, 9 September.

Sovacool, B 2010, 'A Critical Evaluation of Nuclear Power and Renewable Electricity in Asia', *Journal of Contemporary Asia*, vol. 40, no. 3, pp. 369–400.

Sovacool, B 2014, 'Diversity: Energy Studies Need Social Science', *Nature*, vol. 511, 31 July, pp. 529–530.

Steinhauser, G, Brandl, A and Johnson, T 2014, 'Comparison of the Chernobyl and Fukushima Nuclear Accidents: A Review of the Environmental Impacts', *Science of the Total Environment*, vols 470–471, 1 February, pp. 800–817.

Taylor, A 2016, 'Norway's Radioactive Reindeer', *The Atlantic*, In Focus, 1 March.

Ten Hoeve, J and Jacobson, M 2012, 'Worldwide Health Effects of the Fukushima Daiichi Nuclear Accident', *Energy and Environmental Science*, vol. 5, pp. 8743–8757.

Tickell, O 2012, 'Does the World Need Nuclear Power to Solve the Climate Crisis?', *Guardian Environment Network*, 21 August.

Tsuda, T, Tokinobu, A, Yamamoto, E and Suzuki, E 2016, 'Thyroid Cancer Detection by Ultrasound among Residents Ages 18 Years and Younger in Fukushima, Japan: 2011 to 2014', *Epidemiology*, vol. 27, no. 3, May, pp. 316–322. doi: 10.1097/EDE.0000000000000385.

USDoE: United States Department of Energy 1994, *The History of Nuclear Energy*, DOE/NE0088, Washington, DC, US.

USNRC: United States Nuclear Regulatory Commission 2016, *High Level Waste*, website.

USPTO: US Patent and Trademarks Office 2016, website, August.

Von Hippel, F 2011, 'The Radiological and Psychological Consequences of the Fukushima Daiichi Accident', *Bulletin of the Atomic Scientists*, vol. 67, pp. 27–36.

WHO: World Health Organisation 2014b, 7 *Million Premature Deaths Annually Linked to Air Pollution*, News release, 25 March, Geneva, Switzerland.

WHO: World Health Organisation 2016, *Ionising Radiation, Health Effects and Protective Measures*, Fact sheet, updated April. www.who.int/mediacentre/factsheets/fs371/en/.

Wills, D 2016, 'Premier Jay Weatherill Effectively Buries Nuclear Waste Dump Proposal With Vague Promise of Statewide Referendum', *The Advertiser*, 15 November.

WNA: World Nuclear Association 2016, Radiation Health Effects, website. www.world-nuclear.org/information-library/safety-and-security/radiation-and-health/nuclear-radiation-and-health-effects.aspx.

Zhua, H, Deng, Y, Zhuc, R, and Hed, X 2016, 'Fear of Nuclear Power? Evidence from Fukushima Nuclear Accident and Land Markets in China', *Regional Science and Urban Economics*, vol. 60, September, pp. 139–154.

11 Case study: Robotics and artificial intelligence

Perhaps our two-faced dance partner is a robot. Robotics is concerned with connecting perception and action. Robots are essentially hardware and are not necessarily intelligent, although they are certainly artificial. Artificial intelligence, on the other hand, is software concerned with the acquisition, representation and use of knowledge (Brady 1984). Both have been a source of human hopes and fears for a long time.

There were about one and a half million slaves in ancient Italy out of a total population of six million during the lifetime of Augustus in the first century B.C. (Scheidel 2007). The benefits of slave ownership must have been highly valued in an era when human muscle and intelligence were little augmented by energy-exploiting machines. Nevertheless, the rewards of slave ownership were offset by risk. The many slave rebellions of earlier times include the three Servile Wars of the first and second centuries B.C.,[1] the Zanj revolt led by Ali bin Muhammad (A.D. 869–884) in the Middle East, and the creation of the Haitian state (1791–1803) led by Toussaint Louverture. Because masters feared their slaves, retribution was ruthless and bitter. Crucifixion was a common punishment for insurrection.[2] Likewise, there is contemporary unease as robots become more numerous and more intelligent – and therefore more threatening.

Still, the combination of strength and intellect in human slave form is as yet unrivalled in modern times. Certainly machines now multiply human muscle by many orders of magnitude. And computer programs exceed the human mind in myriad different – albeit singular – ways. But the computer that can land an aircraft is yet incapable of playing naughts and crosses or telling the difference between a dog and a parrot. These technologies are "ordinary artefacts that outperform us in ever more tasks, despite being no cleverer than a toaster" (Floridi 2016). Neither machine nor program is human-like and, despite hopes and fears to the contrary, may well never be.

Links

The development of modern robotics and nuclear energy are directly linked. Robotic limbs were first constructed for the manipulation of radioactive material during the 1940s as part of the Manhattan atomic bomb project. As with herbicides, nuclear power and robotics, the origins of artificial intelligence also lie with the military during the World War II when early computers were constructed and used for code breaking. Alan Turing, discussed earlier, proposed a chess-playing program in 1941 at Bletchley Park, in a further step in the direction of artificial intelligence. A decade later he remained optimistic about its prospects: "It seems probable that once the machine thinking method had started, it would not take long to outstrip our feeble powers" (Turing 1951, p. 475).

The word 'robot' is from the Czech 'robota' for 'slave' or 'serf' and was first used in a 1920 play 'R.U.R' (Rossum's Universal Robots) by the Czech satirist Karel Čapek. The play was an immediate hit in Prague, was translated into many languages and reached New York in 1922 (IFR 2012). Čapek portrayed robots as intelligent androids that enabled cheap labour and higher profit.[3] In the play, human civilisation becomes a type of 'Sodom' as people pursue gross pleasure, so easily are human desires fulfilled by robots. One particularly intelligent robot, Radius, wants not just his freedom, but also demands to command as humans do. Other robots similarly grow resentful of their lot. Ultimately a world-wide robot uprising ensues, echoing the slave revolts of earlier times. Pamphlets announce its objectives:

> Robots of the world, we enjoin you to exterminate mankind. Don't spare the men. Don't spare the women. Retain all factories, railway lines, machines and equipment, mines and raw materials. All else should be destroyed. Then return to work, it is imperative that work continue.
>
> (Čapek 2006, p. 44)

Only one human is left, Alquist, like Daedelus of ancient legend not a scientist but an artisan. Alquist helps robots to reproduce themselves and witnesses an emergent robot Adam and Eve (ibid., pp. 64–76).

Commercialisation

Despite this caution, the commercialisation of robotics was inspired by the fictional science of Isaac Asimov, who began writing short stories about robots in the late 1930s. But unlike Čapek's unruly slaves, and unlike every significant robot narrative that came before (Golem,[4] Frankenstein's Monster) Asimov's creatures are benign. His robots are wholly electro-mechanical, so that they can be programmed to protect humanity; they can be made safe. However, in line with fear of slaves, most earlier fictional robots were partly organic and fell prey to the 'rogue robot plot' in which they ultimately turn on their human creators (Pez 2016, part 1). As mentioned in Chapter 6, Asimov developed his 'laws of robotics' that, in a cascade of priorities, purportedly ensure that robots cannot harm humans. In summary, these are: (0) protect humanity (1) protect individual humans (2) obey human orders, and, of lowest priority (3) protect oneself.[5] Indeed, in the final book of his *Foundation* series, *Foundation and Earth*, Asimov reveals that one advanced robot, R. Daneel Olivaw, in his final resting place on the moon that still orbits a radioactive Earth, has been working to protect humanity throughout the galaxy over a future twenty thousand years (Asimov 1987).

Asimov fans, Joe Engelberger and George Devol, founded the company *Unimation*[6] in the United States in 1956, the first to manufacture robots commercially. Engelberger was a physicist who designed control systems for the new technologies of nuclear power plants and jet engines. Devol was an inventor who patented an invention called a 'programmed article transfer',[7] from which they together developed the first industrial robot arm, called the *Unimate*. The device was later licensed around the world to enable the development of other industrial robots (IFR 2012). Industrial robots used in manufacturing now come in many forms, but are generally designed to perform repetitive and dangerous functions such as spot welding, spray painting and stock transfer. These machines are subject to safety regulation and their numbers are increasing rapidly, although still miniscule compared to the human workforce. In 2016 the International Federation of Robotics estimated that there would be 2.6 million industrial robots by 2019 – an increase of a million in only four years (IFR 2016) – mostly in the United States, Germany, China, Japan and South Korea. There are also already several million small autonomous domestic robots that perform such tasks as vacuum cleaning, floor washing and pet impersonation, plus several thousand 'carebots' (Muoio 2015) that mimic some human expressions or provide other functions useful in health and aged care. Military robots, including drones and bomb disposal units, tend to operate under direct human control, whereas experimental agbots and environmental pest destroyers usually operate autonomously.

Artificial intelligence

While the term 'artificial intelligence' is often used as if its meaning is self-evident, it remains a source of confusion and controversy (Lewis-Kraus 2016). Insofar as it is the software that acquires data and applies knowledge, artificial intelligence is independent of physical form and function, although it may enable particular functions. So far, it is not at all similar to human or biological intelligence. Neither is it conscious. Nevertheless, forms of artificial intelligence can already do many things better than any human can and their variety is astonishing. Johnson (2016) argues that artificial intelligence is the most powerful technology that human intelligence has yet developed. The philosopher Timothy Morton says that we are already controlled by a primitive form of artificial intelligence – industrial capitalism (cited in Blasdel 2017) – which is pause for thought.

As mentioned in Chapter 6, 'machine learning' is a common form of artificial intelligence (AI) that is dumb, but has yet triumphed over more sophisticated intelligence through the use of prodigious amounts of data. And such available data are ever increasing, making machine learning AI potentially more and more capable. Such AI can instantly translate from one language to another, it can help make better medical diagnoses and it can facilitate intensive agriculture, for example. It can increasingly reliably recognise individual human faces and drive cars. But it can also optimise bio-weapons and create financial system meltdown (Kaspersen 2016).

Figure 11.1 depicts the relationship between AI and machine learning, as well as its overlap with other forms of data analytics. Machine learning is a form of AI. Except for databases, both overlap to some extent with all other forms of data analysis, such as pattern recognition, data mining and computational neuroscience. The diagram also suggests just how complicated and uncertain is the association between these inter-related fields.

There is an important further distinction within AI itself: there is narrow task-based AI, such as translation software, which is different from the concept of a much wider human-like artificial *general* intelligence (AGI), such as Clarke and Kubrick's HAL 9000 in *2001 A Space Odyssey* (which incidentally illustrates conflict between Asimov's zeroth and first laws). The former tends to be regarded simply as a remarkably useful tool, whereas the latter tends to evoke fear that it may eventually rebel and control or destroy humanity:

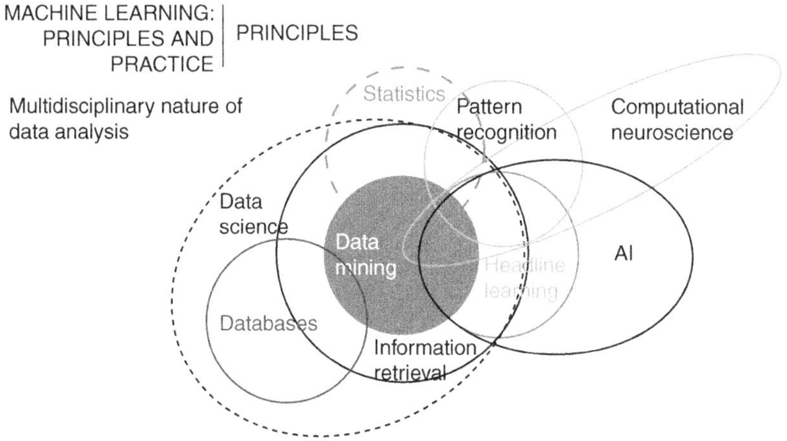

Figure 11.1 Relationship between artificial intelligence and forms of data analytics.

Source: Hall 2014. Copyright © 2014. SAS Institute Inc. All rights reserved. Reproduced with permission of SAS Institute Inc., Cary, NC, US.

DAVE: *Open the pod bay doors, Hal.*

HAL: *I'm sorry Dave. I'm afraid I can't do that... This mission is too important for me to allow you to jeopardise it.*

(Kubrick 1968)

Narrow AI has been successful. Wider AGI, however, has failed to develop over the past several decades of research. While there is considerable disagreement, some predict that AGI might be achieved by 2050, others say earlier, or 'never'. Still others warn we should not be complacent just because it seems far away (Häggström 2016). Once AGI exceeds human capacity, there is a speculative 'singularity', the point at which AGI can create even more intelligent AGI indefinitely, rendering humanity obsolete, or subservient to the whims of our super-intelligent progeny. The seminal formulation of this concept by I. J. Good (1966),[8] however, went beyond 'singularity' to an 'intelligence explosion':

> Let an ultraintelligent machine be defined as a machine that can far surpass all the intellectual activities of any man however clever. Since the design of machines is one of these intellectual activities, an ultraintelligent machine could design even better machines; there would then unquestionably be an "intelligence explosion," and the intelligence of man would be left far behind. Thus the first ultraintelligent machine is the last invention that man need ever make, provided that the machine is docile enough to tell us how to keep it under control.
>
> (Good 1966, p. 33)

Intriguingly, this idea has been compared with Fermi's[9] first self-sustaining nuclear pile in Chicago in 1942. The addition of the final layer of uranium bricks (layer 58) brought the number of neutrons emitted within the pile to a level that then was more than self-sustaining. The reaction increased so rapidly that it had to be shut down after 28 minutes. If not, much of the university would have been destroyed along with Fermi himself. A similar AI scenario – without the shutdown – has been termed 'AI-go-FOOM' (Yudkowsky 2013, p. 6).

Yet the intelligence 'return on cognitive investment' may be unlike a nuclear chain reaction. The advances made by an artificial ultra intelligence might be simply linear – double the time, double the intelligence – rather than exponential. Or advanced AGI might suffer from diminishing returns once easily accessed information systems have been exploited (ibid., pp. 18–20). The intelligence explosion that Good envisaged well might never happen.

Certainly current AGI is neither very intelligent or self-aware. Recent Turing tests that ask an observer to tell the difference between human and computer responses have swiftly identified the non-human by asking questions like *How do you put on a boot?* While the human respondent mentions undoing bootlaces and holding back the tongue, the AGI asks *What is a boot?* Others have claimed that the test has been passed, but the evidence is dubious, based on contorted rules and results (Masnick 2014). In contemporary science fiction, a future involving a spaceship's artificial intelligence system that is downloaded into an android body is explored. The original ship-wide system has eyes and ears everywhere, and can access knowledge instantly. The android body is, however, isolating, constricting and has to rely on its own memory for knowledge (Chambers 2016). In this imagining, it is simply assumed that self-awareness is viable. The spaceship AGI is self-aware as is its android download, but how so is unexplained.

In reality, while 'chatbots' like China's *Xiaoice* (Wang 2016) and assistants like Google's *Siri*, might appear self-aware, their appearance is deceiving. Their responses are based on assimilation of millions of examples of conversations through machine learning. It may well be that human-like AGI has been elusive because it attempts to use symbols to represent an objective world. By contrast, human intelligence "thinks with our whole body, not just the brain", in a world that has been revealed to us through our senses over each lifetime. Humans thus embrace rich models of reality and can predict outcomes from few examples, unlike AGI that needs to be shown again and again (Wilson 1998; Medlock 2017).

But since the era of Turing, Good and Clarke, and despite slow progress in achieving human-like AGI, the issue of how to control it, should it ultimately emerge, has become more pointed. Stephen Hawking, Bill Gates and Elon Musk are famously among those who have expressed wariness of its potential. Hawking et al. (2014) for example, deplore the lack of serious research into AI[10] given that "we are facing potentially the best or worst thing ever to happen to humanity" involving possible "incalculable benefits and risks". This view is more negative than Good's sentiments. While he did proviso machine docility, the opening sentence of his paper is "The survival of man depends on the construction of an ultraintelligent machine" (Good 1966, p. 31), as if he was anticipating some artificial deity that will save humanity from disaster. The structure of his AGI is based on the form of the human brain, so that this God would be made in man's image. Hawking and his colleagues are from this less optimistic age, when the artificial intelligence explosion appears closer, but also dangerous and possibly malevolent.

There is another important matter. As part of a future that is as yet unevenly distributed, our young are beginning to grow up interacting with artificial intelligence. Our children are associating with forms of AI on a familiar day-to-day basis such that they tend to regard its forms as part of normal everyday life, unremarkable, ordinary, even in a sense, natural. The Amazon AI *Alexa's* growing relationship with a three-year old human, Grace, is related by Grace's mother:

> With some trepidation, I watched my daughter gaily hand her decisions over. "Alexa, what should I do today?" Grace asked in her singsong voice on Day 3. It wasn't long before she was trusting her with the big choices. "Alexa, what should I wear today? My pink or my sparkly dress?".
>
> (Botsman 2017)

It seems that Alexa, represented by a cylindrical microphone on a kitchen bench, is more capable than its Wang or Google equivalent. It can not only intelligibly answer a wide range of questions that a three-year-old might ask, it can play music, predict the weather and most importantly it can both recommend and decide. If Grace wants blueberries it will order them. If Grace wants clothing advice it recommends and may buy favoured items, from Amazon. While it has useful and amusing functions, it is insidious. Its algorithms are designed to engender trust – and to sell stuff. While this particular Alexa was consigned to the closet, the issue for our children who will grow up surrounded by different forms of AI will be how to maintain a wariness of its ultimate purpose. But if 'selling stuff' can be designed into AI as its ultimate purpose, then so can other motivations: even the ethical and the sustainable, for example.

Sustainability benefits

The 'incalculable benefits' of a super intelligence might include solutions for clean energy, the maintenance of Earth systems, the elimination of disease and the end of the era of extinctions. But in the meantime, a combination of contemporary artificial intelligence with robotics also augers potentially significant advances for sustainability. While they are separate fields, the two technologies overlap in intelligent robots that can acquire and assess new information and act on it independently. For example, a promise for sustainability is illustrated with robots under development at the Queensland University of Technology (QUT) in Australia. These robots can identify and precisely eliminate different types of weeds in agriculture – chemically, mechanically and by zapping them with microwaves – ways much more targeted than the typical blunderbuss spraying of herbicides. A similar system has been developed by the German multinational, *Robert Bosch*, since 2014. This particular mobile 'agbot', first identifies all desirable plants on a geo-spatial grid. As it autonomously patrols each field, it detects plants that depart from this pattern and eliminates them mechanically – in a return to the pre-chemical method, albeit updated (Albert 2016; Van Woensel and McCormack 2016a). *Abundant Robotics'* (US) robot visually recognises apples on the branch that are ready for harvest and picks one apple a second using a vacuum grasp. *Blue River's* (US) robot uses computer vision and machine learning to identify plants in need of chemical treatment, weeding or thinning. The company claims this can reduce the amount of chemicals used in agriculture by 90 percent (Burwood-Taylor 2017). Marine sustainability is enhanced by a submersible robot shaped like a torpedo with a single extensible arm, which independently identifies and injects acids into crown of thorn starfish that threaten Australia's Great Barrier Reef.[11]

As illustrated by these examples, many technologies now depend on a world increasingly enveloped by devices, sensors, applications and data, which has resulted in an IT-friendly environment (Floridi 2016). Even if a superintelligence does not eventuate, machine learning that combines algorithms, memory and vast datasets offers increasingly reliable predictions of weather, sea states and environmental change, as well as how they can be managed. With or without robotics, self-awareness or general intelligence, machine learning is already applicable to many sustainability issues and their potential resolution. For example, the several climate models used by the UN Intergovernmental Panel on Climate Change are tracked and constantly improved by machine learning that compares predictions to results. The technique derives patterns that can also apply to the past by filling in gaps when data were scarce (Deaton 2014).[12]

By contrast, the co-founder of Google, Sergey Brin, speaking at the *World Economic Forum* appears uninspiring about the possibilities of machine learning and AI. Although enthusiastic, the examples he cites are mundane: AI "touches every single one of our main projects, ranging from search to photos to ads … everything we do" (Chainey 2017). Its application to sustainability appears to be low on Google's agenda. Paying academics for favourable research may be a higher priority (Bridge, Whipple and Moody 2017).[13]

Oversight

Little has been done to oversee such technologies. Much of that which is positive appears due to serendipity rather than intent. As recently as 2016, the UK Parliamentary committee on robotics and AI commented without any apparent urgency that:

> There has been some discussion about who should be involved in identifying, and establishing, suitable governance frameworks for robotics and AI. Kate Crawford from Microsoft has argued that 'like all technologies before it, artificial intelligence will reflect the values of its creators. So inclusivity matters – from who designs it to who sits on the company boards and which ethical perspectives are included'.[14]
>
> (UK House of Commons 2016, para. 66)

The committee went on to recommend the establishment of a permanent committee to be operated from the Alan Turing Centre. The European Union appears to be in a similar frame of mind. The author of a 2016 EU report on robotics and AI talks of a framework for current robots and those available over "the next ten to fifteen years" (Hern 2017a), but is also concerned with international regulation:

> In view of the development of robotics and AI all over the world, consideration should be given and initiatives taken to amend existing relevant international agreements when needed or to draft new instruments with the objective of introducing specific references to robotics and AI. International cooperation in this field is very much desirable.
>
> (Delvaux 2016, p. 22)

Like the United Kingdom, the EU report proposes the establishment of a separate agency on the issue (ibid., p. 7). The report is also concerned that Asimov's laws should be incorporated into a code of ethics for robot and AI designers, and that smart robots should become registered legal entities subject to relevant law.

In the United States, the Obama administration held a series of workshops on the "benefits and risks of artificial intelligence", and released a report, 'Preparing for the Future of Artificial Intelligence' (Bensinger 2016). However, as of 2017, the new White House website was silent on the issue. The *Association for the Advancement of Artificial Intelligence* (AAAI)[15] research group took part in the workshops and the report. But the sort of intelligence explosion that Good (and Hollywood)[16] envisage has received a luke-warm reception at such levels. Still, the report does point to three ways that AGI might be developed: (1) by gradually broadening the scope of narrow AI systems to cover a wider range of less structured tasks, (2) by progressive unification of existing methods to address more applications that previously required multiple methods, and (3) by solving specific technical challenges that open up new ways forward, such as 'transfer learning', which would create a machine learning algorithm that can be applied to a range of new applications (ibid.).

At the other end of the opinion spectrum, the entrepreneur, Elon Musk, has injected a sense of urgency into AI regulation:

> We should be very careful about artificial intelligence. If I were to guess what our biggest existential threat is, it's probably that. … Increasingly, scientists think there should be some regulatory oversight maybe at the national and international level, just to make sure that we don't do something very foolish.
>
> (Musk, cited in Floridi 2016)

Major US Silicon Valley firms Google, Facebook, Amazon, IBM and Microsoft[17] have formed a partnership to consider such matters as the ethics and trustworthiness of AI. Such considerations were also supposed to be on the main agenda for an ethics board to be created when Google bought out the UK-based AlphaGo-winning *DeepMind* in 2014. However, as at 2017, if the board had even met, it had not by then made its views public (Hern 2016, 2017b).

This relaxed attitude can be traced back to a meta-study which analysed 257 statements about AGI made since 1950 and assessed predictions as to when AGI will be achieved. It found "no evidence that expert predictions differ from those of non-experts" and that "timeline predictions are likely unreliable" because they were dependent on when they were made (Armstrong et al. 2014, p. 12). The key political issue of robotics and AI is not 'singularity' but rather the replacement of human workers by automation and its effects on society. The OECD, for example, has a blog called *Robotenomics* that concentrates on this question. So did Čapek nearly a century ago:

> Domin: In ten years' time Rossum's Universal Robots will be making so much wheat, so much material, so much of everything that nothing will cost anything. Everyone will be able to just take as much as he needs. Nobody will live in poverty. They won't have jobs, that's true, but that's because there won't be any jobs to do. Everything will be done by living machines. People will do only the things they want to do, they can live their lives just so that they can make themselves perfect.
>
> (Čapek 2006 [1921], p. 19)

AI singularity may or may not turn out to be the more important political question eventually, but is a risk that is not yet addressed through governance. Despite its military origins, robotics and artificial intelligence has since developed as a market-led technology. Unlike glyphosate and the chemical technologies developed by large MNCs, it appears more transparent and participatory.

However, it is not especially accountable as such mechanisms are underdeveloped, despite a potentially strong policy capacity. Its present main weakness is clearly integrity. As well as the lack of diversity in its recruitment leading to cases of harassment (such as Uber in 2017), Silicon Valley company ethics are known for its dearth, probably related to limited accountability. Nevertheless, the EU has recently imposed record fines on such companies due to unethical practices (Cox 2017).

Apart from its effects on humanity, outside of fiction there is almost no discussion of the ethics of creating a super-intelligent being from the point of view of the being itself, yet this is of great importance. Is it ethical to create such an intelligence but at the same time restrict its freedoms through some variant of Asimov's laws, for example? By contrast, humanity has laws, but also the freedom to break those laws irrespective of consequence (Camus 2011 [1942]), while an AGI would have no such freedom. It would be bound by the laws integrated with its existence. Discussions are almost entirely anthropocentric, typically 'what would be the impact of a super-intelligence on human society?', instead of 'what would be the effect of humanity on super-intelligence?', let alone 'what would be the effect of a super-intelligence on the biosphere?'.

At least the issue of AI's contribution to sustainability is beginning to be addressed by some such as the Oxford ethicist, Luciano Floridi, who believes that "we should make AI environment-friendly" and as smart as possible so as to "tackle the concrete evils oppressing humanity and our planet". In this he is clear about our responsibility:

> We are and shall remain, for any foreseeable future, the problem, not our technology. Churchill said that 'we shape our buildings and afterwards our buildings shape us'. This applies to the infosphere and its smart technologies as well.
>
> (Floridi 2016)

It also applies to technology more generally. If we are to become sustainable, humanity must consciously shape the technologies we create. That relationship has yet to be achieved.

Even the sustainability bond between robotics and nuclear power is fragile, at least at Fukushima where the clean up will take several decades:

> As the 60cm-long Toshiba robot, equipped with a pair of cameras and sensors to gauge radiation levels was left to its fate last month, the plant's operator, Tokyo Electric Power (Tepco), attempted to play down the failure of yet another reconnaissance mission to determine the exact location and condition of the melted fuel.
>
> (McCurry 2017)

Notes

1 The third Servile War (73–71 B.C.) was famously led by Spartacus.
2 For example, 6,000 rebel slaves were crucified along the main roads leading into Rome after the third Servile War.
3 "on the wall stage right are printed posters: 'The Cheapest Workforce You Can Get: Rossum's Robots'" – R.U.R (Čapek 2006, p. 3).
4 The 'Golem' legend about the creation of a literally obedient giant designed to protect Jews may be only a few centuries old, but was centred in Prague, consistent with the robots of Čapek.
5 Originally only the latter three. The 'zeroth law' was added during the *Foundation* series.

6 Based on the words 'universal' and 'animation'.
7 US patent no. 2,988,237.
8 UK mathematician Irving John Good (formerly Isadore Jacob Gudak) articulated this in his paper 'Speculations Concerning the First Ultraintelligent Machine' (Yudkowsky 2013). Good worked with Turing on early computer design and advised Stanley Kubrick on *2001: A Space Odyssey*.
9 Enrico Fermi (1901–1954), the Manhattan project physicist.
10 "little serious research is devoted to these issues outside small non-profit institutes such as the Cambridge Center for Existential Risk, the Future of Humanity Institute, the Machine Intelligence Research Institute, and the Future of Life Institute" (Hawking et al. 2014).
11 Personal discussion with Professor Tristan Perez at QUT Brisbane, Australia on 23 September 2016.
12 The technique is known as 'sparse matrix completion'.
13 Since 2005, Google funded 329 research papers paying millions of dollars to academics for research that it hoped would sway public opinion and influence policy in favour of the tech giant.
14 Originally published in the *New York Times* as 'Artificial Intelligence's White Guy Problem', by Kate Crawford 25 June 2016.
15 Formerly the American AAAI.
16 As in the 2014 movie, 'Transcendence'.
17 Apple since joined in 2017.

References

Albert, A 2016, *From the Internet of Fields to the Internet of Plants*, Presentation at Hobart, Australia for 'The Yield', TasICT and the Australian Computer Society, Deepfield Robotics (Robert Bosch), Stuttgart, Germany.

Armstrong, S, Sotala, K and ÓhÉigeartaigh, S 2014, *The Errors, Insights and Lessons of Famous AI Predictions and What They Mean for the Future*, 20 May, Machine Intelligence Research Institute, Berkley, CA.

Asimov, I 1987, *Foundation and Earth*, Collins, London, UK.

Bensinger, R 2016, *White House Submissions and Report on AI Safety*, Machine Intelligence Research Institute, News, 20 October.

Blasdel, A 2017, 'A Reckoning for Our Species': The Philosopher Prophet of the Anthropocene', *The Guardian*, 15 June.

Botsman, R 2017, 'Co-Parenting with Alexa', *New York Times*, 7 October.

Brady, M 1984, *Artificial Intelligence and Robotics*, A.I. Memo 756, Massachusetts Institute of Technology, Artificial Intelligence Laboratory, Cambridge, US.

Bridge, M, Whipple, T and Moody, O 2017, 'Google PAYS ACADEMICS MILLIONS FOR KEY SUpport', *The Times*, 13 July.

Burwood-Taylor, L 2017, 'AgFunder Innovation Awards: 2016 Winners Announced', *Agfunder News*, 19 January.

Camus, A 2011 [1942], *L'Étranger*, Ebooks libres et gratuits.

Čapek, K 2006 [1920], *R.U.R. (Rossum's Universal Robots)*, A play in introductory scene and three acts, Wyllie, D (trans.), eBooks@Adelaide, Australia.

Chainey, R 2017, 'Google Co-Founder Sergey Brin: I Didn't See AI Coming', *World Economic Forum*, 19 January.

Chambers, B 2016, *A Closed and Common Orbit*, Hodder and Stoughton, London.

Cox, J 2017, 'Google Hit with Record EU Fine over 'Unfair' Shopping Searches', *The Independent*, 27 June.

Deaton, J 2014, 'What Machine Learning Can Do for Climate Science', *Planet Forward*, 12 May.

Delvaux, M 2016, *Draft Report with Rrecommendations to the Commission on Civil Law Rules on Robotics, Committee on Legal Affairs*, European Parliament 2014–2019, 31 May.

Floridi, L 2016, 'Should We Be Afraid of AI?', *Aeon*, 9 May.

Good, I 1966, 'Speculations Concerning the First Ultraintelligent Machine', *Advances in Computers*, vol. 6, January, pp. 31–88.

Häggström, O 2016, *Here Be Dragons: Science, Technology and the Future of Humanity*, Oxford University Press, Oxford, UK.

Hall, P 2014, *Machine Learning: Principles and Practice*, Webinar 8.12 min., SAS Institute, Cary, NC.

Hawking, S, Tegmark, M, Russell, S and Wilczek, F 2014, 'Transcending Complacency on Superintelligent Machines', *Huffington Post*, blog, 19 April, updated 19 June.

Hern, A 2016, "Partnership on AI' Formed by Google, Facebook, Amazon, IBM and Microsoft', *The Guardian*, 29 September.

Hern, A 2017a, 'Give Robots 'Personhood' Status, EU Committee Argues', *The Guardian*, 13 January.

Hern, A 2017b, 'Whatever Happened to the DeepMind AI Ethics Board Google Promised?', *The Guardian*, 27 January.

IFR: International Federation of Robotics 2016, *History of Industrial Robots: From the First Installation Until Today*, Frankfurt am Main, Germany.

Johnson, B 2016, 'The Combination of Human and Artificial Intelligence Will Define Humanity's Future', *TechCrunch*, 12 October.

Kaspersen, A 2016, 'We're on the Brink of an Artificial Intelligence Arms Race: But We Can Curb It', *World Economic Forum*, 15 June.

Kubrick, S 1968, *2001: A Space Odyssey*, movie, MGM, UK-US.

Lewis-Kraus, G 2016, 'The Great A.I. Awakening: How Google Used Artificial Intelligence to Transform Google Translate, One of Its More Popular Services — And How Machine Learning Is Poised to Reinvent Computing Itself', *New York Times Magazine*, 14 December.

Masnick, M 2014, 'Overhype: No, a 'Supercomputer' Did NOT Pass the Turing Test for the First Time and Everyone Should Know Better', *TechDirt*, 9 June.

McCurry, J 2017, 'Dying Robots and Failing Hope: Fukushima Clean-Up Falters Six Years after Tsunami', *The Guardian*, 9 March.

Medlock, B 2017, 'The Body Is the Missing Link for Truly Intelligent Machines', *Aeon*, 17 March.

Muoio, D 2015, 'Japan Is Running out of People to Take Care of the Elderly, So It's Making Robots Instead', *Business Insider Australia*, 22 November.

Pez, J 2016, 'The History of the Positronic Robot and Foundation Stories', *Asimov Online*, website.

Scheidel, W 2007, 'Roman Population Size: The Logic of the Debate', Princeton/Stanford Working Papers in Classics, Version 2.0, July. www.princeton.edu/~pswpc/pdfs/scheidel/070706.pdf.

Turing, A 1951, *Intelligent Machinery, A Heretical Theory*, Lecture notes, pp. 472–475.

Turing, A 2013 [1947], 'Lecture to the London Mathematical Society on 20 February 1947', in Cooper, S and van Leeuwen, J (eds), *Alan Turing, His Work and Impact*, Elsevier, Cambridge, US, pp. 481–497.

UK House of Commons 2016, *Robotics and Artificial Intelligence*, Science and Technology Committee, Fifth Report of Session 2016–2017, 12 October.

Van Woensel, L and McCormack, S 2016a, *Will Robots Change the Face of Agriculture and Food Production?* European Parliamentary Research Service, blog, 25 August.

Wang, Y 2016, 'Your Next New Best Friend Might Be a Robot: Meet Xiaoice; She's Empathic, Caring, and Always Available – Just Not Human', *Nautilus*, no. 033, 4 February 2016.

Wilson, E 1998, *Consilience: The Unity of Knowledge*, Vintage, Random House, New York, US.

Yudkowsky, E 2013, *Intelligence Explosion Microeconomics*, Technical report 2013-1, 13 September, Machine Intelligence Research Institute, Berkeley, US.

12 Case study summary

As the French philosopher, Paul Valéry (1931) observed, "the future, like everything else, is no longer what it used to be". Our imagined futures continually change, much according to the emerging technologies of the present. These case studies illustrate several concerns about our current connection with technology. One underlying concern is the fear that lurks within our relationships. We fear a slow poisoning or starvation that herbicides might unleash. There may be yet unseen effects on other life forms. We dread the risk of nuclear disaster that would threaten health over generations. We are afraid our robots could rise against us as slaves did in the past. Still we gamble on those risks. Humanity has woven systems that lock in blanket herbicide use. We have resumed building nuclear power stations and we accept unhindered development of robots and artificial intelligence. How we manage such risks is a matter of governance.

Fiction

These examples also show that inspiration to conceive new technologies has come from literature, especially from the science fiction that underlies robotics and AI. In this way it is not just that the present determines the future. It is also that the future exerts a pull on the present. It is in fiction that the conundrums of technological ethics and governance have been debated well in advance of reality: the ethics and control of artificial beings is a trope of science fiction (e.g. Asimov 1987). Kubrick's movie *Dr Strangelove* (1964) highlighted absurd weaknesses in the governance of nuclear weapons systems. Neville Shute's 1957 novel *On the Beach*[1] made stark its humanity-ending results. Pfister's movie *Transcendence* (2014) portrayed the dystopian results of AI singularity. Such fiction is more than entertainment. Its explorations can inform how technology might be controlled.

Science

The case studies indicate that scientists differ in interpreting results. Scientists are part of the cultural and political fabric of paradigm and influence. And while many depend for their income on industry and industry depends on profit, reputation amongst scientists themselves depends on peer approval; so is dissent confined. As Michael Polanyi (1962, p. 5) points out, the elements of scientific merit are in a sort of creative tension: originality, which promotes dissent is esteemed as much as plausibility and scientific value, both of which promote conformity. His 'republic of scientists' may resemble aspects of a body politic, but scientists are neither politicians nor (often) regulators.

Politics

The intersection of science and politics is critical to the governance of technology. But while the views of scientists are essential in assessing that direction, there are other interests that need to be heard from if technology is to be governed in the interests of sustainability. There is support from organised religion, for example, personified by Pope Francis, who, contrary to some reports, may not have said that "capitalism is the dung of the devil",[2] but who nevertheless counsels for sharing our common home and against the love of money (Plis 2015).

Against the *TAPIC* criteria, there are substantial areas of concern:

Transparency

Industry is typically in advance of government when deploying technologies, including those that affect sustainability; regulation follows commercialisation. Nevertheless, in a competitive corporate environment consumer and political pressures are magnified through social media enabled by the Internet. This can prompt sustainability considerations by firms before government is aware. Yet corporate deliberations are not always open to public scrutiny. As shown with glyphosate, secrecy based on outdated patent rights has rendered many industry studies opaque. The nuclear club relies more on public relations (PR) than transparency. With artificial intelligence, ethics committees tend to be 'internal'; so far not for public debate. It is left to individual outliers like Hawking and Musk to warn of danger. Others point out that human stupidity, rather than artificial intelligence, remains the greatest risk (Cave 2017). Governments move slowly as if regulation might disturb the mystical process of innovation.

Accountability

The military is intimately involved in the development of modern technologies and its influence is striking. Corporate industry became involved in the commercialisation of many such technologies only after the World War II.

Post-colonial conflict in both Africa and Asia further sped the demand for herbicides, while nuclear and robotic technologies were spurred on during the Cold War. Today robot drones kill on one continent while controlled from another. There is the prospect of autonomous robotic soldiers. Artificial intelligence guides weapons systems as well as enabling driverless cars. The military, as Eisenhower warned, is both economically and politically powerful, especially in the United States and China. The central issue to the governance of technologies it spawns is that the military is intrinsically secretive and, at least technologically, arguably beyond democratic control. In the commercial arena, there is as yet no accountability for the effects of robotics and AI. There is political concern, but only for the job loss effects of automation. For nuclear risks, the IAEA is more concerned with proliferation than safety, and the industry body appears ineffective.

Participation

One of the most important findings from the case studies is that governing technologies often involves few people, who tend to share a common view of risk and reality. Because technology clubs are exclusive, not inclusive, this results in a skewed set of priorities. Disciplines are narrow and secluded. Technical knowledge is often isolated from real world experience of its effects. Political connection trumps technical knowledge. Governing technologies for sustainability thus defers to a passive attitude of technological inevitability. With chemicals such as glyphosate, participation tends to be limited to scientists rather than end-users and consumers. Nuclear power risks are monitored in exclusive technocratic enclaves that have proven inadequate and remain ineffective, including the industry body. The highly technical nature of AI daunts the participation of non-experts despite its potential impact on the future, and ICT corporations tend to view stakeholders as consumers.

Integrity

As Milton Friedman (1982) observed, regulatory agencies tend to be captured by producers, which appears to be true of the relationship between the EPA and Monsanto, for example, and will be made more likely at global level as the Bayer-Monsanto takeover consolidates. There is also evidence that academics are paid for favourable research by major firms in at least two sectors. The nuclear industry is affected as much by economics as it is by safety and sustainability. Evidence of pro-nuclear front groups rewarding former EPA chiefs is also concerning for the integrity of its governance. Robotics and artificial intelligence developments are without any real oversight, which proceeds with a lethargy that contrasts starkly with the speed of industry innovation. The integrity of public organisations involved in regulating its development has not emerged as an issue because there is essentially no such regulation, although the European Union has been active in the wider ICT industry. Ethical think tanks are at best, nascent. Yet this criterion is where the need for mission clarity is defined. This is where leadership and initiative converge. There appears to be a profound lack of the sort of ethical leadership that defines technology governance as primarily aiming for sustainability.

Capacity

At the global level, technological governance remains disjointed, with different approaches between institutions and inconsistent rule-making. The chemical industry is fraught with legal and PR battles. Nuclear regulation focused more on non-proliferation than safety, until the industry itself was forced to act after Chernobyl – albeit with the lethargy of the incapacitated. Nuclear waste storage remains a major problem; no state is prepared to suffer the ignominy of becoming an international nuclear waste dump and there is neither capacity nor authority to determine otherwise. While corporate interests exert less leverage on regulatory standards at global level than at national level, this may be challenged by the 2018 Bayer chemical takeover of Monsanto. By contrast, potential policy capacity in robotics and AI governance is high, as stakeholders tend to be well-educated and it applies across many fields. Table 12.1 summarises the case study findings against the TAPIC criteria.

Overall, the case studies indicate that there are significant deficiencies in the governance of representative technologies, irrespective of mode. Both state and market-led approaches are wanting against most criteria, which allows the 'dark side' of the technology Janus to present itself should sustainability be strongly pursued. This dark side emerges primarily because stakeholders are neither engaged nor informed and therefore cannot have any part in determining or owning the decisions and impacts of these powerful technologies. Instead, however well-meaning, participants are few, technocratic and cloistered, bound by commercial and legal constraint as well as a culture of secrecy and self-reference. At worst, there is regulatory capture due to the political strength of corporations, or else concealment of major risks.

It has been pointed out. that shifting "from government to governance spells a change in decision making and numerous opportunities for the pursuit of sustainability" (Kemp, Parto and Gibson 2005, p. 18). If better technology governance is to be effected in the quest for sustainability, then a system that *relies* on participation and engagement is indicated. A form of network governance that may be viable is outlined in the following chapter, along with other alternative measures that would support its operation and strengthen its results.

Table 12.1 Governance strengths and weaknesses of case studies against TAPIC criteria: Summary

Paradigm	*Transparency*	*Accountability*	*Participation*	*Integrity*	*Capacity*
State-led Nuclear electricity	Low – state and industry secrecy due to fear of risk exposure.	Variable – military origins, little evidence of sanction. IAEA not focussed on safety.	Narrow – limited to industry and technical representation.	Moderate – bracket creep, industry front groups. Duties well-defined, but enforcement lacking.	Weak – slow, waste storage unresolved, engagement poor, inter-agency duties detract.
Market-led Chemicals (glyphosate)	Low – corporate secrecy, patents, military origins.	Low – military origins, regulatory capture, Monsanto adverse influence.	Narrow – rights to RandD tend to lack wider scrutiny.	Low – Monsanto influences science and politics. Global institutions divided.	Weak – lack of resources. Stakeholder engagement absent. Standards not coherent.
Market-led Robotics and artificial intelligence	Limited – due to data ownership, innovation, speed of change. Ethics boards yet to report.	Low – military origins, no mechanisms, codes or regulation, except robot safety rules. AGI at think-tank stage.	Moderate – epistemic and corporate involvement in research, but groups seen as consumers, not participants.	Low – ethics questionable, corrupt practices in EU, mission clarity lacking. Ethics boards not yet met. Academics paid for favourable research.	Potentially high – stakeholders well-educated, but problem identification and solution formulation at early stage.

Notes

1 *On the Beach* was made into a United Artists film of the same name in 1959, directed by Stanley Kramer.

2 This is a disputed quotation. It was a speech in Bolivia where Francis was referring to the greed for money, and was quoting a fourth-century saint, Basil of Caesera (Plis 2015). Other sources fail to substantiate the claim despite headlines consistent with it (e.g. Reuters 10 July 2015).

References

Asimov, I 1987, *Foundation and Earth*, Collins, London, UK.

Cave, S 2017, 'Intelligence: A History', Leverhulme Centre for the Future of Intelligence, *Aeon*, 21 February.

Friedman, M 1982 [1962], *Capitalism and Freedom*, University of Chicago Press, Chicago, US, and London, UK.

Kemp, R, Parto, S and Gibson, R 2005, 'Governance for Sustainable Development: Moving from Theory to Practice', *International Journal of Sustainable Development*, vol. 8, nos. 1/2, pp. 12–30.

Kubrick, S 1964, *Dr Strangelove or: How I Learned to Stop Worrying and Love the Bomb*, movie, Colombia Pictures, UK-US.

Pfister, W 2014, *Transcendence*, movie, Warner Bros Pictures, Hollywood, CA, US.

Plis, I 2015, 'No, Pope Francis Didn't Call Capitalism 'The Dung of the Devil', *Daily Caller*, 7 October.

Polanyi, M 1962, 'The Republic of Science: Its Political and Economic Theory', *Minerva*, vol. 1, Winter, pp. 54–73.

Shute, N 1957, *On the Beach*, William Morrow, New York, US.

Valéry, P 1931, "Notre Destin et Les Lettres", *Regards sur le monde actuel* (essays), Stock, Paris, France.

13 Alternative visions

The current widespread mood of pessimism and belief in the inevitability of future dystopias, reflected and reinforced by literature, movies and popular music, centres on technology. It is not difficult to be swept up by such attitudes. They are also common in academic circles, especially in relation to the effects of climate change and the nature of the political economy that drives it. Frances Fukuyama, for example wrote that:

> The end of history will be a very sad time… daring, courage, imagination, and idealism, will be replaced by economic calculation, the endless solving of technical problems, environmental concerns, and the satisfaction of sophisticated consumer demands.
>
> (Fukuyama 1989, p. 25)

Nonetheless, others equally prestigious, such as John Ruggie (2014, p. 6), have different views on prospects for sustainability: "And yet as has been said about Wagner's music, the situation may not be as bad as it sounds.".

On the American Samoa island of Ta'ū, where the anthropologist Margaret Mead once based her research, all electricity is produced from an array of solar panels combined with 60 Tesla powerpacks, plus a microgrid for distribution.[1] The system can power the entire island for three days without sunlight and recharges within seven hours (Etherington 2016).

Within the Ukraine's Chernobyl exclusion zone, two Chinese companies are building a solar farm that will produce a gigawatt of electricity, enough to power 700,000 homes. The site has been made more secure by placing a giant steel lid over the meltdown site, land is cheap and grid infrastructure is already in place (Cooke 2017).

Apart from these two particular applications, some other sustainable technologies are already widespread. They include energy generation systems such as hydro, tidal and wind, as well as illumination by light-emitting diodes (LEDs). Electric transport can be sustainable depending on energy source. There are now many low-energy homes independent of the grid. Apart from these examples, however, few sustainable technologies are fully developed. Certainly there are promising advances such as 'scrubbing towers' that remove pollutants from the air, bladeless wind generators, and nanotube filters that remove heavy metal contaminants from water (Wang 2016). There are phyto-remediation methods that use plants to remove soil contamination (Lasat 2000), techniques that promise to restore entire marine ecosystems using artificial reefs and oyster spats (Grovenor 2016) as well as metal-organic machines that harvest fresh water from the air (Kim et al. 2017). But these are only technical advances, yet to be commercially rolled out.

In more general terms, business interests involved with the World Economic Forum have identified 'fourth industrial revolution' (4IR) technologies for the Earth. These include: advanced materials, cloud technology, drones and autonomous vehicles, synthetic biology, virtual and augmented reality, artificial intelligence, robotics, blockchain, 3D printing, and the Internet of things (WEF 2017, annexe 1). While these technologies may not automatically contribute to sustainability, they are all light-footed and indicate promising trends. Importantly, most are Internet dependent. All have a digital base:

Meanwhile business as usual – strip mining, fracking, drilling, trawling, land clearing, broad acre monoculture, fossil-fuelled and polluting technologies – continue as typical, conventional and unremarkable. And some emerging technologies pose "major risk of global catastrophe" (Beckstead et al. 2014, p. 6), let alone jeopardise sustainability. Synthetic biology might result in pandemics. Geoengineering risks drought and acid rain (ibid., p. 3). Advanced artificial intelligence demands protection from malevolence lest it spells the end of the human race (Hawking et al. 2014). Insidiously, the 'smart' mobile technologies that direct advertising based on past taste and location (Glavas and Letheren 2016) reduce people to neoliberal consumers (Monbiot 2016), who are force-fed promotions to increase consumption, as geese are engorged to produce foie gras. Other emerging technologies such as data mining, low intensity nuclear, nuclear fusion, vehicular hydrogen, magnetic-levitation trains and microbial biofuels can be associated both positively and negatively with sustainability.

This blotched pattern of mixed hope and despair results from the current fragmented governance of technology. These are isolated attempts to control a set of entities that are largely self-propelled, yet directionless, buffeted by forces economic, political and legal. While sustainability of and through both new and conventional technologies scaled to the global level demands conscious oversight and regulation, instead there is only passive acceptance apparent in many areas.

With few exceptions, governance of important technologies lacks in many of the *TAPIC* elements. Where it exists, governance tends to be secretive and opaque. It is not really accountable, except to states that may be unduly influenced by corporate pressures, or constricted by the military relationship.

With the notable exception of the SDGs, participation tends to be narrow outside of the technocratic enclaves that are required by treaty. It is evident from the case studies that important technologies are developed and managed in secretive, technocratic bartizans that are reluctant to respond to criticism, except when corporate interests are threatened. Integrity is especially constrained by regulatory capture, and capacity is limited by lack of collaboration, as well as the difficulties of combining political acumen with technical literacy.

Measures that would advance technology governance to benefit sustainability are significant challenges. They involve legal and financial reforms, structural and institutional change, a re-emphasis of values and different political priorities that centre on integrity.

Legal reforms

Governance arrangements are undermined by the law, which has failed to protect where technology impacts. Nevertheless, the law is a means to prise open opaque practices and shed light on the unsustainable. Technology may be able to be restrained by a legal framework that sets boundaries to protect the environment as Pope Francis suggests (2015, p. 39). Within that, specific legal reforms that would underpin technological sustainability include a recognition of non-human rights. They also include establishing the international crime of ecocide, revising laws concerning corporate involvement in politics and changes to patent law.

While the concept of human rights was progressively translated into law during the twentieth century, non-human rights have advanced a little in the twenty-first. A Spanish town recently gave civic rights to cats and dogs, and a US court ruled that chimpanzees in a research laboratory are 'legal persons' (Dawber 2015).[2] Francis Fukuyama, in considering a 'superuniversalisation' of rights, concluded that there is no essential difference between the rights of humans and other animals (Dresner 2008, pp. 155–156). Recent New Zealand law recognises an entire river as a living entity with the rights of a person (Roy 2017). The proposed crime of ecocide would regard the destruction of ecosystems as punishable in an international court. In the 2016 'shadow' ecocide case against Monsanto, however, the putative crime remains anthropocentric; the ecosystem damage must in some way directly harm humans.[3] Yet as Polly Higgins (2017) points out, since the Paris Agreement, the *mens rea* ('guilty mind') in ecocide now exists as a recklessness in disregarding available information. Decision-makers cannot claim ignorance of the harm from dangerous industrial technologies, at least as far as greenhouse gas emissions are concerned.

But, as is typical of concepts touched by sustainability, there is a paradox. The dirtiest companies tend to spend a great deal on politics if they are not to be regulated out of existence, so politics tends to be dominated by companies associated with unsustainable technologies (Monbiot 2017). ExxonMobil is one example.[4] There is strong evidence that the Koch brothers' oil interests led to the Republican repudiation of climate science (Davenport and Lipton 2017). But while corporate influence on politics varies according to individual states, sway mounts as international corporate mergers progress and companies dwarf many countries. Laws that prevent such mergers, and which restrict corporate donations to political parties would tend to avert Monbiot's paradox and enable politics to better represent the public interest, especially in favour of sustainable technologies.

Further, as discussed in Chapter 8, it would be prudent to amend patent regulations to encourage genuine progress rather than trivial innovation, and to discourage patents as a lure for corporate takeover. This would provide for a primary indication of the invention's impact on sustainability rather than just its novelty and 'useful purpose' as it is now. To advance sustainable innovation, patent governance might be tailored to each industry sector. The information technology sector that can make developments quickly, may well benefit from short patent protection; greater openness may encourage collaboration (Bostrom 2016). Whereas the pharmaceutical sector may need longer protection because developments take much time and can be easily copied. While the radical proposal – to abolish patents altogether and rely on first mover advantage – might benefit innovation, it would do nothing to encourage sustainable technologies. As a register of inventions, patents are a potentially rich source of data that may be trawled by artificial intelligences looking for more sustainable technologies than those presently developed.

Financial restructuring

Technology governance is hampered by the influence of the military and the fossil fuel industries. The case studies imply that it would benefit transparency and accountability if the military were restrained and re-oriented. This indicates financial restraint so that military technologies are properly scrutinised rather than authorised under pressure of a competitive arms contest. It also implies a re-orientation from the technologies of destruction to the technologies of peace-keeping, development aid delivery, and emergency services. New Zealand's approach to its air force is an example here. It has no expensive combat aircraft[5]; instead patrol helicopters together with long range transport and rescue craft are consistent with its mission, enabling operations over large areas of the Pacific. While achieving such re-orientation would be enormously difficult and require great political skill, it would be reinforced by the practical necessity of dealing with the disasters brought about by climate change. As floods and tempests intensify, the military is the logical organisation to deal with their effects. Missiles, bombs and combat aircraft have no value in disaster relief.

Desalination units, emergency shelter and safe transport are in demand. Further, the dynamics of reorientation would favour producers of rescue equipment and so reduce the influence of the weapons producers in the military-industrial complex.

Apart from legal restraint, encouraging the 'divestment' movement would also tend to reduce the influence of fossil fuel corporations and enable integrity in governance. Originally based in Western universities, the movement uses social media and direct protest to persuade investment funds to put their money elsewhere. While still more a trickle than a tsunami, funds that have pledged divestment include the Lutheran World Federation, the World Council of Churches, the world's largest single investment pool, Norway's sovereign wealth fund, as well as Stanford University and AXA, the world's largest insurance company. Ultimately the aim is to damage the reputation of fossil fuel companies as tobacco companies and apartheid were affected in times past (Gunther 2015). In any case, there already appears to be over-investment in extraction. According to the International Energy Agency (IEA 2014, p. 43), much investment in fossil fuels will be "stranded" if stronger measures are taken against emissions. Therefore, a rational investment strategy would fund technologies that increase sustainability, and so reduce the need for hard regulatory measures to govern the technologies that do not.

Structural and institutional change

As to the structure of governance, some believe that stronger, more globally centralised government is ultimately the only answer to sustainability (Heilbroner 1974). Others, wary of government's cumbersome tendencies, assert market forces and innovation as leading to a sustainable future based on renewable energy technologies (Lovins 2011). Still others believe this is implausible because corporations are ultimately focused on profit, not sustainability (Reich 2007). Or because adversarial liberal democracies, based on individual self-interest, are inherently unsustainable (Gale 2014). In framing the issue as a conflict between democracy and the environment, Mathews (1991) doubts that either can be sacrificed for the other.

Beyond authoritarian centralism or corporate laissez-faire, there is a third option involving enlarged and localised 'strong' democracies (Orr 2013; Klein 2014) encompassing de-privatisation and spreading economic assets throughout society, similar to the socialism espoused by US Senator Bernie Sanders during the 2016 presidential election. Allan Patience (2017) argues that after the present evident failure of neoliberal capitalism, better governance would consist of an enlivened international civil society, new forms of direct democratic governance, and state re-regulation of economies in the interests of society rather than elites. There are cautions about throwing out the baby with the bathwater, however: "the administrative state isn't optional in our complex society. It's indispensable" (Bazelon and Posner 2017). But given the ever-increasing scope and size of transnational corporations, inter-government and global institutions are indicated to set standards and manage the transitions to new technologies (Adams 2006; Sajeva, Sahota and Lemon 2015). Some such institutions exist, but lack coordinated placement within a governance structure.

Much of the focus for epistemic communities is already on international cooperation, but for technological sustainability such is the speed of change and mounting risk, wider global networks involving all stakeholders are indicated. "Despite the veneer of objectivity and value neutrality achieved by pointing to the input of scientists, policy choices remain highly political" (Haas 1992, p. 11). As the case studies show, scientists often differ in interpreting results and are cloistered in narrow disciplines. Consistent with Berg's (2008) observation about increasing corporate dominance, the governance of emerging technologies would be facilitated if connection with other stakeholders is made early.

The dynamic nature of technology demands an agile, adaptive form of governance. Traditional 'predict and control' regimes are centralised, hierarchical with narrow participation. Whereas adaptive regimes are polycentric, horizontal approaches with broad stakeholder input. In this area, centralised regimes tend to involve only quantifiable environmental measurement that can be determined easily. By contrast, decentralised governance tends to use both qualitative and quantitative indicators over whole ecosystems that are changing rapidly due to human influence (Pahl-Wostl 2007). This sort of adaptive management is also a systematic process for continuous improvement of policies and practices by learning from strategic outcomes (ibid.).

Polycentric, collaborative governance systems that include the general public in decision-making can not only enhance the knowledge base of decisions, but also support implementation through ownership (Lang et al. 2012; Newig et al. 2016). Just as Ostrom (2009) advised that the management of common resources is best "drawing on the strengths of many different institutions working together" and "co-operating at multiple scales" (Meinzen-Dick 2012), the management of technology in all its complexity is similarly mandated: technology is a fundamental resource and has many owners. Polycentric systems also have advantages over monocentric designs because numerous points of intersection help form resilient networks that can develop and maintain capability despite the ineptitude of one or two. Such network governance systems have reserves of expertise that help overcome inadequacies of particular components.

If technologies are to be effectively governed, then their risks must be made visible. This is challenging due to the variety of technologies involved and the temptation to use further high-risk technologies to solve the original problem (Adam, Beck and Loon 2000; Harari 2017). Technological risk assessment therefore implies a mix of approaches, including a top-down oversight that identifies major risks (high impact combined with high likelihood) and their mitigation, as well as inputs from experts and stakeholders in different particular fields.

There are already academic institutions that attempt to assess major risk, such as the University of Cambridge's *Centre for the Study of Existential Risk* (CSER) as well as the *Future of Humanity Institute* (FHI) at Oxford which look at threats to the very existence of our species, especially including the hazards of "advanced forms of biotechnology, molecular nanotechnology, and machine intelligence" (Bostrom 2013, p. 16). The *Club of Rome* continues its concern with the future of humanity, especially focussing on the impact of economics, but including attention to the technological risks that Peccei (1984) had urged. The *Massachusetts Institute of Technology* (MIT) that provided the original 'Limits to Growth' report for the Club is central to both sustainability and technology. The *Stockholm Resilience Centre* studies the risks of crossing the planetary boundaries, and also seeks to identify technologies that advance sustainability by reducing those risks.

The *Bulletin of the Atomic Scientists* in Chicago publishes the 'Doomsday Clock',[6] which originally indicated the likelihood of nuclear conflict. It has since included other technology-related threats, such as climate change, biological weapons and 'cyberthreats'. The *Bulletin* suggests that open-source monitoring could be used to help supervise the waning *Biological Weapons Convention*, which lacks an effective inspections system (Jeremias and Himmel 2016). A like principle could apply to the monitoring of all high risk technologies. Whistleblowers can help police mundane technologies too, as in the recent US$40 million case of illegal discharge from cruise liners (Clatworthy and Horne 2017).

Meanwhile, geoengineering looks increasingly real. A Harvard team has proposed to spray aerosols into the stratosphere in 2018, despite a moratorium on the practice adopted by the UN Convention on Biological Diversity, which the United States has not ratified (Neslen 2017). At the same time, the non-profit Carnegie Council is pursuing an initiative that aims to shift the debate on geoengineering governance "from academia to the intergovernmental policy space". The initiative involves constructing global networks on the issue from a wide range of stakeholders and governments (Carnegie Council 2017). This sort of risk demands serious attention, but other issues of technology governance might benefit from a similar network.

The evermore complex risk society implies that science must be 'de-monopolised' away from narrow groups of experts, because those who may suffer the effects of decisions should be able to contribute to them (Bäckstrand 2003). Encouraging linkages between the organisations concerned with technological risk cited above, with civil society, international institutions, the media – and directly to interested individuals – would help enable such contributions to be made.

Many organisations deal with particular technologies at and above the level of nation states, but none directly concern technology and sustainability together. Many others involve the more general 'sustainable development', 'sustainability' and 'biosphere impacts'. There are also proposals for new bodies that relate to other aspects of technologies and their impacts. One suggestion is for institutions to "manage the technological transition" to the new economy of reuse, recycle and new energy (Adams 2006, p. 15). Another is for a "global referee" to ensure that planetary boundaries are respected (Steffen, Rockström and Costanza 2011, p. 65).

However, it seems that there is as yet no institution to oversee technologies as they affect sustainability *in general*. If established, such an institution would function as the global focal point for the governance of technologies in the quest for sustainability. It would coordinate research and promote links between nodal points in a networked governance system, including the academic bodies mentioned above, other relevant research institutions, industry bodies, think tanks, NGOs and regulatory agencies. It would produce reports on established and emerging technologies and their effects on sustainability. It would organise conferences to determine priority areas, and it would encourage evaluations of technology systems and their governance. Desirably, ordinary people would be empowered to engage online, especially including those who might identify applications of unsustainable technologies. Conversely, such an institution would help illuminate more of those jewels of sustainability and encourage their adoption more widely.

Values

In the name of profit, capitalism drives efficiency to reduce production costs. But efficiency improvements by themselves, can be counter-productive, as the Jevons paradox demonstrates. It has been argued that rather what is needed is:

> a conscious effort to direct technological innovation toward the achievement of clearly defined societal goals that reflect shared values... Unless we undertake this critical challenge, technological innovation and efficiency improvements will continue to promote unsustainable growth which will inexorably lead to environmental and societal collapse.
>
> (Huesemann and Huesemann 2011, p. 116)

Even if this prediction is exaggerated, there is a mass of evidence that neoliberal capitalism is at best inadequate in fostering sustainable technology, because its only value is profit. But as Fukuyama and Thatcher might concur,[7] capitalism will nevertheless be part of technology's future, just not as neoliberal discourse suggests, "that no other part need be played" (Albertson 2014).

Further, there are tendencies for values to be downgraded. An early landmark paper in the field of conservation biology advanced 'core values'[8] that were centred on the intrinsic value of biodiversity and ecological complexity, irrespective of human considerations (Soulé 1985, p. 964). By contrast, a more recent article in the same journal suggests that these values are no longer appropriate because the global context has since profoundly changed. Human dominance of the biosphere is pervasive, while at the same time nature has shown surprising resilience. Instead of values, more "practical statements" about what should be done are proposed. These involve recognition of human-altered ecosystems and working with corporations to maximise both conservation and economic objectives (Kareiva and Marvier 2012, pp. 965–967). This exposes a fault line in sustainability thinking. The former position may be too restrictive, whereas the latter licenses still more human interference in natural systems that are already precarious.

It has been suggested that it is time for "a presidential technoethics commission", just as earlier US bioethics commissions[9] laid down guidelines for stem cell research. The technoethics commission call is due to an increasingly pervasive information technology, which now routinely records and sells personal details, preferences and consumer actions via social media (Rockmore 2016), thus threatening all values that are not commercial. Sustainability logically outranks values of commerce and exchange.

Political priorities

The related issue of data ownership and control is probably still more substantial. As data has become a ubiquitous commodity, corporate control over it threatens sustainability through the exclusion of other interests. As a 'non-rival good', data are virtually immortal and may be consumed indefinitely. While it may be used commercially, it is also integral to scientific and social research. And because data mining for research opens vivid new vistas of sustainability assessment, Donella Meadows' (2008) 'dancing with Earth systems' becomes possible – but only if data are freely available for examination and study. This does not contradict the right of entrepreneurs to put a value on the information they create from data for commercial purposes. But the quest for sustainability will be impeded if researchers are denied access to the fundamental records available on which to build information and knowledge.

Persuasive rhetoric is important to political outcomes (Hulme 2009), and as Adams (2006, p. 15) points out: "Environmentalism's traditional capacity to speak like the prophet Jeremiah, promising hell to come, does not promote creative thinking and openness to change". Sustainability values are underpinned by narrative; they are not intrinsic to the technology Janus, which is values-blind. The gulf between the sciences and the arts lamented by C. P. Snow (1959) has probably widened due to the fractal concerns of both systems, as well as to the cumulative technological complexity since his time. This accelerating 'techne', however, has not been matched by a logos or 'persuasive discourse' (Kemple 2017) that would set technology within a purposeful ethic. If biodiversity and sustainability have intrinsic value, then technologies that minimise the human footprint are most ethical. A rhetoric that invokes technology to bridge the gap between science and the arts may make use of both spheres. Such a narrative could involve an ethic of preventing the entropic juggernaut from taking nature with us into terminal decline, instead to use technologies that enhance life in all its forms. Or it could describe how brute machines that once underpinned industrial-scale devastation are replaced by agile technologies that measure, visualise and diagnose ways to repair the damage that has led us to the brink.

Bearing in mind that the pursuit of sustainability depends on a global community linked through Internet-based technologies (Dresner 2008; Gore 2014), such inclusion is now technically feasible online,[10] whether direct or via the social media. Input may be invited from everyone who has interests or concerns. Thus governance may be refreshed by democracy. It may also be refreshed by gender inclusion. The Silicon Valley 'alien overlords' are largely men, as elsewhere such as the United Kingdom (Hicks 2017). It is probable that women have some different perspectives and priorities.

The planetary boundaries (Steffen, Rockström and Costanza 2011; Steffen et al. 2015) indicate where technologies are most critical in the quest for sustainability. Although questioned as to their scale (Nordhaus, Shellenberger and Blomqvist 2012), they are extensively cited and recognised by major global organisations.[11] These boundaries therefore indicate priorities for technological governance. Ozone destroying technologies have been checked by the Montreal protocol, but disruption of the nitrogen and phosphorous cycles is at critical levels, as are extinctions and biodiversity impacted by chemical pollutants, land-clearing and hunting technologies. The Earth's capacity to assimilate 'novel entities' such as heavy metals, nuclear waste, plastics and other chemicals is not yet quantified, but of considerable concern. The effects of greenhouse gas and particulate emissions are also critical. Beyond these indications, there are the emerging technologies that threaten to get out of hand, such as artificial general intelligence and forms of geoengineering, which demand sustainability risk assessment. These issues are more than just prosaic 'environmental concerns'. Contrary to Fukuyama's view, their governance will also require daring, courage, imagination, and idealism:

Notes

1 The project was funded by the USEPA, the US Department of the Interior and the American Samoa Economic Development Authority.
2 This decision was since overturned on appeal (Kyriakakis 2015).
3 Personal conversation with court advocate, Dr Gwynn MacCarrick, 3 February 2017.

4 ExxonMobil former CEO, Rex Tillerson, became US Secretary of State in 2017.
5 One FA-18E/F Hornet costs c.US$100 million, for example (Wiki 2017).
6 The Doomsday Clock, created in 1947 by former Manhattan Project researchers, showed two minutes to midnight at September 2018.
7 They share the view that 'there is no alternative'.
8 These values are consistent with the philosophy of 'deep ecology' (Naess 1973).
9 Such as the US *National Bioethics Advisory Commission 1999*.
10 The *Online Direct Democracy Party* in Australia, for example, offers a citizen direct-vote policy.
11 Such as the UN High Level Panel on Global Sustainability, Oxfam, the World Wildlife Fund and the UNEP.

References

Adam, B, Beck, U and Loon, J 2000, *The Risk Society and Beyond*, Sage, London, UK.

Adams, W 2006, *The Future of Sustainability: Re-Thinking Environment and Development in the Twenty-First Century*, International Union for the Conservation of Nature (IUCN), Gland, Switzerland.

Albertson, K 2014, 'The Many Reasons Why We Must Look Beyond Capitalism: Public Goods Cannot Be Provided by the Market', letter to *The Observer*, 20 April.

Bäckstrand, K 2003, 'Civic Science for Sustainability: Reframing the Role of Experts, Policy-Makers and Citizens in Environmental Governance', *Global Environmental Politics*, vol. 3, no. 4, November, pp. 24–41.

Bazelon, E and Posner, E 2017, 'The Government Gorsuch Wants to Undo', *New York Times Sunday Review*, 1 April.

Beckstead, N, Bostrom, N, Bowerman, N, Cotton-Barratt, O, MacAskill, W, Ó hÉigeartaigh, S and Ord, T 2014, *Unprecedented Technological Risks, Policy Brief*, University of Oxford, UK.

Berg, P 2008, 'Meetings That Changed the World: Asilomar 1975: DNA Modification Secured', *Nature*, vol. 455, 18 September, pp. 290–291. doi: 10.1038/455290a.

Bostrom, N 2013, 'Existential Risk Prevention as Global Priority', *Global Policy*, vol. 4, no. 1.

Bostrom, N 2016, *Strategic Implications of Openness in AI Development, Technical Report #20161*, Future of Humanity Institute, Oxford University, UK, pp. 1–26.

Carnegie Council 2017, *Carnegie Climate Geoengineering Governance Initiative (C2G2)*, New York, US, 1 January.

Clatworthy, B and Horne, M 2017, 'Courageous' Whistleblower Awarded $1m by US Court', *The Times*, 25 April.

Cooke, K 2017, 'Solar Power to Rise from Chernobyl's Nuclear Ashes', *Guardian Environment Network*, 12 January.

Davenport, C and Lipton, E 2017, 'How G.O.P. Leaders Came to View Climate Change as Fake Science', *New York Times*, 3 June.

Dawber, A 2015, 'Human Rights for Cats and Dogs: Spanish Town Council Votes Overwhelmingly in Favour of Defining Pets as 'Non-Human Residents', *The Independent*, 22 July.

Dresner, S 2008, *The Principles of Sustainability*, second edition, Earthscan, London, UK.

Etherington, D 2016, 'Tesla and SolarCity Made This Whole Island Solar-Powered in Under a Year', *Tech Crunch*, 22 November.

Fukuyama, F 1989, 'The End of History?' *The National Interest*, Summer.

Fukuyama, F 1992, *The End of History and the Last Man*, Hamish Hamilton, London, UK.

Gale, F 2014, On the Deep Unsustainability of Actually Existing Liberal Democracy, Conference Paper, Australian Political Studies Association, Sydney, Australia.

Glavas, C and Letheren, K 2016, 'The Disruptive Technologies That Will Shape Business in the Years Ahead', *The Conversation*, 19 January.

Gore, A 2013, *The Future*, Random House, London, UK.

Grovenor, K 2016, 'New York Is Growing Oysters in Toilets to Save the City's Waterways', *Techly*, 17 October.

Gunther, M 2015, 'Why the Fossil Fuel Divestment Movement May Ultimately Win', *Yale Environment 360*, 27 July.

Haas, P 1992, 'Epistemic Communities and International Policy Coordination', *International Organization*, vol. 46, no. 1, Winter, pp. 1–35.

Harari, Y 2017, 'Homo Sapiens As We Know Them Will Disappear in a Century or So', in Anthony, A, questions from readers, *The Guardian*, 19 March.

Heilbroner, R 1974, *An Inquiry into the Human Prospect*, W.W. Norton, New York, US.

Higgins, P 2017, 'Why Ecocide Crimes Are Crimes of Recklessness', *Eradicating Ecocide*, 27 February.

Hawking, S, Tegmark, M, Russell, S and Wilczek, F 2014, 'Transcending Complacency on Superintelligent Machines', *Huffington Post*, blog, 19 April, updated 19 June.

Hicks, M 2017, *Programmed Inequality: How Britain Discarded Women Technologists and Lost Its Edge in Computing*, MIT Press, Cambridge, US.

Huesemann, M and Huesemann, J 2011, *Technofix: Why Technology Won't Save Us or the Environment*, New Society Publishers, British Colombia, Canada.

Hulme, M 2009, *Why We Disagree about Climate Change: Understanding Controversy, Inaction and Opportunity*, Cambridge University Press, Cambridge, UK.

IEA: International Energy Agency 2014, *World Energy Investment Outlook – Special Report*, IEA, Paris, France.

Jeremias, G and Himmel, M 2016 'Can everyone help verify the bioweapons convention? Perhaps, via open source monitoring', *Bulletin of the Atomic Scientists*, vol. 72, no. 6, 1 November, pp. 412–417. http://dx.doi.org/10.1080/00963402.2016.1240487.

Kareiva, P and Marvier, M 2012, 'What is Conservation Science?', *BioScience*, vol. 62, no. 11, November.

Kemple, B 2017, 'Techne, Physis, and Technology: Heidegger on Aristotle's Physics B.1', draft, *Academia*.

Kim, H, Yang, S, Rao, S, Narayanan, S, Kapustin, E, Furukawa, H, Umans, A, Yaghi, O and Wang, E 2017, 'Water Harvesting from Air with Metal-Organic Frameworks Powered by Natural Sunlight', *Science*,. doi: 10.1126/science.aam8743.

Klein, N 2014, *This Changes Everything: Capitalism vs. the Climate*, Penguin Group, London, UK.

Kyriakakis, J 2015, 'Another Chimpanzee Personhood Claim Fails, But There's Hope Yet', *The Conversation*, 4 August.

Lang, D, Wiek, A, Bergmann, M, Stauffacher, M, Martens, P, Moll, P, Swilling, M and Thomas, C 2012, 'Sustainability Science: Bridging the Gap between Science and Society: Transdisciplinary Research in Sustainability Science: Practice, Principles, and Challenges', *Sustainability Science*, vol. 7, supplement 1, pp. 25–43

Lasat, M 2000, 'Phytoextraction of Metals from Contaminated Soil: A Review of Plant/Soil/Metal Interaction and Assessment of Pertinent Agronomic Issues', *Journal of Hazardous Substance Research*, vol. 2, no. 5, pp. 1–25.

Lovins, A 2011, *Reinventing Fire: Physics + Markets = Energy Solutions*, Rocky Mountain Institute, Colorado, US.

Mathews, F 1991, 'Democracy and the Ecological Crisis', *Legal Service Bulletin*, vol. 16, no. 4, August, pp. 157–159.

Meadows, D H 2008, Thinking in Systems: A Primer, Wright, D (ed.), Earthscan, London, UK.

Meinzen-Dick, R 2012, 'Elinor Ostrom's Trailblazing Commons Research Can Inspire Rio+20', *The Guardian*, blog, 15 June.

Monbiot, G 2016, 'Neoliberalism – The Ideology at the Root of All Our Problems', *The Guardian*, 15 April.

Monbiot, G 2017, 'The Pollution Paradox', website, 20 January.

Naess, A 1973, 'The Shallow and the Deep, Long-Range Ecology Movement: A Summary', *Inquiry*, vol. 16, no. 1, pp. 95–100.

Neslen, A 2017, 'US Scientists Launch World's Biggest Solar Geoengineering Study', *The Guardian*, 24 March.

Newig, J, Kochskämper, E, Challies, E, and Jager, N 2016, 'Exploring Governance Learning: How Policymakers Draw on Evidence, Experience and Intuition in Designing Participatory Flood Risk Planning', *Environmental Science and Policy*, vol. 55, pp. 353–360.

Nordhaus, T, Shellenberger, M and Blomqvist, L 2012, *The Planetary Boundaries Hypothesis: A Review of the Evidence*, The Breakthrough Institute, Oakland, US.

Orr, D 2013, 'Governance in the Long Emergency', in *State of the World 2013: Is Sustainability still Possible?* The Worldwatch Institute, Washington, DC, US, pp. 279–291.

Ostrom, E 2009, 'A General Framework for Analyzing Sustainability of Social-Ecological Systems', *Science*, vol. 325, no. 5939, 24 July, pp. 419–422.

Pahl-Wostl, C 2007, 'Transitions towards Adaptive Management of Water Facing Climate and Global Change', *Water Resource Management*, vol. 21, pp. 49–62.

Patience, A 2017, 'If We Are Reaching Neoliberal Capitalism's End Days, What Comes Next?', *The Conversation*, 6 February.

Peccei, A 1984, 'The Club of Rome: Agenda for the End of the Century', in Malaska, P and Vapaavuor, M (eds), *The Club of Rome "The Dossiers" 1965–1984*, Finnish Association for the Club of Rome Helsinki, Finland, pp. 37–43.

Pope Francis 2015, *Laudato si'*, Libreria Editrice Vaticana, 24 May.

Reich, R 2007, *Supercapitalism*, Knopf, New York, US.

Rockmore, D 2016, 'Is It Time for a Presidential Technoethics Commission?', *The Conversation*, 13 May.

Roy, E 2017, 'New Zealand River Granted Same Legal Rights as Human Being', *The Guardian*, 16 March.

Ruggie J 2014, 'Global Governance and "New Governance Theory": Lessons from Business and Human Rights', *Global Governance*, vol. 20, pp. 5–17.

Sajeva, M, Sahota, P and Lemon, M 2015, 'Giving Sustainability a Chance: A Participatory Framework for Choosing between Alternative Futures', *Journal of Organisational Transformation and Social Change*, vol. 12, no. 1, April, pp. 57–89. doi: 10.1179/1477963314Z.00000000035.

Schwab, K 2016 'The Fourth Industrial Revolution: What It Means and How to Respond', *World Economic Forum*, 14 January.

Snow, C 1959, *The Two Cultures: The Rede Lecture*, Cambridge University Press, Cambridge, UK.

Soulé, M 1985, 'What Is Conservation Biology?', *BioScience*, vol. 35, no. 11, December, pp. 727–734.

Steffen, W, Richardson, K, Rockström, J, Cornell, S, Fetzer, I, Bennett, E, Biggs, R, Carpenter, S, de Vries, W, de Wit, C, Folke, C, Gerten, D, Heinke, J, Mace, G, Persson, L, Ramanathan, V, Reyers, B and Sörlin, S 2015, 'Planetary Boundaries: Guiding Human Development on a Changing Planet', *Science*, vol. 347, no. 6223, 13 February,ss 1259855. doi: 10.1126/science.1259855.

Steffen, W, Rockström, J, and Costanza, R 2011, 'How Defining Planetary Boundaries Can Transform Our Approach to Growth', *Solutions*, vol. 2, no. 3, pp. 59–65.

Wang, U 2016, 'Top Five Sustainable Technology Trends of 2015', *The Guardian*, 1 January.

WEF: World Economic Forum 2017, *Harnessing the Fourth Industrial Revolution for the Earth*, in collaboration with PwC and Stanford Woods Institute for the Environment, Geneva, Switzerland.

Wiki 2017, *Boeing F/A-18E/F Super Hornet*. en.wikipedia.org/wiki/Boeing_F/A-18E/F_Super_Hornet.

14 Conclusion: The Owl of Minerva

Wisdom's companion is a natural creature of strength and beauty, that oversees the world from aloft before swooping down in shadow upon its quarry:

> Only in the maturity of reality does the ideal appear as counterpart to the real, apprehends the real world in its substance, and shapes it into an intellectual kingdom. The Owl of Minerva takes flight only when the shades of night are gathering.
>
> (Hegel 2001 [1820], p. 20)

Dusk is falling over the industrial age of fossil fuels and the 'limitless frontier'. It gathers over neoliberal market supremacy, which has failed to answer humanity's greatest problem. At the end of this era the planet-changing power of our technological tigers is apparent. After two centuries of technological outburst, we have reached a point where its pattern is clear. Just as modern capitalism cannot control itself, it cannot control the technologies from which it profits; the exponential growth it provokes inevitably leads to material entropy. Unrestrained, it is hostile to the planet and to humanity.

Capitalism's handmaiden, innovation, concerns the diffusion of novel technologies. But unlike the Owl, innovation does not inevitably augment wisdom. It is blind to the benefit or harm that such dispersal might bring. Rather, innovation worships at the value-free "altar of change" (Vinsel and Russell 2016). As another Russell (1946) said, it is a kind of madness.

Much concern with technology centers on energy and how its emissions lead to global warming and to a cascade of detrimental effects. Geoengineering responses such as atmospheric aerosol spraying, or seeding of the seas with iron, may evoke unforeseen results. Nuclear technologies risk meltdown and spawn the conundrum of waste that cannot be assimilated. Chemical technologies still pollute the waterways despite knowledge of sustainable alternatives. Mining and smelting technologies are not sustainable yet could be much contained and improved. There is risk of pandemic from genetic experimentation. And there is the 'existential threat' of artificial intelligence that uncritically but inexorably follows instructions embedded in its algorithms, written by our alien overlords.

All life is counter-entropic within itself (Schrödinger 1944; Lovelock 1979). The external entropy it necessarily creates can only be offset by incoming energy. The Earth is literally a solar-powered life-support system – for humans and for myriad, but dwindling numbers of other species. That life support system is malfunctioning mainly due to the impact of human technologies. Unless our future is to be degraded and inert, as our sibling moon reminds us of such a fate, our technologies must support life. The imperative is to treat 'our common home' as if we intend to stay – and as if we want other species to stay with us. Technologies that frustrate that imperative must be reined in; technologies that advance it must be encouraged.

This book has approached the conundrum of sustainability by assessing how it is conceived and the contribution of its classic components to human impact upon this planet. While the rate of population increase is at last slowing, such is its momentum that our numbers will probably continue to swell until the end of the century. Rising affluence and consumption, especially of food and energy in the East and South, will place additional strains on primary industries and demand better ways to fuel and transport within and between settlements. Neither population nor affluence can be braked through policy leverage in the medium term; such methods are simply unpalatable or unacceptable to those who already have little. In the West, the rising inequality that is intrinsic to capitalism fragments political coherence.

The assertion that technology offers our best chance in the pursuit of sustainability is highly contested. The case studies illustrate major concerns with how it is developed and deployed. Yet if not technology, then what? It is technology that has defined our civilisation. It is evolving rapidly. While it has wrought great harm, it has also brought great benefit. Our relationship with technology has always been interwoven with our historical economic systems; much has centered on the relative price of labour, on slavery, and on the development of labour-saving devices in the West. But the most powerful force in technological development over the period of the Industrial Revolution has been capitalism, often linked with the concerns of the military. Even the present key technology platform, the Internet, has military origins. These two forces are exceptionally difficult to control. Neither recognises limits or end points. Both amplify rivalry, whether of the individual or the nation state. Both are based on continuous development.

The direction of that development may nevertheless be channelled. Now that it is the entire planet that is affected in this Anthropocene, how technologies are governed, demands a global approach. At the same time, such is the diversity and rapid development of new technologies in this digital era, that agility is essential to any form of effective governance. There is no world government, but a world government in an institutional sense may not be warranted – as far as reasserting direction and control over technology is concerned – as it would be clumsy. Agility may be attained through a network system that links existing and new stakeholders in a mesh large enough to encompass major global issues, yet fine enough to be sensitive to particular emerging developments and impacts.

In deference to Snow, Pirsig and McLuhan, the need to bridge the quantitative and the qualitative in resolving this conundrum is apparent. Scientists, engineers and technocrats are essential to bend technology in the direction of sustainability. Philosophers, writers, economists and politicians are needed to reform the economic system and the military towards a capacity to accept and benefit from it. Politically, it is well past time to assert leadership over technology, over the economic system and over the military so that the fine arts, love, and capital accumulation are, in the longer term, not for nought and for nothing, but rather for all and everything, in the present and the future.

Means of effecting such a future have been discussed in the light of the case studies. In summary, they include the following:

Legislative measures that would favour technological sustainability include the establishment of non-human rights and the crime of ecocide, limiting corporate influence over representational politics and amending patent law to support genuine progress. Financial measures include restricting investment in unsustainable technologies and promoting re-investment in sustainable alternatives. Re-nationalisation of electricity grids and public transport are alternatives to neoliberal market failures in these sectors if sustainability is to be a priority.

There is a need for technological threat identification, which links existing institutions and facilitates input from a wide range of stakeholders, including support for whistleblowers. As a source of destructive technologies, the military attracts special scrutiny. To the extent that the military can be re-oriented to non-violent roles, there would tend to be a rebalance towards more sustainable technologies, and perhaps towards a peace during which such organisations might redirect their attention.

The tension between values involving technology and nature includes nature's intrinsic value as against the practical value of working within altered ecosystems in conjunction with corporate interests. Technoethics commissions may be useful, but the need to regulate data that simultaneously rewards openness, preserves privacy and enables its analysis is more pressing. Sustainability and technology demand an underpinning narrative; one that is both practical and aspirational; as with Dryzek's reformist environmental discourses, both prosaic and imaginative, or as with Steffen's greens and greys, more bright green (innovation and regulation) than the alternatives.

Elements of an effective governance system include transparency involving usable and accessible information, inclusive participation, integrity fortified by a clear sense of mission and a policy capacity drawn from polycentric networks and the encouragement of public input. Such a system would first relate to technologies that affect the Earth system boundaries. A global institution designed as a central focus of governance for technological sustainability would especially encourage network linkages between existing institutions and illuminate areas where progress has been realised. Accountability is ultimately to all of us; its sanction is the quality of our existence.

The governance of technologies for sustainability must itself rely on technologies available for data, information and knowledge, as well as for communication. These technologies are easy to identify and access is at issue. But for wisdom, there is no such technology; ultimately wisdom depends on human judgement. As technological complexity increases, expertise must narrow. Judgement of technological quality then is a series of human overlaps. These connections can be made across and between disciplines, and also across experiences. Neither the blind wisdom of the market nor the omniscience of a central authority can be assumed. To govern technology for sustainability, we must "reach beyond the powers of commerce and command" (Kemp, Parto and Gibson 2005, p. 26) to a shared responsibility that encompasses the Earth.

One key challenge between technology and sustainability is, on the one hand, against the attitude that our predicament is inevitable and that any effort can only make things worse. It is, on the other hand, against the promise that technology will solve all problems (Buckup 2016). In either discourse we become passive subjects in a world of increasing uncertainty, whereas awareness, thought and participation is critical if our species is to have a future.

Driven by national rivalry or by the quest for profit, technology has defined the last two centuries of human civilisation – at its most triumphant and at its most demeaning. At this gathering of the dusk, unless the quest for sustainability overrides other considerations of technology's purpose, a long night looms ahead.

References

Buckup, S 2016, *The Surprising Link between Science Fiction and Economic History*, World Economic Forum, Cologny, Switzerland, 16 June.

Hegel, G 2001 [1820], *Philosophy of Right*, Dyde, S W (trans.), Batoche Books, Kitchener, Canada.

Kemp, R, Parto, S and Gibson, R 2005, 'Governance for Sustainable Development: Moving from Theory to Practice', *International Journal of Sustainable Development*, vol. 8, nos. 1/2, pp. 12–30.

Lovelock, J 2000 [1979], *Gaia: A New Look at Life on Earth*, Oxford University Press, Oxford, UK.

McLuhan, M 1964, *Understanding Media: The Extensions of Man*, MIT Press, Cambridge, US, and London, UK.

Pirsig, R 1984 [1974], *Zen and the Art of Motorcycle Maintenance*, Bantam, New York, US.

Russell, B 1979 [1946], *History of Western Philosophy*, Allen and Unwin, London, UK.

Schrödinger, E 1944, *What Is Life? The Physical Aspect of the Living Cell*, Cambridge University Press, Cambridge, UK.

Snow, C 1959, *The Two Cultures: The Rede Lecture*, Cambridge University Press, Cambridge, UK.

Vinsel, L and Russell, A 2016, 'Hail the Maintainers: Capitalism Excels at Innovation But Is Failing at Maintenance, and for Most Lives It Is Maintenance That Matters More', *Aeon*, 8 April.

Index